Faces of
Mathematics

Faces of Mathematics

An Introductory Course for College Students

SECOND EDITION

A. Wayne Roberts
Macalester College

Dale E. Varberg
Hamline University

HARPER & ROW, PUBLISHERS, New York
Cambridge, Philadelphia, San Francisco,
London, Mexico City, São Paulo, Sydney

1817

Cover photo: Michel Craig

The problem on page 11 is from *The Moscow Puzzles* by Boris Kordemsky, copyright © 1972 Charles Scribner's Sons. Reprinted by permission of Charles Scribner's Sons.

The picture of Leonhard Euler on page 91 was provided courtesy of The Bettmann Archive Inc.

Reptiles (page 268) and *Birds* (page 274) by M. C. Escher are reprinted courtesy of the Escher Foundation. Collection Haags Gemeentemuseum — The Hague.

Kepler's Model is reprinted courtesy of The Princeton University Press (Source: Hermann Weyl, *Symmetry,* p. 76, copyright 1952 by The Princeton University Press).

The photograph of Albert Einstein on page 289 was provided courtesy of Wide World Photos, Inc.

"From the Minutes of a Borough Council Meeting" was reprinted by permission of Curtis Brown, Ltd. Copyright 1943 by Robert Braves and Alan Hodge.

Sponsoring Editor: Fred Henry
Project Editor: Eleanor Castellano
Designer: Michel Craig
Manager of New Book Production: Kewal K. Sharma
Production Assistant: Jacqui Brownstein
Compositor: Ruttle, Shaw & Wetherill, Inc.
Printer and Binder: R. R. Donnelley & Sons Company
Art Studio: J & R Art Services, Inc.

Library of Congress Cataloging in Publication Data

Roberts, A. Wayne (Arthur Wayne), 1934-
  Faces of mathematics.

  Includes indexes.
  1. Mathematics—1961-   I. Varberg, Dale E.
QA39.2.R6 1982   510   81-23779
ISBN 0-06-045471-7   AACR2

Contents

PREFACE   *ix*

DEPENDENCE CHART   *xv*

I Solving Problems   2

1 Strategies   4
1.1 Clarify the Question   *5*
1.2 Organize the Given Information   *10*
1.3 Experiment, Guess, Demonstrate   *16*
1.4 Transform the Problem   *23*

2 Numeric Methods   29
2.1 The Problem of Measuring and Counting   *30*
2.2 Need for Negative Numbers   *38*
2.3 Fractions   *43*
2.4 Decimals and Percents   *52*
2.5 Exponents and Scientific Notation   *59*

3 Algebraic Methods   67
3.1 Solving Equations   *68*
3.2 Problems That Lead to One Equation   *74*
3.3 Problems That Lead to Two Equations   *82*

II Finding Order   90

4 Numerical Patterns   92
4.1 Number Sequences   *93*
4.2 Arithmetic Sequences   *100*
4.3 Geometric Sequences   *105*
4.4 Population Growth   *113*
4.5 Compound Interest   *120*
4.6 Fibonacci Sequences   *127*

5 Programming a Computer   134
5.1 Binary Arithmetic   *135*
5.2 Computing Machines   *142*
5.3 Step-by-Step Directions   *148*
5.4 Flowcharting for a Computer   *153*
5.5 BASIC Programming   *159*

6 Systematic Counting   166
6.1 Fundamental Counting Principles   *167*
6.2 Permutations   *173*
6.3 Combinations   *178*
6.4 The Binomial Theorem   *184*

7 The Laws of Chance   191
7.1 Equally Likely Outcomes   *192*
7.2 Independent Events   *200*
7.3 The Binomial Distribution   *206*
7.4 Some Surprising Examples   *213*

8 Organizing Data   220
8.1 Getting the Picture   *221*
8.2 On the Average   *228*
8.3 The Spread   *235*
8.4 Sigma Notation   *242*
8.5 Correlation   *248*

9 Geometric Paths   254
9.1 Networks   *255*
9.2 Trees   *263*
9.3 The Platonic Solids   *268*
9.4 Mosaics   *274*
9.5 Map Coloring   *281*

III Reasoning and Modeling   288

10 Methods of Proof   290
10.1 Evidence but Not Proof   *291*
10.2 Deduction   *298*
10.3 Difficulties in Deductive Thinking   *307*
10.4 Deduction in Mathematics   *314*

11 From Rules to Models   321
11.1 The Consequences of Given Rules   *322*
11.2 Finite Geometries   *332*
11.3 The Axiomatic Method   *340*
11.4 Models   *345*

12 Geometries as Models   355

12.1 Euclid's Work   *356*
12.2 Non-Euclidean Geometry   *365*
12.3 Lessons from Non-Euclidean Geometry   *373*
12.4 Lessons from Euclidean Geometry   *378*

IV Abstracting from the Familiar   384

13 Number Systems   386

13.1 The Counting Numbers   *387*
13.2 Extending the Number System   *395*
13.3 Modular Number Systems   *402*
13.4 Equations with Integer Answers   *410*

14 The System of Matrices   417

14.1 Boxes of Numbers   *418*
14.2 Properties of Matrix Multiplication   *424*
14.3 Some Applications   *430*

15 Algebraic Structures   436

15.1 Basic Concepts of Algebra   *437*
15.2 Mathematical Rings   *444*
15.3 Mathematical Fields   *450*
15.4 Solving Equations   *456*

ANSWERS   *461*
NAMES AND FACES INDEX   *489*
SUBJECT INDEX   *491*

Preface

It has long been held that anyone who aspires to be educated must study mathematics. We still believe it, and this book is intended to be a source book for those who want to see what mathematics can contribute to a liberal education. In particular, we have in mind those college students who plan to take just one or two semesters of mathematics. Perhaps they want to satisfy a distributive requirement, or perhaps they are prospective elementary school teachers who need a broadened and deepened perspective on mathematics.

A number of books have addressed themselves to this audience. We think they generally miss the mark for either of two reasons. Some try to survey the content of mathematics, offering a smorgasbord from which users may choose according to their taste. Such books are often superficial, although this is not our principal objection. The availability of interesting and potentially practical topics is not the only reason—or perhaps even the main reason—great thinkers insisted that educated people should study mathematics. They believed, as we believe, that the study of mathematics can help us to learn something about thinking itself: how to state our problems clearly, sort out the relevant from the irrelevant, argue coherently, and abstract some common properties from many individual situations. It is toward these goals that we wish to move.

This brings us to the second kind of book available for the purposes we have in mind. This type of book emphasizes the methodology rather than the content of mathematics. In some cases, the focus is on the foundations of mathematics and the rigor of mathematical proof. In others, it is on how mathematicians approach problems, or on the historic development of great mathematical ideas. We have been greatly impressed with many of these books, and wish to acknowledge here the particular influence of such writers as Polya, Wilder, and Richardson. Books of this type, however, have one drawback: they are too difficult for the audience we have in mind.

We have tried to steer a middle course. Insofar as it was consistent with maintaining a light, readable, often humorous style that would appeal to our audience, we have selected topics that can be presented in some depth. Moreover, we have continually addressed ourselves to the larger contention that mathematics is the ideal arena in which to develop skill in the areas of organizing information, analyzing a problem, and presenting an argument.

AN EMPHASIS ON PROBLEM SOLVING

To a large extent, we have tried to achieve our goals by getting the reader involved in solving problems. The great mathematicians were problem solvers, and insight into their work requires, we believe, some involvement in their activity. But beyond the confines of mathematics, every person—painter or scientist, carpenter or homemaker—must solve problems. We believe it is possible to describe principles of problem solving that can carry over into many areas of our lives.

Our book begins, therefore, with a discussion of strategies for problem solving. These strategies are illustrated with puzzle type problems. Some of these problems are a well-established part of mathematical lore; their origin has long since been lost. Most of them appear in one guise or another in popular books of problems. We make no effort to identify the sources of problems we have used; they have found their way into our notes over many years of teaching. However, we are pleased to identify several puzzle books that have been favorites of ours. If we have succeeded in what we tried to do, then these books may well include some of our readers among the audiences they have served so well. The books are H. E. Dudeney, *Amusements in Mathematics,* New York: Dover, 1970. (reprint of a book first published in 1917); H. E. Dudeney, *536 Puzzles and Curious Problems,* New York: Scribners, 1967; E. R. Emmet, *Puzzles for Pleasure,* Buchanan, N.Y.: Emerson Books, 1972; Martin Gardner, *Mathematical Puzzles,* New York: Harper & Row, 1961. (see also, Gardner's regular monthly column in *Scientific American*); J. F. Hurley, *Litton's Problematical Recreations,* New York: Van Nostrand-Rinehold, 1971; B. A. Kordemsky, *The Moscow Puzzles,* New York: Scribners, 1972; and C. F. Linn, *Puzzles, Patterns and Pastimes,* Garden City, N.Y.: Doubleday, 1969.

PREPARATION OF ELEMENTARY SCHOOL TEACHERS

We said in the Preface to the first edition of this book

It is our belief, not shared (we are sad to say) by all educators, that a course developed along the lines of this book would be an excellent preparation for an elementary school teacher. We feel that our text, in its emphasis on lively problems and its attention to those mathematical concepts judged to be essential knowledge for all educated people, offers an attractive alternative to the dreary routine involving sets, distinguishing between numbers and numerals, and the associative law of addition—the usual fare in texts designed for teachers.

Since that paragraph was written, the mood has shifted. Problem solving has been elevated to a place of central importance in the teaching of mathematics, and some schools have adopted our text for their teacher education course. May their numbers increase!

A WORD ABOUT THE TITLE

We chose the title *Faces of Mathematics* for two reasons. First, we wanted to emphasize the fact that mathematics was developed by human beings, real people with real faces. True, they may have had special talents, but on the whole they lived their lives subject to the same constraints as anyone else. Results in mathematics do not arise through divine revelation; they represent the hard work of individual men and women. The faces and brief biographies of many of the most significant contributors to this field appear in this book.

Second, we wanted to suggest the analogy that mathematics is like a finely cut diamond; it must be seen from several sides to be fully appreciated. Each view exposes a new face with its own distinctive features. Four of these faces—solving problems, finding order, building models, and creating abstractions—reflect those activities most characteristic of mathematicians. We have organized our book around these four faces.

A SPECIAL WORD TO STUDENTS

Many years of teaching have convinced us that most students who fall within this book's intended audience approach mathematics with fear and trembling. We have made every effort to ease this anxiety by using simple examples, clear explanations, and a limited technical vocabulary. Our aim is to demonstrate that mathematics is interesting, relevant, and learnable.

We believe that problem solving is the heart of mathematics, and that you must try problems—many problems—if you are to understand the subject. Most sections begin with a problem; every section ends with a host of problems for you to try. They are carefully arranged in order of increasing difficulty, the most challenging being identified with an asterisk. Be sure to work at the problems; it is the only way to learn mathematics. It is also the activity most likely to help you later in life.

ADVICE TO TEACHERS

This text can be used in a variety of ways. The book contains sufficient material for a full-year (two-semester) course. It is also easy to make selections for the typical semester course offered at many colleges. Both of us have used a preliminary version of the book in one-semester courses. Professor Varberg's course,

which emphasized problem solving, was based on Chapters 1,
2, 3, 4, 5, 8, 9, and 13. Professor Roberts's course was more
philosophical, with special attention given to clear thinking and
precise writing. He used most of Chapters 1, 2, 3, 10, 11, 12, 13,
14, and 15. There are many other possibilities. The Dependence
Chart, that follows the Preface, will help you design a course to
your liking.

CHANGES IN THE SECOND EDITION

Second editions give authors a second chance at trying to
explain things that readers found difficult in the first edition. We
have made a few changes of this type, though users of the first
edition were kind to us in this regard.

The most obvious change is that much of the review material
on fractions and decimals has been transferred from the appendix
of the first edition into a new Chapter 2 in the second edition.
Also incorporated in Chapter 2 is material on the number system
which had been in Chapter 12 of the first edition. The new
Chapter 2 thus provides a comprehensive review of numeric
methods. In responding to the users who felt that this review
should be in the body of the text, we tried still to avoid having
our readers feel that they've seen all this before. We retained the
emphasis on problems and puzzles, and we decided to use
calculators as a tool for teaching. No teacher or text is so
unforgiving as a calculator in requiring that a student learn to
perform arithmetic calculations in the correct order, and students
—along with other people—do love to punch buttons.

We believe that the introduction of material on calculators is
consistent with the goal of using something with natural interest
to teach underlying concepts. We have been careful, however,
to keep the book organized so that the teacher of a well prepared
class can skip Chapter 2.

There are new problems following many sections of the text.
Many of them, identified with a [c] in the margin, require that the
student use a calculator to get a solution. In Part III of the text,
where students need to digest more mathematical prose than
they are accustomed to reading, Self-Tests have been provided
at the end of each section.

ACKNOWLEDGMENTS

Most of the art work done by Stan Olson for the first edition has
been retained, and it seems appropriate that we once again
acknowledge his work. We also pointed out in our first edition
that in typing our entire manuscript, Idella Varberg demonstrated
that some marriages can withstand, in a spirit of continuing good

humor, the strain of working closely on a project of this scope. It is a pleasure four years later to report that the marriage endures, and her work as typist has continued.

This edition has afforded us the pleasant opportunity to work with Ted Ricks, who encouraged this second edition at all the crucial places, Eleanor Castellano, who has been very diligent in attending to all the details of production, and the entire staff of Harper & Row. Our thanks go to them.

A. Wayne Roberts
Dale E. Varberg

Dependence Chart

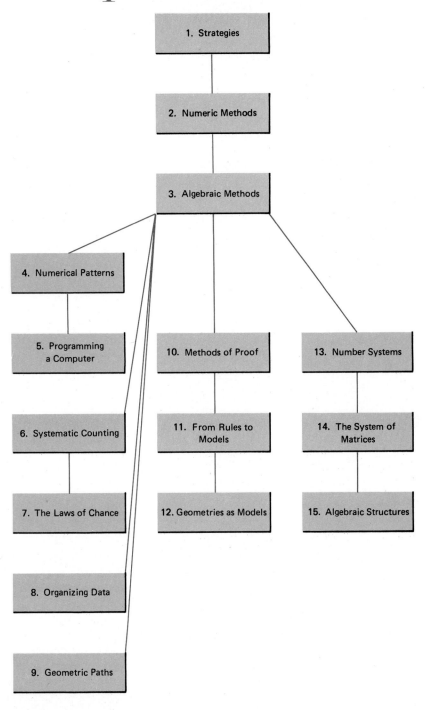

A great discovery solves a great problem but there is a grain of discovery in the solution of any problem. Your problem may be modest; but if it brings into play your inventive faculties, and if you solve it by your own means, you may experience the tension and enjoy the triumph of discovery.
George Polya

Part I
Solving Problems

PROBLEMS

There was a time when almost everyone associated elementary mathematics with long lists of problems to be solved: theoretical problems and practical problems, problems requiring long computations and problems beautiful in their simplicity, many monotonously simple drill problems and a few utterly baffling problems. To some, mathematics seemed nothing more than a collection of memorized tricks which could, with luck, be matched to the problems they were designed to solve.

In an effort to get away from this view, the so-called new math was developed to emphasize the structure and unity of mathematics. The intention was laudable, and in certain ways successful, but when pushed too far, this approach too became pedantic. One feels the need to learn abstract principles only when one has worked on concrete problems. Skeletons are wonderfully useful, but it is easier to sell pictures of those that are covered with meat. Problems are the meat of mathematics and the focus of this book.

Every subject has its problems. In contrast, however, to many useful areas of human inquiry (medicine, psychology, economics, etc.) where a clear and enduring answer is seldom expected, mathematical problems admit the possibility of uncontestably correct answers. They therefore afford us an excellent medium in which we can focus attention not on the answers but on how they are obtained. In Part I, we undertake such a study, suggesting that there are principles applicable to solving a host of common problems.

It is not essential to our purposes to consider only practical problems. What is essential is that our problems illustrate the principles we have in mind, that they be interesting, that they pose a challenge—yet seem enough within grasp to be tantalizing—and that they draw out from our imagination creative ideas about which we are pleased to say, "I thought of that."

George Polya (1887–   )

George Polya was born in Hungary and educated at the universities of Budapest, Vienna, Göttingen, and Paris. After teaching for 26 years at the Swiss Federal Institute of Technology in Zürich, he became affiliated with Stanford University where he has continued his research in advanced mathematics, research that has resulted in over 200 papers and several books.

Polya's research in pure mathematics has earned him a place of honor among the world's leading contemporary mathematicians. But he is also famous for the research and writing he has done on the nature of problem solving. His books, *How to Solve It,* 2nd ed. (Garden City, N.Y.: Doubleday, 1959), *Mathematics and Plausible Reasoning* (2 vols.) (Princeton, N.J.: Princeton University Press, 1954), and *Mathematical Discovery* (2 vols.) (New York: Wiley, 1962) are widely read expositions of the art of solving problems. We are pleased to acknowledge that the ideas we express in Part I have been profoundly influenced by reading Polya's books.

Chapter 1
Strategies

Solving a problem is similar to building a house. We must collect the right material, but collecting the material is not enough; a heap of stones is not yet a house. To construct the house or the solution, we must put together the parts and organize them into a purposeful whole.
GEORGE POLYA

Whose clock keeps the best time?

Mine does. It doesn't run at all.

Mine does. It loses 20 seconds an hour.

Mine does. It gains 8 minutes per day.

HOMER

HUGO

HORACE

1.1 / Clarify the Question

Before you try to answer a question, you should have some idea of why the question was asked. Before you try to work out a solution, you should be sure that you're not working on the wrong problem. Before you try to master the complexities of a problem, you should discard facts that are irrelevant to the question to be answered. We would hesitate to mention things obvious, were it not for the fact that they are so easily and so often overlooked.

WHAT IS THE REAL QUESTION?

Sometimes a question is phrased in an ambiguous way. Then our first job is to try to make sense out of it. Take the question above, for example. Does the questioner wish to know which clock is correct most often? That's Homer's—right twice a day, with no maintenance. If the questioner wishes to know which clock runs closest to the correct rate, then either Hugo's or Horace's will do; both vary from the correct time by 8 minutes a day. But if the goal is to have a clock that, when correctly set in the morning, will most reliably get its owner to appointments on time, then Hugo's is ahead in the running.

DON'T IMPOSE CONDITIONS THAT AREN'T THERE

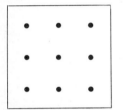

There was a blind beggar who had a brother, but this brother had no brothers. What was the relationship between the two? If you have any trouble answering this question, it is because you are imposing a condition that isn't there. Try again.

Here is another problem: Without taking your pencil off the paper, draw four straight line segments that will pass through the nine dots at the left. Again, if you have trouble, it is because you are unconsciously assuming something not stated in the problem.

Three points can be arranged so that the three distances they determine are equal. Can four points be arranged so that the six distances they determine are equal? If your answer is no, you have once again imposed a limiting assumption.

In a regulation nine-inning baseball game which was not cut short prior to completion, the pitcher for the Tooterville Toads threw the minimum number of pitches possible for a pitcher who goes all the way. How many pitches did she throw? You struck out if you thought 27 was the correct answer.

It is quite common for us to make unconscious assumptions which then lead us to work on a problem quite different from the one given to us. The problems above are interesting because people often assume, in turn, that all blind beggars are male, that line segments cannot extend beyond the square region containing the given points, that the four points all have to be in the same plane, and that such an outstanding pitching performance would surely be rewarded with a game won. (No, the answer is not 24, either, though that is closer.)

Sex makes the difference.

Sex makes no difference.

STRIP THE PROBLEM OF IRRELEVANT DETAILS

Problems, especially those from real life, often come to us cluttered by details which may have little to do with finding a solution. Then our first job is to pare away the extraneous trappings so as to expose the core of the matter. What is the problem in its peeled, undisguised form? This is a question we must always ask.

A fast 10-car express train left sunny New York at precisely 12:01 P.M. on Monday, January 2, traveling at 100 kilometers per hour and heading for Boston. One hour later, in perfect weather, a slow-moving freight train loaded with coal left Boston for New York at 40 kilometers per hour. At 5:19 P.M., in a blinding snowstorm, the two trains crashed head on and both engineers were killed instantly. Which train was farther from New York at the time of the crash?

There is hardly a piece of information in the whole paragraph that has anything to do with the question asked. Peeled right down to the core, the question is this: If two objects (trains) are at the same spot, which of them is farther from another spot (New York)? And the answer of course is that they are equally far away.

That was really too easy. See if you can get to the bottom of this brainteaser: Consider two cylindrical glass containers, one holding 2 liters of water and the other 1 liter of red wine. One milliliter of water is transferred to the container of wine and mixed in thoroughly. Then 1 milliliter of the mixture is transferred back to the container of water. Is there now more wine in the water or water in the wine?

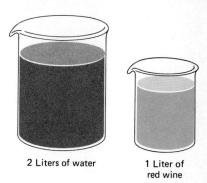

2 Liters of water   1 Liter of red wine

Everyone sees that the redness of the wine and the shape of the containers are irrelevant. But it may take more thought to realize that the amounts of water and wine we start with, the amounts transferred (provided they are equal), and the thoroughness of the mixing are also irrelevant. The only thing that really counts is that we end up with as much liquid in each container as we had at the beginning. Then it is clear that the water removed from the container of water has been replaced by wine, and the wine removed from the container of wine has been replaced by water. Thus there is exactly as much wine in the water as water in the wine.

Now ask yourself the same question if 100 two-step transfers were made.

SUMMARY

In this chapter, we are not trying to teach a mathematical technique applicable to a certain class of problems; we are trying to suggest patterns of thought that can be useful in any problem-solving situation. It follows that the problems we consider here are different from the ones usually encountered in mathematics texts (and the ones you will encounter in subsequent chapters of this text). They do not yield to a single method. They do call for imagination, perseverance, and clear thinking.

Don't expect to solve all the problems in Problem Set 1.1, or maybe even a majority of them. We didn't get them all the first time we tried either. Do take satisfaction in being able to solve some of them. Do express appreciation for the insight of those who "see" solutions that you didn't. Do let the overall impact of the problems be this: When correctly approached, difficult problems sometimes turn out to be easy.

Speaking of problems that most people don't get right off the bat reminds us to tell you that the pitcher of a losing team needs to pitch only eight innings, so she must throw 24 balls to get the necessary "outs," and one more to account for the run by which she loses.

Can we give any advice to the person who finds himself or herself getting stuck on almost everything? The best advice is probably this: Don't try to solve a problem too quickly. Force yourself to go through the steps, simple as they seem, discussed in this section. Do read the question over and over until you are certain that you understand what is being asked. Do consider carefully whether or not you are imposing restrictions not stated in the problem. Do try to list the essential information, omitting what is irrelevant.

If, after a real effort, you are still stumped, note that hints for solving a good many of the problems are given following the list of problems. Since it is hard to give hints that do not immediately ruin the problem, we encourage you to refer to the hints only as a desperation measure. Rather than solving many problems by using the hints, you should, in the spirit of our opening comments, preserve for yourself in as many cases as possible the right to say of a solution, "I thought of that."

PROBLEM SET 1.1

1. If there are 12 one-cent stamps in a dozen, how many two-cent stamps are there in a dozen?
2. Homer had seven apples and ate all but three. How many were left?
3. A chemist discovered that a certain reaction took 1 hour and 20 minutes when he wore a blue tie, but only 80 minutes when he wore a red tie. Why the difference?
4. Homer went to bed at eight o'clock one evening after setting the alarm to go off at nine o'clock the next morning. How many hours of sleep did he get?
5. Five apples are in a basket. How can you divide them among five girls so that each gets an apple but one apple remains in the basket?
6. Which is worth more: 1 pound of $10 gold pieces or $\frac{1}{2}$ pound of $20 gold pieces?
7. A ship stands anchored offshore with a rope ladder hanging over its side. The ladder has 12 rungs, and the distance between the rungs is 12 inches. At 4:00 P.M. the ocean is calm, and the bottom rung just touches the water. But then the tide begins to come in, raising the water level 3 inches every hour. When will the water just cover the third rung from the bottom?
8. An apartment house has six stories, each the same height. How many times as long will it take me to run up the stairs to the sixth floor as it does to run up to the third floor? Assume that I start at the first floor.
9. Homer always buys the same style of socks, choosing either green or blue. He knows that there are 28 pairs of socks in his drawer, but in

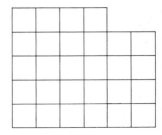

the dark he cannot tell colors. How many socks should he take out to the light to be sure he has a match?

10. Three kinds of apples are mixed in a basket. How many apples must you take out to be sure of getting at least two of one kind? At least three of one kind?

11. If 3 hens lay 3 eggs in 3 days, how many eggs will 300 hens lay in 300 days?

12. If 3 cats catch 3 rats in 3 minutes, how many cats will catch 100 rats in 100 minutes?

13. Grandma Uphill makes tasty French toast in a pan which holds only two slices. After browning one side of a slice, she turns it over. Each side takes 30 seconds. How can she brown both sides of three slices in 1½ instead of 2 minutes?

14. Trace and then cut the figure in the margin to make two identical pieces. Make your cuts along the lines.

15. Without raising your pencil from the paper, draw five straight line segments connecting all 12 of the displayed dots and make sure you end where you started.

16. Without raising your pencil from the paper, draw six line segments connecting the 16 dots shown in the margin. Can you do this in such a way that you end where you started?

17. In a rectangular room, how can you place 10 small tables so there are 3 on each wall?

18* Zippo is an intercity messenger service which provides 1-day delivery for any box, the maximum dimension (length, width, or height) of which does not exceed 15 inches. The architectural firm of Woff and Toppel wishes to send an acetate drawing which cannot be folded but can be rolled. The drawing is 40 inches by 24 inches. How can they use the Zippo service?

19* A man walks 1 mile south, 1 mile west, and then 1 mile north, ending where he began. Where did he start his journey? Is there more than one answer?

20* Three pennies can all be placed in contact as indicated in the diagram. Four pennies can be placed so that each touches all the others by laying the fourth on top of the other three. How can you place five pennies so that each penny touches the other four?

21* Homer has five pieces of chain which he wants to join into a long chain (see diagram below). He could open ring 3 (first operation), link it to ring 4 (second operation), then open ring 6 and link it to 7, etc. In this way he would complete the job in eight operations, but it can be done in six. How?

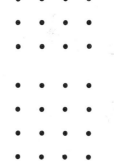

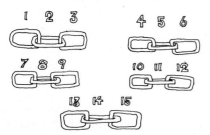

22.* Hugo has six pieces of chain, each with four links. It costs one cent to cut a link and two cents to weld one closed. How can the six pieces be fashioned into one chain for as little money as possible?

Hints to Selected Problems

2. Four is wrong. Read the question again—carefully.
4. Thirteen is an unlucky guess.
5. Don't add conditions not stated.
7. Ships float.
9. How many socks can he take out *without* getting a match?
10. How many apples can you take out that are all different?
11. How many eggs do three hens lay in 300 days?
12. How many rats can three cats catch in 1 minute?
15. Start drawing horizontally from a point 2 units to the left of the upper left dot.
17. Can a table be on two walls?
18. This is a cockeyed problem.
19. If you thought of the North Pole, fine. But don't stop with that. There are many other answers.
21. It's as easy as 1, 2, 3.

1.2 / Organize the Given Information

A lone goose was flying in the opposite direction from a flock of geese. He cried: "Hello, 100 geese!" The leader of the flock answered: "We aren't 100! If you take twice our number and add half our number, and add a quarter of our number, and finally add you, the result is 100, but . . . well, you figure it out."

The one goose flew on, but could not find the answer. Then he saw a stork on the bank of a pond looking for frogs. Now among the birds the stork is the best mathematician. He often stands on one leg for hours, solving problems. The goose descended and told his story.

The stork drew a line with his beak to represent the flock. Then he drew a second line of the same length, a third line half as long, another line a fourth as long, and a very small line, rather like a dot, to represent the goose.

"Do you understand?" the stork asked.

"Not yet."

The stork explained the meaning of the lines: The first and the second represented the flock, the third half the flock, the next a quarter of the flock, and the dot stood for the goose. He rubbed out the dot, leaving lines that now represented 99 geese. "Since a flock contains four quarters, how many quarters do the four lines represent?"

Slowly the goose added $4 + 4 + 2 + 1$. "Eleven," he replied.

"And if 11 quarters make 99 geese, how many geese are in a quarter?"

"Nine."

"And how many in the entire flock?"

The goose multiplied 9 by 4 and said:

"Thirty-six."

"Correct! But you couldn't get the answer yourself, could you? You . . . goose!"

This delightful story comes from Russia where it has been told in homes and schools for over 50 years. (Boris Kordemsky, *The Moscow Puzzles,* Scribner, New York, 1972, p. 95.) We tell it to make a point. Always try to organize the given information in a systematic graphic way.

DRAW A PICTURE

Hugo Hardback, the librarian, has in his collection of rare books a Greek lexicon which was published in four volumes, each having exactly the same number of pages. Unfortunately, the volumes have stood lo these many years, side by side, volumes I, II, III and IV, without ever having been used. In fact, a bookworm has bored a hole straight through the set, from the first to the very last page, taking care, however, not to betray its presence by boring through the outside covers of the first and last volumes. The pages in each volume measure 2 inches from first to last, and each cover is $\frac{1}{8}$ inch. Assuming that the bookworm has accomplished its feat with the shortest tunnel possible, how long is the tunnel?

This problem is tricky, and without a picture we could easily dig in too far. Even with a picture, you should hold up a book to convince yourself that, when shelved, the first page is adjacent to the right cover of volume I and the last page is next to the left cover of volume IV. Once we have the picture shown at the left, our problem is easy. You finish it.

MAKE A DIAGRAM

Each day at noon a ship leaves New York for LeHavre, France, and another ship leaves LeHavre for New York. The trip across the Atlantic Ocean takes six full days. How many ships will a ship leaving New York today meet during the journey to LeHavre?

If you answer 6, you are forgetting about the ships already enroute. The diagram below clarifies matters. A ship leaving New York today will encounter 11 ships on the high seas. If we include the ships it meets in each of the harbors, the answer is 13.

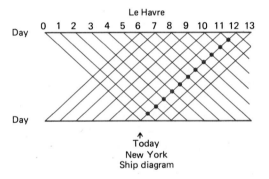

Ship diagram

KEEP A SYSTEMATIC RECORD

In how many ways can you add eight odd numbers together to get 20? Numbers may be repeated, but changes in order are not to be counted as new solutions.

This type of problem demands a system. If one looks for solutions in a random way, he or she is almost bound to miss some.

As in packing a suitcase, it is wise to start with the largest items and work down. Now, we can't use 19, 17, or 15 in the solution. They use up too much space—it's impossible to add seven more odd numbers and still get 20. But 13 works:

$$13 + 1 + 1 + 1 + 1 + 1 + 1 + 1 = 20$$

Next we try 11, then 9, etc.:

$$11 + 3 + 1 + 1 + 1 + 1 + 1 + 1 = 20$$
$$9 + 5 + 1 + 1 + 1 + 1 + 1 + 1 = 20$$
$$9 + 3 + 3 + 1 + 1 + 1 + 1 + 1 = 20$$
$$7 + 7 + 1 + 1 + 1 + 1 + 1 + 1 = 20$$
$$7 + 5 + 3 + 1 + 1 + 1 + 1 + 1 = 20$$
$$7 + 3 + 3 + 3 + 1 + 1 + 1 + 1 = 20$$
$$5 + 5 + 5 + 1 + 1 + 1 + 1 + 1 = 20$$
$$5 + 5 + 3 + 3 + 1 + 1 + 1 + 1 = 20$$
$$5 + 3 + 3 + 3 + 3 + 1 + 1 + 1 = 20$$
$$3 + 3 + 3 + 3 + 3 + 3 + 1 + 1 = 20$$

Note that we have placed the bigger numbers to the left. It gives us a systematic way of writing down all 11 solutions.

SUMMARY

When a problem has many details, it is hard to keep track of them in our minds. A list, chart, or diagram may bring these details into clearer focus and help us perceive relations between them. And if a picture can be drawn that somehow represents the problem, draw it. This is a maxim that great problem solvers follow religiously.

PROBLEM SET 1.2

1. Fourteen clothespins are placed on a line at 7-foot intervals. How far is it from the first to the last?

2. At six o'clock the wall clock struck six times. Checking my watch, I noticed that the time between the first and last strokes was 30 seconds. How long will the clock take to strike twelve midnight?

3. Each hour, on the hour, a bus leaves Dallas for Houston and another bus leaves Houston for Dallas. The trip takes 5 hours. How many buses will a bus leaving Dallas at 10:00 A.M. meet on its way to Houston?

4. After hearing a gloomy Sunday sermon on the danger of smoking, Hugo promised himself to cut back 2 cigarettes a day, smoking 2 less on Monday than he had on Sunday, etc. He kept his promise for a week, that is, through Saturday. During the whole week he smoked 63 cigarettes. How many did he smoke the day of the sermon?

5. A lazy and rather careless frog fell into a cistern 21 feet deep. It was the morning of July 4, and it may have been a firecracker that did it. After thinking over its plight, the frog started to climb and made 3 feet by nightfall. Next morning it discovered that it had slid back 1 foot during its sleep. Satisfied with this pattern (3 feet upward progress during the day, 1 foot downward slide at night), the frog worked its way up the side of the cistern. On what date did it get out?

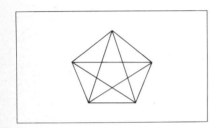

6. I live in a place where the temperature goes up sharply during the day and down at night. This affects my watch. I notice that it gains $\frac{1}{2}$ minute during the day but then loses $\frac{1}{3}$ minute during the night; thus it is $\frac{1}{6}$ minute fast for a 24-hour period. One morning—July 1—my watch showed the right time. On what date did it first show 5 minutes fast?

7. How many triangles are there in the picture at the left?

8. Amy plans to invite six of her friends (call them A, B, C, D, E, and F) to lunch on five different days. She has three small tables at which only two people can sit. How can she arrange it so that no two people ever sit together twice? You are to assume that Amy never sits down.

9. A detachment of 10 soldiers must cross a river. The bridge is broken, and the river is deep. What should they do? Suddenly the sergeant in charge spots two boys playing in a little rowboat near shore. The boat is so tiny that it can hold only two boys or one soldier. Still the soldiers manage to cross the river. How?

10. A man has to take a wolf, a goat, and some cabbage across a river. His boat has room enough for the man (who must row) plus either the wolf or the goat or the cabbage. If he takes the cabbage with him, the wolf will eat the goat. If he takes the wolf, the goat will eat the cabbage. Only when the man is present are the goat and the cabbage safe from their enemies. Nevertheless, the man carries the wolf, goat, and cabbage across the river. How?

11. A chemist has two identical test tubes, each able to hold 50 milliliters of liquid. In a beaker she has 48 milliliters of solution, and she wishes to measure off exactly 42 milliliters. How should she proceed?

12. In how many ways can you add six even numbers (bigger than zero) to get 20? Changes in order are not counted as new solutions.

13. In how many ways can you add five positive whole numbers to get a sum of 11? Changes in order are not counted as new solutions.

14. If there are nine people in a room and every person shakes hands exactly once with each of the other people, how many handshakes will there be?

15. A railroad line runs straight from Posthole to Podunk. Center City is on this line, exactly halfway in between. Klondike is just as far from Posthole as it is from Center City, and Center City is as far from Klondike as it is from Podunk. All cities are connected by direct railroad lines. If it is 6 miles from Posthole to Klondike, how far is it from Klondike to Podunk?

16. I have already covered one-third of the distance from Podunk to Boondocks, and after I walk another kilometer I'll be halfway there. How far is it from Podunk to Boondocks?

17. The new pastor at First Lutheran Church plans to post the page numbers for three hymns at each Sunday service. To do it, he must buy plastic cards each with one large digit on it, but obviously he wishes to minimize the number of cards to be bought. The hymnal has hymns numbered from 1 to 632. How many cards must he buy to make sure that any selection of the three hymns is possible?

18. We expected you to assume that 6's and 9's required different cards

in Problem 17. Suppose 6's can be turned upside down to form 9's. Then how many cards are required?

19. What is a plausible basis for the order in which the following 10 digits are arranged?

$$8 - 5 - 4 - 9 - 1 - 7 - 6 - 3 - 2 - 0$$

20.* What is the sum of all the counting numbers from 1 to 100? What is the sum of the *digits* in all the counting numbers from 1 to 100?

21.* Of the members of three athletic teams at a certain school, 21 are on the basketball team, 26 on the baseball team, and 29 on the football team. Fourteen play baseball and basketball, 15 play baseball and football, and 12 play football and basketball. Eight are on all three teams. How many team members are there altogether?

22.* A pet store offered a baby monkey for sale at $1.25. The monkey grew, and the next week it was offered at $1.89. Not one to monkey around, the shopkeeper subsequently raised the price to $5.13, then to $5.94, and next to $9.18. Finally, during the sixth week, an organ grinder bought the monkey at $12.42. How were the prices figured?

23.* Andrew Algaard was born on April 1, 1863, and died on May 25, 1950. How many days did he live? (Don't forget leap years.)

Hints to Selected Problems

1. Draw a picture; try the problem first with just three clothespins.
3. See the LeHavre–New York ship problem in the text.
5. Make a diagram showing the position of the frog each morning.
7. You will have to find a systematic way of counting. How many triangles are not subdivided into smaller triangles? How many are subdivided into exactly two smaller triangles, etc.?
9. A picture may help you get started.
11. As a first step, divide the liquid equally between the two test tubes (most test tubes are transparent, so this is usually possible). Keep track of each step.

	Beaker	Tube A	Tube B
Start	48	0	0
Step 1	0	24	24
Step 2			

14. Label the people A, B, C, . . . , I. Let A shake hands with B, C, . . . , I. Then let B shake hands with C, D, . . . , I, and so on.
15. Draw a good picture.
17. First figure out how many 0's are needed, then how many 1's, then 2's, and so on.
19. Eight, five, four, etc.
21. Draw three overlapping circles, each circle representing the players from one of the teams as in the margin. Put appropriate numbers in each of the seven regions.
22. Count pennies; the increases were 64, 324, 81, 324, and 324. Make a chart showing prices and increases.

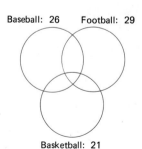

Baseball: 26   Football: 29

Basketball: 21

Teams A, B, and C participated in a track-and-field meet in which each team entered one contestant in each event. Points were awarded in each event for a third-place finish, more points for second-place finish, and still more for finishing first. At least three events were held. A won the meet with a total score of 14 points, while B and C tied with 7 points each. B won first place in the high jump. Who won the pole vault, assuming no ties occurred in any of the events?

1.3 / Experiment, Guess, Demonstrate

Our first hurdle is to get over the shock of being expected to answer such a question. To get off the starting block, suppose we assume that there are exactly three events, and that first, second, and third places are rewarded with 3, 2, and 1 points, respectively.

Now let's take the advice in Section 1.2 and organize the meager information we have.

| | Team | | |
Event	A	B	C
High jump		3	
Pole vault			
Mystery event			
TOTAL	14	7	7

We notice almost immediately that our guess must be wrong, since there is no way that team A could have accumulated 14 points; the most possible would have been 8.

We could just guess again—trying four events or five events; or we could assign more points in each event. And if nothing else occurred to us, this is the course of action we would probably follow.

But actually, there is more to observe from our table than the mere fact that we are wrong. The teams earned a total of $14 + 7 + 7 = 28$ points. We asked, How many points were awarded after each event? Our first guess was $3 + 2 + 1 = 6$. That would mean there were $\frac{28}{6} = 4\frac{2}{3}$ events—which is ridiculous. So we are now conscious of something that should have occurred to us in the first place:

$$\binom{\text{Number of points awarded}}{\text{after each event}} \text{ times } \binom{\text{number of}}{\text{events}} \text{ is } 28$$

If we had noticed this at the beginning, we never would have tried guesses of 6 and 3.

We know that a certain number of points was awarded for third place, more for second, and still more for first. The total therefore must be *at least* $1 + 2 + 3 = 6$, but we know that 6 does not work. Let's try 7, which would mean there were four events. Moreover, a little experimenting with groups of three positive integers that add up to 7 quickly convinces us that $7 = 4 + 2 + 1$ is the only possibility; 4 points were awarded for first place, 2 for second, and 1 for third. Our table now takes a slightly different form.

| | Team | | |
Event	A	B	C
High jump		4	
Pole vault			
Event 3			
Event 4			
TOTAL	14	7	7

Again we ask if there is any way that A could have obtained 14 points. This time there is—just one. Team A would have had to take first in all events save the high jump, and there it would have had to take second. But that hurdle we didn't have to jump. Team A won the pole vault.

Since we have reached our conclusion on the basis of an experiment, two questions naturally occur. Is our present guess consistent with the rest of the information (can we make assignments of the points available so that B and C each have 7 points)? Second, is there any other set of trial values that is consistent with the given information? When you have checked these out, you will most certainly agree that team A won the pole vault.

ELIMINATE WRONG ANSWERS

If the number of possible answers is reasonably small, one can often find the correct one by a systematic process of eliminating the wrong ones. (This procedure is a favorite of students taking multiple-choice exams.) Here is a well-known puzzle in which this process ultimately yields the solution:

A farmer's wife drove to town to sell a basket of eggs. To her first customer she sold half her eggs and half an egg. To the second she sold half of what she had left plus half an egg. And to a third she sold half of what she then had and half an egg. Three eggs remained. How many did she start with? She did not break any eggs.

Now, two things are clear. The answer is an integer considerably greater than 3, and yet it can't be very large (baskets hold a finite number of eggs). We could start by trying some reasonable number, say 17, and then 18, 19, 20, etc., eliminating wrong answers along the way. Eventually, we reach 31 and, lo, it works. To her first customer she sold $\frac{31}{2} + \frac{1}{2} = 16$, to her second she sold $\frac{15}{2} + \frac{1}{2} = 8$, and to her third $\frac{7}{2} + \frac{1}{2} = 4$, leaving 3 eggs unsold.

BEGIN WITH A SIMPLE CASE

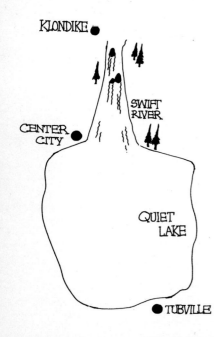

Center City is right at the mouth of Swift River where it flows into Quiet Lake. Upstream a few miles is the thriving town of Klondike, while across the lake and the same distance away is the village of Tubville. Horace proposes a race with his good friend Hugo. Horace will row upstream to Klondike and return. Hugo will row across Quiet Lake to Tubville and return. Assuming that they are equally good oarsmen and that they start from Center City at the same time, who will win?

Since Horace will be helped by the current in one direction and hindered in the other, it is tempting to guess that he and Hugo will arrive back at Center City in a dead tie. That's a good guess, but it happens to be wrong. It gives us a chance to emphasize the third word in our section title: *demonstrate*.

Our problem is notable for its lack of information. We don't know how far it is between the cities, how fast the men can row, or the speed of the current in Swift River. Can it be that none of these things make any difference?

Let's begin with a concrete, simple case. Suppose that it is 12 miles from Center City to Klondike, that the men can row 6 miles per hour (mph), and that Swift River flows at 2 mph. Then Horace will make $6 - 2 = 4$ mph upstream, and he will reach Klondike in 3 hours. Downstream he will make $6 + 2 = 8$ mph. It will take him $1\frac{1}{2}$ hours, or a total of $4\frac{1}{2}$ hours for the roundtrip. In the meantime, Hugo, rowing at a steady 6 mph in Quiet Lake, will make the 24-mile roundtrip in 4 hours flat.

This is enough to eliminate our original guess. Is it enough to declare Hugo the winner? No. Let's try a different set of data. Suppose that the two oarsmen can row at only 4 mph and that the other information is as before. Now Horace can go only 2 mph upstream. It will take him 6 hours to reach Klondike. On the return he will make 6 mph, and it will take 2 hours. This gives a total of 8 hours altogether. Meanwhile, Hugo will need only $\frac{24}{4} = 6$ hours to make the whole trip. He will be back at Center City by the time Horace gets to Klondike.

Now we begin to see what happens in general. It will certainly take Horace longer to row upstream than downstream. Thus the current will hinder him for a longer time than it will help him. He can't possibly compensate for the time lost going upstream on the return trip, and so he will always lose.

Of course, there is a nice algebraic demonstration of this fact. But that is the subject of Chapter 3.

SUMMARY

When faced with a problem that has many potential solutions, experiment a little. If you work in a systematic way, you may find the answer rather quickly. At the very least, you may observe a pattern that suggests the correct solution. If so, make a conjecture. State it boldly. But don't forget to check it carefully. Mathematicians are not satisfied until they have demonstrated that a solution is correct.

PROBLEM SET 1.3

1. In the barnyard is an assortment of chickens and pigs. Counting heads I get 13; counting legs I get 46. How many pigs and how many chickens are there?

2. In a certain family, each boy has as many sisters as brothers, but each sister has only half as many sisters as brothers. How many brothers and sisters are there in the family?

3. On a 24-item true-false test, Professor Witquick gave 5 points for each correct response but took off 7 points for each wrong one.

	1	
3		
4		2

Homer answered all the questions and came up with a big fat zero for his score. How many did he get right?

4. Place the numbers 5 through 9 in the vacant squares at the left so that every row, horizontal, vertical, or diagonal, has the same sum.

5. With a pile of about 1000 identical cubical bricks, a cubical monument is to be built, and it is to stand on top of a square plaza constructed from the same pile of bricks. There are to be equal numbers of bricks in the monument and in the plaza. What is the largest monument that can be built?

6. Move only three coins from the first arrangement to produce the second.

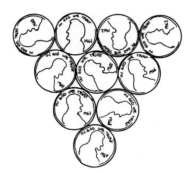

7. The odometer of the family car showed 15,951 kilometers. The driver noticed that this number was palindromic; that is, it read the same backward as forward. "Curious," the driver said to herself. "I imagine it will be a long time before that happens again." But 2 hours later, the odometer showed a new palindromic number. How far had the car gone during the 2 hours?

8. Notice that $(1 + 2)(3 + 4 + 5 + 6 + 7 + 8 + 9) = 126$. Place mathematical signs between all the following digits to make the equation correct:

$$1\ 2\ 3\ 4\ 5\ 6\ 7\ 8\ 9 = 100$$

9. Assume the earth to be a perfect sphere, and that a rope has been tightly tied around the equator. Snip the rope, insert a 6-foot section of extra rope, and take up the slack by pulling the now lengthened rope into a circle concentric with (but slightly larger than) the equator. Is the extra room so obtained enough for an ant to slip through? A mouse? A cat? How about you, good reader, could you slip through? (The circumference of the earth is approximately 25,000 miles.)

10. A ball of yarn of a certain size will make one mitten. If its diameter were doubled, how many mittens would it make?

11. Homer and Hugo started together to go from Pitstop to Posthole. Homer ran half of the time and walked half of the time. Hugo ran half of the distance and walked half of the distance. Assuming they ran and walked at the same rate, who got to Posthole first?

12. Rodney Roller drives 20 miles to work each morning. The first 10 miles are through heavy traffic and he averages 20 mph. The second 10 miles are further out of the city where he averages 40 mph. What is his average speed for the whole trip?

13. Three hockey teams, A, B, and C, play a three-game tournament. We have the following information:
    a. Each team played two games.
    b. A won two games; B tied one game.
    c. B scored a total of two goals; C scored three goals.
    d. One goal was scored against A, four against B, and seven against C. Find the scores of all three games.

14. For a series of games involving five hockey teams, we have the following partial information. Fill in the rest of the table, figure out who played whom, and determine what the score was for each game.

	Played	Won	Lost	Tied	Goals for	Goals against
A	3	2	0		7	0
B	2	2	0		4	1
C	3	0	1		2	4
D	3	1	1		4	4
E	3	0			0	

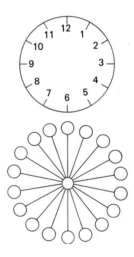

15. By drawing straight lines across the face of a clock, can you divide it into three regions so that the sums of the numbers in the three regions are equal? Four regions? Six regions?

16. Write the numbers 1 through 19 in the circles shown so that the numbers in any three circles on a straight line total 30.

17. A Hamline graduate attended a recent class reunion and sent in the following account. "I met 15 former classmates. More than half were nurses, the rest were lawyers. Of the nurses, most were females and there were still more female lawyers. Both of the latter facts were true even if you included me. My friend, a noted lawyer, left his wife at home. What conclusion can you draw about me?"

18. Consider the arrangement of coins shown in the margin. We wish to interchange the position of the penny and the nickel by moving one coin at a time to an adjoining open space. Can it be done and if so how?

19.* Show that, if every student at a certain college has at least one friend there, at least two students have the same number of friends.

20.* Discover a pattern in the diagram below. Make a conjecture. State it clearly.

$$
\begin{aligned}
1 &= 1 \\
1+3 &= 4 \\
1+3+5 &= 9 \\
1+3+5+7 &= 16
\end{aligned}
$$

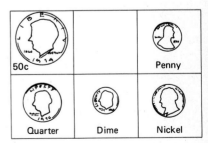

21.* A coin collector has 12 coins that look alike, but one is counterfeit and has a weight different from that of the other 11. Her scale is not sensitive enough to detect the slight difference, but she does have an excellent two-pan balance. Show how she can find the counterfeit coin in the least number of weighings.

22.* Do Problem 21 assuming there are 13 coins, one of which is counterfeit.

23.* We have 10 piles of 10 quarters each. One pile consists entirely of counterfeits, and each coin in this pile weighs 4.6 grams. A good quarter weighs 5.6 grams. You have a single-pan scale which is accurate to hundredths of a gram. What is the minimum number of weighings needed to identify the bad pile?

24.* There are 1 million dots on a sheet of paper. Can a straight line be drawn, no matter how the dots are situated, so that no point is on the line and exactly one-half of the dots are on each side of the line?

25.* The firm that designed the streetcar line from Podunk to East Podunk goofed. It forgot to allow for expansion of the rails due to the heat of the summer sun. When laid in February, the rails were perfectly straight and exactly 1 mile (5280 feet) long. By mid-July, they had expanded in length by 2 feet, causing them to buckle up in the middle, thus forming the shape of an isosceles triangle. How high above the earth did the middle point rise? First guess, and then figure it out mathematically.

26.* If a two digit number is multiplied by the product of its digits, a three digit number is formed, each digit of which is the unit digit of the original number. Find the original number.

Hints to Selected Problems

1. Make a chart listing chickens, pigs, heads, and legs. The first entry could be 6, 7, 13, and 40, which is wrong. Try other possibilities.

2. There must be one more boy than girl in the family.

3. Try various possibilities.

5. If the cube is $4 \times 4 \times 4$, then the plaza must be 8×8 (since they are to use the same number of bricks). We cannot use a cube of $5 \times 5 \times 5$, since 125 is not a perfect square. Is there any number larger than 4 that works?

7. 16,961 is a new palindromic number, but it's not the first one after 15,951. Find the first one.

9. $C = 2\pi r$, where π is a little more than 3. What happens to C when r is increased by 1?

10. $V = 4\pi r^3/3$, where V is the volume and r is the radius.

11. Did Hugo run half of the time?

12. Thirty is wrong. What does average speed mean?

14. Do things in the order mentioned. To fill in the table, first ask yourself how many games were played, then how many ties there were, and then which teams played to ties.

20. $5 = 2n - 1$, where $n = 3$; $7 = 2n - 1$, where $n = 4$.

21. You don't have to put all the coins on the balance the first time.

24. Try the same question with 10 dots, noting that the 10 dots determine $(19)(9)/2 = 45$ slopes. Use a line with a slope different from these 45.

1.4 / Transform the Problem

It is exactly 100 miles from Posthole to Pitstop. At 8 A.M. Horace sets out from Posthole to visit his friend Hugo in Pitstop. Unbeknown to Horace, Hugo sets out at exactly the same time to visit Horace. Both ride at 10 mph. As Hugo leaves, his tireless dog Trot runs on ahead at 30 mph. When the dog meets Horace, he immediately turns tail and heads back to his master, and when he meets his master he turns again and races back to Horace. Trot persists in this behavior until the two friends meet midway between the two towns. How far did Trot run?

Often a problem that sounds very complicated becomes extremely simple when looked at the right way. Just as one turns a jewel to examine every facet, so one should look at a problem from every angle. Often an elegant solution will appear when the problem is seen in the right light.

Take the problem above, for example. If one begins with the first idea that occurs to him, he will probably try to find the length of the first leg of Trot's journey. Then he will add on the length of the return trip, which is of course somewhat shorter. Continuing this way, he will have to compute and add together the lengths of all of Trot's trips. But how many back-and-forth trips does Trot make? And how long is each trip? Just thinking about Trot's trips tires most of us.

But wait. This problem is one of those jewels that ought to be seen from another angle. Ask yourself how long Trot ran. As long as Hugo (or Horace) rode. And since Hugo rode 50 miles at 10 mph, that is exactly 5 hours. Now, Trot ran all this time at 30 mph, and so he covered 150 miles. There it is—a simple, elegant solution to what appeared to be a hard problem.

Note how we obtained this solution. We changed the original question, How far did Trot run? to one that is much easier to answer, How long did Trot run? The answer to the second question allowed us to answer the first.

REFORMULATE THE QUESTION

> You turn the problem over and over in your mind; try to turn it so that it appears simpler. The aspect of the problem you are facing at this moment may not be the most favorable: Is the problem as simply, as clearly, as suggestively expressed as possible? Could you restate the problem?
> *George Polya*

One hundred and seventeen tennis players enter a park district tournament. It is an elimination tournament in which the loser of a match is out. In the first round, 1 player draws a bye and 58 matches are played, so that 59 players go into the second round. How many matches will have to be played in order to determine the champion?

To get a feel for the problem, let's start with a simple case, say 5 players, and work out the results of a typical tournament as shown below. Note that four matches are played. But is this the way to approach a problem with 117 players? And what if there were 1017 players? Surely none of us would want to draw a diagram in that case.

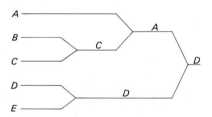

Let's try to reformulate the question. What happens in a tennis match? Well, one person wins and one person loses. But if a person loses, she is out. So every match has a loser, and there are exactly as many matches as there are losers. So our question becomes, How many losers are there? Answer: All but one of the players are losers. So, how many matches will be played if we start with 117 players? It must be 116; and if we start with 1017 players, there will be 1016 matches.

Suddenly the whole problem has become a triviality. It was simply a matter of restating the question. By the way, each match also has a winner. Why doesn't it help to look at the problem from this viewpoint?

FIND A NEW PROBLEM EQUIVALENT TO THE OLD ONE

A Tibetan monk, starting at daybreak and pausing often to rest, slowly made his way up the winding trail to the peak of Mt. Arazza, arriving just at sunset. There at a pagoda he meditated all night. The next morning he started back down over the same trail, arriving at the base exactly at noon. Demonstrate that he was at some point along the trail at exactly the same time on both days.

The Monk's Path

To do this, let us imagine (mathematicians must be able to pretend) a second monk who has exactly the same habits as the first one. Let him start from the base of the mountain on the second day and follow exactly the same pattern the first monk followed on the first day. That is, let him pause for rest, hobble along, and do everything in exactly the same way and at the same speed. This second monk must meet the first monk coming down, and their meeting point is the solution to the original problem.

SUMMARY

Sometimes it is possible, for a given question, to ask a different question which, if answered, will yield the answer to the original one. What is it that you want to find? Is there something else directly related to it that would be easier to find? Can the whole problem be reformulated? These ideas are worth thinking about even if you can clearly see your way through the problem. A chisel that cuts a stone cleanly, exposing its colored layers, is much to be preferred over a sledge hammer that smashes it beyond recognition.

PROBLEM SET 1.4

1. How many matches will be required to determine the champion in a tennis tournament that starts with 89 players?
2. In a certain state, the champion basketball team is determined as follows. All 508 teams enter district tournaments, the winners go to regional tournaments, and finally the regional winners go to the state tournament. How many games must be scheduled to determine the champion team?
3. Thirteen teams enter a double-elimination baseball tournament (in which a team must lose twice to be eliminated). How many games must be played to determine a champion if the winner never loses? If the winner loses one game?
4. Homer has a pile of old lead shot which he plans to melt and pour into a form requiring 1 quart of lead. How can he tell if he has enough? All he has to measure with is a 1-gallon container which happens to be about half full of water.
5. Henry has a 12-inch ruler marked only in inches. He wants to divide a $7\frac{1}{4}$-inch-wide board into six strips of equal width. How should he mark the board for sawing?

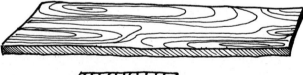

6. Amy and Susan, looking into a cylindrical barrel which appears to be about half full of liquid, argue about whether it is in fact a little more or a little less than half full. How can they settle the argument without a measuring device?
7. The distance from Chicago to New York is 800 miles. A nonstop train leaves Chicago for New York at 60 mph. Another train leaves New York for Chicago at 40 mph. How far apart are the trains 1 hour before they meet?
8. Two college students were traveling from Evanston to Chicago on an electric train. "I notice," said one of them, "that trains coming in the opposite direction pass us every 10 minutes. What do you think

—how many trains from Chicago arrive in Evanston every hour, given equal speeds in both directions?" "Six, of course," the other answered, "because 60 divided by 10 is 6." The first student did not agree. What is your answer?

9. In the opening problem about Trot and the two bicyclists, suppose Horace races along at 15 mph while Hugo pedals at a leisurely 10 mph. If Trot races back and forth at 30 mph, how far will he run before the cyclists meet?

10. The rollers in the diagram at the right have circumferences of 1 foot. How far does the slab move each time the rollers go through a complete revolution?

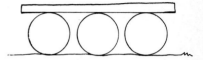

11. Farmer Brown has followed the same routine for years now. Each morning he milks the cow at the barn, carries the milk can to the river to cool, and then walks on to the house for breakfast. Long ago he figured out exactly where he should hit the river in order to minimize the length of his path. How did he do it?

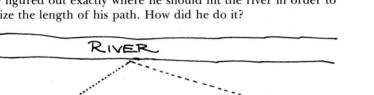

Farmer Brown's milk route

12. If the spider at the right wants to walk the shortest path to the hole on the opposite side, tell it how this path can be found.

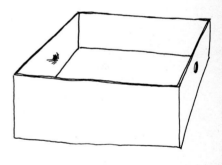

13. A farmer's wife with a carving knife has a cube of cheese 3 inches on a side. She wishes to slice the cube into 27 small cubes 1 inch on a side. She can do this with the six cuts shown in the figure below. Can she reduce the number of cuts if she rearranges the pieces after each cut?

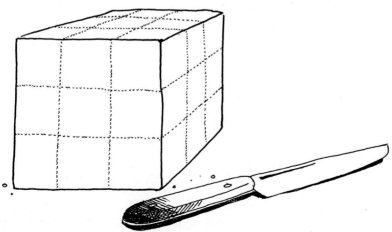

14. With a large power mower, it takes 3 hours to mow a lawn. Using a smaller mower, it takes 6 hours. If Amy and Homer work together using the two mowers, how long will it take?

15. Two opposite corners have been removed from a standard chessboard. Can this mutilated chessboard be covered with 31 dominoes, each of which can cover two squares?

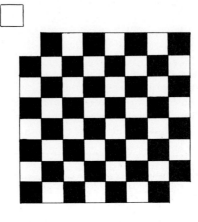

16.* If the squares removed in Problem 15 are of opposite colors (but anywhere on the board), can the remaining squares be covered with 31 dominoes?

17.* My only timepiece is a wall clock. One day I forgot to wind it, and it stopped for I don't know how long. Anyway, I wound it up again and went to visit a friend across town whose clock is always correct. I stayed awhile and then returned home. Then I made a simple calculation and set my clock correctly. How did I do it?

18.* Sadie entered life at 7 pounds and died exactly 100 years later weighing 79 pounds. Show that, at some time in her life, her age was equal to her weight.

19.* A rubber string is laid out along a line. Then it is stretched along this line by pulling both ends, but not necessarily the same amount. Show that at least one point remains stationary.

Hints to Selected Problems

5. Laying the ruler squarely across the board won't help.
6. Tip the barrel, but how far?
7. How fast are they closing the distance between them?
10. Think of the motion of the slab as being due to two factors: (a) the movement of the centers of the rollers, and (b) the rotation of the rollers about their centers.
11. Try a reflection and then remember that a line provides the shortest distance between two points.
13. Study the center cube.
15. One domino always covers two squares—but of what kind?

Chapter 2
Numeric Methods

Through and through the world is infested with quantity. To talk sense is to talk quantities. It is no use saying the nation is large—how large? It is no use saying that radium is scarce—how scarce? You cannot evade quantity. You may fly to poetry and music, and quantity and number will face you in your rhythms and your octaves.
ALFRED NORTH WHITEHEAD

To familiarize her students with the Celsius scale, Homer's teacher asked everyone to record tomorrow's Celsius temperature at noon. Not wanting to take his thermometer to the beach, Homer decided to take his reading at 11:00 A.M. when he found it to be 36°C. It was just his luck, however, to have a cold front move through, bringing a storm and (he overheard on someone's portable radio) dropping the temperature 22° between 11:00 A.M. and noon. That drop was measured in Fahrenheit, of course. What should Homer report as the noon temperature?

2.1 / The Problem of Measuring and Counting

We have on our desk a questionnaire in which our congressman asks us to assign a 1, 2, or 3 to each of a list of problems confronting the country, according as we believe the problem to be of major, moderate, or minor importance. Next to it is a reference form from a graduate school in which we are asked to rate a student's potential for leadership on a scale from 1 to 10. In very subjective problems (where the answer depends very much on the observer) as well as in objective problems (where almost everyone agrees that there is a correct answer), people often try in some way to assign numbers to possible answers.

There are several reasons for associating numbers with answers. Descriptive answers (she has natural leadership qualities, she could be a leader) are often hard to compare; numeric answers are easy to compare (even, let us point out, when they should not be so used). Descriptive answers are hard to compile into a composite answer; numeric answers can be manipulated in a variety of ways.

COMMON MEASURING UNITS

The two cousins in the accompanying story would have been greatly helped if they had realized that there are several commonly used scales that enable us to rank sensations of coldness. In each of

WHO NEEDS NUMBERS?

Brent and his cousin David, living in distant cities, are having a telephone conversation on Christmas morning. David is speaking.

"I bet it's colder here than it is there."

"Uh uh; it's twice as cold as it was last Christmas."

"It's always colder here than it is there. My mother always says she wishes she lived back where she grew up, 'cause it's not so cold there."

"But my mother says this is the coldest weather ever. How can it be colder than that?"

Unable to respond to that, David is happy that Brent changes the subject.

"I got a new bag of marbles. Now I've got more than you."

"Bet you haven't," says David. "I've got really lots now, cause after I saw you last summer, I played for keeps with a kid, and I got a whole bunch from him."

"Well, I've got way more than last summer. I got a whole bunch, no I mean even more than a whole bunch, a real lot from my grandma that used to be my dad's. And now I got more today."

At that moment David was saved from a second humiliation by the fact that his mother Dolores and Brent's mother Phyllis took over the phone. After exchanging seasonal pleasantries, the conversation turned to Dolores' plans to paint her living room.

"How do you know how much paint to buy?"

"Just tell the clerk in the store how many square feet you have to cover. They can tell the amount of paint from charts."

"How do you figure square feet?"

Phyllis volunteered the information that she had a calculator and would figure the square feet if Dolores would quickly measure the room. Somewhat skeptical of Phyllis' ability, but willing to take a calculated risk, Dolores soon returned to the phone with the required measurements and her own calculator.

"It's 24 feet long and 14 feet wide with a 7 foot ceiling. But don't forget that the fireplace wall is all stone."

"Right," said Phyllis, "so we multiply 14 × 7 for the one short wall, then 24 × 7 for the long wall, and multiply by 2 because there's two of them." They both wrote down the calculation to be made.

$$14 \times 7 + 2 \times 24 \times 7$$

Quick as the wink of a digital display, they had their answers. "It's 16,800," said Phyllis.

"That's funny. I get 21,168," said Dolores.

By the time Dolores went to the store, she had forgotten the dimensions of her living room, but the clerk didn't need them anyhow.

"Do you have an average or a big living room?" was the question put to her.

"I think it's kind of big" was her reply.

"Better take two gallons then. That way you'll probably have some left over if you need to do some touching up later on."

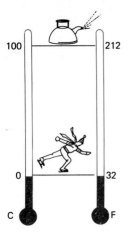

their cities, rising temperature will cause mercury in a tube to expand. If the height of the mercury is marked when water freezes and again when it boils, one has two measures, one of cold and the other of hot, which will also be the same in each city. Now numbers can be assigned arbitrarily to these two points, and the space in between can be partitioned into an arbitrary number of equal units. One logical way to do this is to assign 0 to the lower point, 100 to the upper, and use 100 equal units in between. Another way to do it (which may seem less logical) is to assign 32 to the lower point, 212 to the upper, and use 180 equal units in between. Either system, the Celsius or the Fahrenheit, will do for settling arguments as to the relative coldness experienced in two different places.

We often lose sight of the arbitrariness with which numbers are placed on a scale (where should 0 go?) or how the unit sizes are chosen. By the time we came along, the centimeter (so named because 100 of them make a meter) or the inch (so named because---) were firmly established as measuring units. Dolores and Phyllis both relied on a common understanding of a foot. Many of the problems that we confront are made easier by the fact that numbers have been assigned to commonly accepted units of measurement. We are greatly helped if we become familiar with the various measuring systems (see, for example, the chart on the next page).

We are also helped greatly if we know how to manipulate these numbers. It is to this business that we now turn.

SOME IMPORTANT LAWS

Suppose that Brent and David had spoiled their argument over marbles by using some numeric descriptions. Brent might have said that to the 25 marbles he had last summer, he had now added 23 from his grandmother and the 12 he received for Christmas. And David might have said that last summer's count of 41 had been augmented by winnings of 15. Given each boy's interest in obtaining as large a sum as possible, it is probable that they might have figured their respective totals in several ways.

$$\text{Brent: } (25 + 23) + 12 = 48 + 12 = 60$$
$$25 + (23 + 12) = 25 + 35 = 60$$

$$\text{David: } 41 + 15 = 56$$
$$15 + 41 = 56$$

Brent's equal results illustrate the associative law for addition; David's results illustrate the commutative law. These are two of several laws that are natural consequences of what we mean by addition and multiplication.

Commutative laws	$a + b = b + a$
	$a \cdot b = b \cdot a$
Associative laws	$a + (b + c) = (a + b) + c$
	$a \cdot (b \cdot c) = (a \cdot b) \cdot c$
Distributive laws	$a \cdot (b + c) = a \cdot b + a \cdot c$
	$(b + c) \cdot a = b \cdot a + c \cdot a$

English System   **Metric System**

Length

inch   centimeter
foot (12 inches)   decimeter (10 centimeters)
yard (3 feet)   meter (10 decimeters)
mile (1760 yards)   kilometer (1000 meters)

1 inch = 2.54 centimeters
1 mile = 1.61 kilometers

Liquid Volume

quart   liter

1 quart = .946 liter

Mass or Weight

ounce   gram
pound (16 ounces)   kilogram (1000 grams)

1 ounce = 28.35 grams
1 pound = 0.454 kilogram

Temperature

1°F (Fahrenheit)   1°C (Celsius)

$F = \frac{9}{5}C + 32$ $C = \frac{5}{9}(F - 32)$

Thou shalt not have in thine house divers measures, a great and a small. But thou shalt have a perfect and just weight, a perfect and just measure shalt thou have.
Deuteronomy 25:14-15

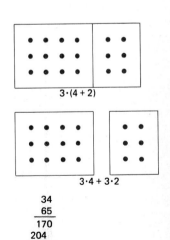

$3 \cdot (4 + 2)$

$3 \cdot 4 + 3 \cdot 2$

Of these, we shall choose the first distributive law for additional comment. That it must hold is almost obvious from the diagram in the margin which can be easily modified to accommodate any particular counting numbers.

We frequently use the distributive law without realizing it. The familiar calculation in the margin is really a shorthand version of

$$65 \cdot 34 = (5 + 60)(34) = 5 \cdot 34 + 60 \cdot 34 = 170 + 2040 = 2210$$

Calculations performed "in one's head" are often done with the aid of the distributive law. To multiply $15 \cdot 28$, think,

$$15 \cdot 28 = (10 + 5) \cdot 28 = 280 + 140 = 420$$

| 34 |
| 65 |
| 170 |
| 204 |
| 2210 |

ORDER OF OPERATIONS

Unfortunately, not all of the rules essential to successful numeric computation follow from such compelling rules. As it turns out, both Phyllis and Dolores got wrong answers to their problem because they forgot a fundamental convention about the order in which operations are to be performed in evaluating

$$14 \times 7 + 2 \times 24 \times 7$$

Phyllis just took the operations as they came. She first multiplied 14×7 to get 98, then added 2 to get 100, then multiplied by 24 to get 2400, and finally multiplied by 7 to get 16,800. (The story says that she did this on her calculator, and there are calculators on the market that will operate just this way; we will say more about that shortly.)

Dolores also used a calculator, but after entering the 14, she noticed that the $7 + 2$ could easily be done in her head, so she multiplied by 9 instead of 7; $14 \times 9 = 126$. She then proceeded with the next multiplications: $126 \times 24 = 3024$ and $3024 \times 7 = 21,168$.

Both answers would be treated with suspicion in a paint store, since both imply that the paint be bought not by the gallon, but by the barrel. The difficulty is that both women have forgotten the rules that govern the order in which operations are to be performed.

---

1. First perform any operations enclosed in parentheses, working from the inner parentheses outward, using rules 2 and 3 below if needed.
2. Next, working from left to right, perform multiplications and divisions as they are encountered.
3. Finally, again working from left to right, perform additions and subtractions as they are encountered.

A mathematician would write the calculation actually performed by Phyllis as

$$(14 \times 7 + 2) \times 24 \times 7$$

The correct answer to this is 16,800. Dolores' actual calculation was

$$14 \times (7 + 2) \times 24 \times 7$$

which does give 21,168. But if the rules stated above are followed in evaluating $14 \times 7 + 2 \times 24 \times 7$, we first perform the multiplications to get $98 + 336$ (the area of the one short wall plus the area of the two long walls), then the addition to get the correct answer of 434 for the problem at hand.

Here are some more examples for you to study.

$$5 + 8 \div 2 \times 6 = 5 + 4 \times 6 = 5 + 24 = 29$$
$$4 + [5 \times (3 + 6) - 13] = 4 + [5 \times 9 - 13] = 4 + 32 = 36$$
$$4 + [5 \times 3 + 6 - 13] = 4 + [15 + 6 - 13] = 4 + 8 = 12$$

CALCULATORS

As has already been mentioned, not all calculators follow the rules we have just stated. Some of the cheapest models do the operations in the order that they come, moving from left to right (Phyllis style). If you have not already purchased a calculator, we suggest that you get a model having at least the following features.

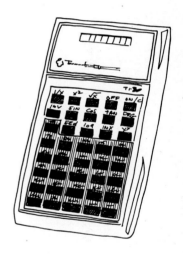

1. *Algebraic logic.* This means that it should follow the rules for order of operations given above. To check, press in order the keys

$$2 \; \boxed{\times} \; 3 \; \boxed{+} \; 4 \; \boxed{\times} \; 5 \; \boxed{=}$$

If you don't get 26, try another model.
2. *Parentheses.* This is self-explanatory. Just look for the keys $\boxed{(}$ and $\boxed{)}$.
3. *A* $\boxed{y^x}$ *key.* This key will be extremely useful later on.

One popular model (the TI 30) with these features is pictured in the margin; there are many others. Some fine calculators use what is called Reverse Polish logic. If you choose one of these, you will have to study your instruction book carefully to see how it works.

We can give one warning, applicable whatever model of calculator you select, which addresses a mistake frequently made by new owners of calculators. It is true that in written work, the product of 3 and 7 may be indicated by 3×7, $3 \cdot 7$ or $(3)(7)$. A calculator will understand only the first expression, however. You must push the key $\boxed{\times}$ to multiply.

SUMMARY

Many problems are most easily attacked by using common measuring units, or otherwise associating numbers with possible solutions. This allows us to compare or to aggregate our answers, but to do so successfully requires that we know the rules that govern calculating. Hand-held calculators can be a great help, but they too depend upon the user knowing the rules. In this section we have reviewed some important laws of arithmetic, and the rules that are used to guide the order in which arithmetic operations are performed.

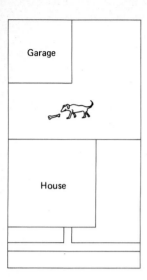

PROBLEM SET 2.1

1. The room discussed by Phyllis and Dolores is 14 feet wide, 24 feet long (often written $14' \times 24'$) and has a 7 foot ceiling. In addition to the one short wall of stone, it has two windows: a $3' \times 4'$ window on the other short wall and a $4' \times 8'$ picture window on one of the long walls. Finally, on the other long wall, there is a floor to ceiling doorway that is 3 feet wide. Using these additional facts, how many square feet are to be painted? Suggestion: Draw a picture of the three walls that are to be painted.

2. Suppose the room described in Problem 1 was $15' \times 28'$ with 8 foot ceilings, that one short wall was still all stone, that the windows were as described, and that the doorway was $3' \times 7'$. Then how many square feet would there be to paint?

3. To confine his hound dog, Librarian Hardback wishes to enclose his backyard with a picket fence. The yard is 50 feet wide and 72 feet deep, but his 42 foot wide house will form a part of one short side, and on two other sides the fence only needs to meet his $24' \times 24'$ garage (see diagram in the margin). If he decides to use three pickets for each foot of fencing, how many pickets are needed to confine Hardback's hound dog?

4. The lot on which Librarian Hardback's home stands is $50' \times 180'$, his house is $42' \times 65'$, and his garage is $24' \times 24'$ (see diagram above). Except for the house, the garage, a 4 foot wide sidewalk across the front of his lot, and a 3 foot wide, 12 foot long walkway from the front door to the sidewalk, the rest of the yard is to be grass. How many square feet of sod should Hardback order?

5. Calculate (without the aid of a calculator) each of the following. Then check by using your calculator.
   (a) $3 \times 4 + 6 \times 2$
   (b) $3 \times 4 + 3 \times 5 + 3 \times 6$
   (c) $5 + 2 \times 6 - 9$
   (d) $12 - 3 \times 2 + 10$
   (e) $3(12 - 2 \times 4 + 1)$
   (f) $9 + 4(3 + 2 \times 6) - 7$
   (g) $3[(12 - 2) \times 4 + 1]$
   (h) $(9 + 4)(3 + 2 \times 6 - 7)$

6. Calculate (without the aid of a calculator) each of the following. Then check by using your calculator.
   (a) $6 \times 2 - 3 \times 2 + 2$
   (b) $4 \times 3 + 7 \times 3 + 3$
   (c) $11 + 2 \times 3 \times 4$
   (d) $16 - 4 \times 3 + 9 \times 2$
   (e) $4[3 + (5 + 4)] - 1$
   (f) $14 + 3 \times 4(16 - 2 \times 3)$
   (g) $4 \times 3 + (5 + 4) - 1$
   (h) $(14 + 3) \times 4(16 - 2 \times 3)$

7. Using parentheses as necessary, express the following operations on the numbers w, x, y, and z.
   (a) the product of x and y added to the product of w and z
   (b) the product of x with the sum of y and z, decreased by w
   (c) the product of x with the sum of y and z decreased by w

8. As in Problem 7, express the following.
   (a) the difference between w and the product of x and y, all multiplied by z
   (b) w, decreased by the sum of y and z, is multiplied by x
   (c) w, decreased by x, is multiplied by the sum of y and z

   [c] 9. Along one wall of my study, there are three shelves, each 8 feet in

length. On another wall, there is a floor to ceiling case with seven shelves, each 3 feet long. And in my bedroom there are six shelves, each 4 feet long. All the shelves are full, and a few sample counts indicate an average of 15 books per foot on the shelves. About how many books do I own?

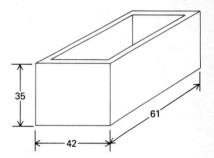

[c] 10. If your heart beats an average of 71 times per minute, how many times will it beat in your lifetime of 70 years (assume every year to have 365 days)?

[c] 11. I bought 27 pencils at 13 cents each, 14 pads of paper at 93 cents each but returned 6 ballpoint pens I'd bought earlier at 34 cents each. What was my bill?

[c] 12. Find the inside volume of an open box whose outer dimensions in inches are 35 by 42 by 61 if the walls and bottom are 1 inch thick. Note: For a box, volume = length × width × height.

[c] 13. Homer's calculations to find the noon temperature on the Celsius scale (See the opening problem to this section) proceed as follows:

(a) Using $F = \frac{9}{5}C + 32$, find the Fahrenheit temperature at 11:00 A.M. when C = 36°.

(b) Subtract the known drop of 22°F to get the Fahrenheit temperature at noon.

(c) Use $C = \frac{5}{9}(F - 32)$ to find the noon Celsius temperature from the known Fahrenheit temperature.

(If you have trouble with this problem because of the division involved, try it again after Section 2.4.)

"And how many hours a day did you do lessons?" said Alice, in a hurry to change the subject.

"Ten hours the first day," said the Mock Turtle; "nine the next, and so on."

"What a curious plan!" exclaimed Alice.

"That's the reason they're called lessons," the Gryphon remarked, "because they lessen from day to day."

This was quite a new idea to Alice, and she thought it over a little before she made her next remark. "Then the eleventh day must have been a holiday?"

"Of course it was," said the Mock Turtle.

"And how did you manage on the twelfth?" Alice went on eagerly.

"That's enough about lessons," the Gryphon interrupted in a very decided tone; "tell her something about the games now." (Lewis Carroll, *Alice's Adventures in Wonderland,* Crown Publishers, New York, pp. (145–146.)

Lewis Carroll (1832–1898), creator of *Alice in Wonderland* and other stories, was actually a mathematics professor at Oxford University. His writings are full of allusions to mathematics.

2.2 / Need for Negative Numbers

Ten, nine, eight, . . ., two, one, zero. What comes next? The answer to this question is of interest to more people than the children who stray into Wonderland. Children in Minnesota, for instance, learn rather early about the need for numbers below zero and the need of bookkeepers for numbers to indicate that an account is "in the red" surely brings us back to earth. We can go even deeper by asking how the buttons in an elevator should be marked to indicate floors below ground level.

The notation used for such numbers varies. The button on the elevator may say B3; a deficit of $3 on a balance sheet may be indicated by parentheses (3); the temperature reading will probably be reported as -3. Whatever the notation, it is clear that we need a whole new class of numbers to supplement the familiar counting numbers 1, 2, 3, 4, ...

THE INTEGERS

The **integers** are the numbers . . ., $-5, -4, -3, -2, -1, 0, 1, 2, 3, 4, 5, ...$ This array is familiar to anyone who has looked at a thermometer lying on its side and that is perhaps the best way to visualize the integers. There they label points equally spaced along a line, as in the diagram below.

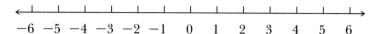

$$-6 \quad -5 \quad -4 \quad -3 \quad -2 \quad -1 \quad 0 \quad 1 \quad 2 \quad 3 \quad 4 \quad 5 \quad 6$$

This line should be thought of as extending infinitely far in both directions; that is the significance of the arrows at each end.

Though our introduction referred to situations where B3 or (3) might be used to indicate a point three units below 0, we have used -3 on our number line. It must be noted that in this notation, the minus sign $(-)$ does not indicate the operation of subtraction. It really should be thought of as a part of the name (-3) of the numeral signifying three units to the left of (or below) 0.

This dual use of the minus sign has its advantages. The numbers 3 and -3 neutralize each other under addition; that is, $-3 + 3 = 0$; and more generally

$$-a + a = 0 = a + (-a)$$

Therefore, it makes little difference whether we think of $3 - 3$ as a subtraction operation or as an addition of 3 and -3. When subtraction is discussed later in this section, we shall note that this dual role of the minus sign can also lead to some confusion.

There is actually a third use of the minus sign; it can be attached to a number that is itself negative. What do we mean by $-(-3)$? The boxed formula above suggests that $-(-3) + (-3) = 0$, which implies that $-(-3) = 3$. In fact,

$$-(-a) = a$$

for any number a.

The absolute value of a signed number n, written $|n|$, is the magnitude of that number apart from its sign. More precisely, if n is positive, $|n| = n$. If n is negative so that $-n$ is positive, $|n| = -n$. The absolute value of n may be thought of as the distance of n from 0.

ADDITION

The addition of signed numbers having the same sign poses little difficulty. Just add in the usual way and affix the common sign. To see that this makes sense in the case of two negative numbers, imagine that you owe $70 to one person (have a balance of -70 in one account), that you owe $50 to another (have a balance of -50 in another account), then you owe $120 altogether; $-70 + (-50) = -120$.

Addition of numbers of different signs is another story. The numbers tend to offset one another, with the one farthest from zero dominating. That sounds easy enough and, if we keep our eyes on the number scale we shouldn't have any trouble. To perform the addition $-4 + 7$, think of starting at -4 and moving 7 units up; you get 3. Similarly $-7 + 4 = (-3)$. Stated as formal rules, things sound more complicated, but here they are.

RULES FOR ADDITION

1. To add two numbers of the same sign, add their absolute values and affix the common sign.
2. To add two numbers of different sign, subtract the smaller absolute value from the larger and affix the sign of the number that had the larger absolute value.

SUBTRACTION

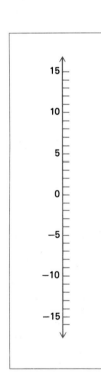

To find $m - n$ is to find the number which, when added to n, gives m. Consider the table below, together with the number scale at the left.

TABLE 2-1			
m	n	The Number We Add to n to Get m	Summary
7	3	4	$7 - 3 = 4$
7	-3	10	$7 - (-3) = 10$
-7	3	-10	$-7 - 3 = -10$
-7	-3	-4	$-7 - (-3) = -4$

The process of subtraction can be described in terms of addition.

RULE FOR SUBTRACTION

Change the sign of the **subtrahend** (that is, the number to be subtracted) and add.

This rule accounts for the previously mentioned possibility of confusion over the minus sign. We have seen papers intended as drill work for elementary school children which read as follows.

Subtract

$$\begin{array}{cccc} 19 & 23 & 15 & 21 \\ -\ 7 & -15 & -\ 8 & -13 \end{array}$$

There is nothing wrong with that, of course, so long as the gold stars get awarded for answers of 26, 38, 23, and 34. But for the answers generally expected at that level, the problem should be stated with the instruction *subtract,* or with the symbol (−) preceding the *subtrahend,* but not both. Study the correct examples in the margin.

This same confusion occurs with some frequency at the high school level in the following context. Consider the student who puzzles for a moment over the equation $x + 4 = -3$ before being struck with the inspiration of subtracting 4 from each side. That is, in fact, a good idea, but if he says "subtract 4" while writing

$$x + 4 = -3$$
$$\underline{ - 4 = -4}$$

then the possibility of trouble arises. For if he now carries through and subtracts on the right side, he just may announce a wrong answer of $x = 1$. (There is far less chance of error if $x + 4 = -3$ is solved by adding -4 to both sides, giving the correct answer $x = -7$; the solution of such equations will be taken up in Chapter 3.)

MULTIPLICATION

The associative and distributive laws discussed in the last section, together with the properties of 0 (namely that $0 + a = a$ and $0 \times a = 0$ for any a) can be used to prove that for any a and b,

$$(a)(-b) = (-a)(b) = -(ab)$$
$$(-a)(-b) = ab$$

The proofs of these propositions are given in Section 15.2, but there are intuitive explanations of these two laws. If I am flat broke today and am going behind at $4 per day, then in three days I'll be worth −$12; $(3)(-4) = -12$. Moreover, three days ago, I must have been worth $12; $(-3)(-4) = 12$.

Calculators automatically handle signed numbers if they are entered correctly. To perform the calculation $3 \cdot (-4 + 9)$, press in sequence the keys

$$3 \;\boxed{\times}\; \boxed{(}\; 4 \;\boxed{\diagup/}\; \boxed{+}\; 9 \;\boxed{)}\; \boxed{=}$$

Here the key $\boxed{\diagup/}$ changes the sign of 4. Note that it follows 4 rather than precedes it. To calculate $-3 \cdot [-4 + 5 \cdot (-3)]$, press

$$3 \;\boxed{\diagup/}\; \boxed{\times}\; \boxed{(}\; 4 \;\boxed{\diagup/}\; \boxed{+}\; 5 \;\boxed{\times}\; 3 \;\boxed{\diagup/}\; \boxed{)}\; \boxed{=}$$

Margin examples:

Subtract		Add	
12	15	12	15
7	11	−7	−11
5	4	5	4

Subtract		Subtract	
12	15	−12	−15
−7	−11	−7	−11
19	26	−5	−4

Perform the indicated operation

12	15	12	15
− 7	−11	+ 7	+11
5	4	19	26

SUMMARY

Negative 3 is not just a positive 3 with a minus sign in front of it. It is a new number, located below 0, 6 units from 3 on the number line. The counting numbers 1, 2, 3, ... together with 0 and the negatives of the counting numbers, form a system called the integers. Rules for calculating with these numbers must be clearly understood.

PROBLEM SET 2.2

1. Add
   (a) $\begin{array}{r} 12 \\ -8 \\ \hline \end{array}$ (b) $\begin{array}{r} -6 \\ -9 \\ \hline \end{array}$ (c) $\begin{array}{r} -14 \\ 5 \\ \hline \end{array}$ (d) $\begin{array}{r} 8 \\ 17 \\ \hline \end{array}$

2. Add
   (a) $\begin{array}{r} 14 \\ 23 \\ \hline \end{array}$ (b) $\begin{array}{r} -6 \\ 9 \\ \hline \end{array}$ (c) $\begin{array}{r} -12 \\ -15 \\ \hline \end{array}$ (d) $\begin{array}{r} 9 \\ -15 \\ \hline \end{array}$

3. For the exercises in Problem 1, subtract rather than add.
4. For the exercises in Problem 2, subtract rather than add.
5. If a ball is thrown upward with a velocity of 88 feet per second from the top of a building that is 104 feet above street level, the height h of the ball at time t is $h = -16t^2 + 88t + 104$. For example, when $t = 4$ seconds,
$$h = -16(4)^2 + 88(4) + 104$$
$$= -256 + 352 + 104 = 200$$
   Find h for the following values of t.
   (a) $t = 1$ (b) $t = 2$ (c) $t = 3$
   (d) $t = -1$ (e) $t = -2$ (f) $t = -3$
   What meaning do you attach to the use of negative values of t? To obtaining a negative value for h?
6. If a ball is thrown upward with a velocity of 88 feet per second from the top of a building that is 104 feet above street level, the velocity v of the ball at time t is $v = -32t + 88$. For example, when $t = 4$ seconds,
$$v = -32(4) + 88 = -40$$
   Find v for the following values of t.
   (a) $t = 1$ (b) $t = 2$ (c) $t = 3$
   (d) $t = -1$ (e) $t = -2$ (f) $t = -3$
   What meaning do you attach to the use of negative values of t? To obtaining a negative value of v?
7. Problem 5 illustrates the general formula: if a ball is thrown upward with a velocity of v_0 feet per second from a spot h_0 feet above ground level, the height of the ball at time t is $h = -16t^2 + v_0 t + h_0$.
   (a) What would it mean to have v_0 negative?
   (b) What would it mean to have h_0 negative?

8. Problem 6 illustrates the general formula: if a ball is thrown upward with a velocity of v_0 feet per second from a spot h_0 feet above ground level, the velocity v of the ball at time t is $v = -32t + v_0$.
   (a) What would it mean to have v_0 negative?
   (b) Is it true that, no matter what the value of v_0, v will eventually be negative as t grows larger? What is the significance of your answer?

9. Use your calculator to evaluate the following.
   (a) $19(-7) + (-14)(-13)$
   (b) $371 + [9(-13) - 56](3)$
   (c) $(-14)[57 - 9(-17)] - (23)(-13)$

10. Use your calculator to evaluate the following.
    (a) $(-11)(19) - (14)(-23)$
    (b) $7[-47 - 11(-16)] + 9(-17)$
    (c) $4[19 + 7(11 - 29)] + (-8)(-17)$

To make a big splash with his friends, Homer invited them to an evening party at his family's pool. A bit of a drip himself, it didn't dawn on him until 4:30 that afternoon that the pool needed to be filled. By the time he had prevailed on a volunteer fireman friend to hook up a hose to a nearby hydrant (capable, they knew, of filling the pool in 5 hours), it was 5:00 P.M. At the same time, they turned on the main faucet of the pool (capable of filling the pool in 12 hours), and they hooked up a garden hose capable of filling the pool in 30 hours. At what time will Homer's friends be able to drop in?

2.3 / Fractions

A straight ahead dive into the problem above turns into a real headache. We are well advised to consider that method of problem solving in which we begin with a related question. In this case, the correct springboard is the question, "How much of the pool will be filled in one hour?"

Let's see. The faucet must fill $\frac{1}{12}$ of the pool in an hour. Similarly, the garden hose and the fire hose will fill $\frac{1}{30}$ and $\frac{1}{5}$ of the pool in an hour. Working together, they will fill $\frac{1}{12} + \frac{1}{30} + \frac{1}{5}$ of the pool in one hour. The next obvious step is to combine these fractions.

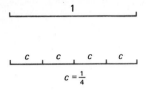

Many problems require, in the process of solution, that we add fractions. Since that subject is itself deep water for many people, we will temporarily set aside our concern over Homer's party in order to offer a generous review of fractions.

QUOTIENTS OF INTEGERS

In Section 2.1 where we discussed the order of operations, we confined our examples to the counting numbers. This meant, for the most part, that we had to avoid division, since division often gives a result that is not a counting number. Thus, while $\frac{10}{2}$, $\frac{18}{3}$, and $\frac{24}{6}$ yield the counting numbers 5, 6, and 4, the divisions $\frac{1}{4}$, $\frac{11}{8}$, and $\frac{25}{50}$ are impossible within the system of counting numbers. To be able to divide any counting number by any other counting number, we need to allow a whole new class of numbers into our computations.

Let us see what meaning these new numbers might have. When division of one counting number a by another counting number b does give yet another counting number c, we write $a/b = c$; it means that $a = bc$. Thus, $\frac{10}{2} = 5$ since $10 = 2 \cdot 5$ and $\frac{18}{6} = 3$ since $18 = 6 \cdot 3$. Accordingly, if $\frac{1}{4}$ is to be a number c, then we would want $1 = 4 \cdot c$. Now c is obviously not a counting number. What is it?

Consider a string of length 1 unit. Cut it into four pieces of equal length. Call this length c. Then $4c = c + c + c + c = 1$. We have our interpretation of $\frac{1}{4}$. The next rational thing to do is to interpret $\frac{3}{4}$ by writing

$$\frac{3}{4} = 3 \cdot \frac{1}{4} = \frac{1}{4} + \frac{1}{4} + \frac{1}{4}$$

Reasoning in the same way gives us an interpretation of $1/b$, and then for a/b, the understanding that

$$\frac{a}{b} = a \cdot \frac{1}{b}$$

for any counting numbers a and b. Moreover, $-a/b$ can be defined to be the number which, when added to a/b, gives 0;

$$-\frac{a}{b} + \frac{a}{b} = 0$$

Carrying our rules for the multiplication of signed numbers over to these new numbers gives us the rules

$$\frac{-a}{-b} = \frac{a}{-b} = -\frac{-a}{b} = \frac{a}{b}$$

In this way, a/b has meaning not only when a and b are counting numbers (positive integers), but when a and b are any integers—with the exception that $b \neq 0$. The collection of all such numbers is called the fractions, or the **rational numbers.**

Consider again the string of length 1, divided into four equal parts. If it were divided into two equal parts, each would have length $\frac{1}{2}$. Thus, $\frac{2}{4}$ and $\frac{1}{2}$ measure strings of the same length. That is,

$$\frac{1}{2} = \frac{2 \cdot 1}{2 \cdot 2} = \frac{2}{4}$$

Considerations like this suggest an important fact. The equality

$$\frac{a}{b} = \frac{k \cdot a}{k \cdot b}$$

holds for any nonzero number k. Thus $\frac{1}{2}, \frac{2}{4}, \frac{3}{6},$ and $\frac{-4}{-8}$ are symbols for the same number. Accept this in the same way that you accept Mark Twain and Samuel Clemens as different names for the same person.

We should learn to read an equality backward as well as forward. Read backward, the last boxed equality says that we can divide the **numerator** and the **denominator** (the top and the bottom) by the same number k. This is commonly referred to as canceling a common factor:

$$\frac{24}{30} = \frac{\not{6} \cdot 4}{\not{6} \cdot 5} = \frac{4}{5}$$

Among the many symbols for the same fraction, one is given a special name, the so-called **reduced form.** If the numerator a and the denominator b of the fraction a/b have no common integer factors greater than 1, and if b is positive, we say a/b is in reduced form. Thus, the reduced form of $\frac{24}{30}$ is $\frac{4}{5}$, and the reduced form of $\frac{6}{-9}$ is $\frac{-2}{3}$.

We call attention to another familiar fact. Notice that $\frac{3}{1}$ is the reduced form of $\frac{6}{2}, \frac{9}{3}$, and so on. We almost never write $\frac{3}{1}$, however, since it is equal in value to 3. In fact for every integer a,

$$\frac{a}{1} = a$$

On the calibrated line introduced in the previous section, we now have a host of new numbers. A few of them are shown on the diagram in the margin.

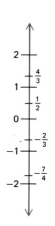

ADDITION AND SUBTRACTION

To add two fractions with the same denominator, we add the two numerators and assign the common denominator. Thus,

$$\frac{a}{b} + \frac{c}{b} = \frac{a+c}{b}$$

You may think of this as an application of the distributive law

$$\frac{a}{b} + \frac{c}{b} = a \cdot \frac{1}{b} + c \cdot \frac{1}{b} = (a+c) \cdot \frac{1}{b} = \frac{a+c}{b}$$

This clearly depends on having a common denominator.

If we wish to add fractions with different denominators such as a/b and c/d, we might multiply top and bottom of the first by d and of the second by b. This gives

$$\frac{a}{b} + \frac{c}{d} = \frac{ad}{bd} + \frac{bc}{bd}$$

Now the two fractions have the same denominator and can be added by the rule above.

$$\frac{a}{b} + \frac{c}{d} = \frac{ad + bc}{bd}$$

$$\frac{5}{12} + \frac{3}{10} = \frac{5 \cdot 10}{12 \cdot 10} + \frac{3 \cdot 12}{10 \cdot 12}$$

$$= \frac{50 + 36}{120}$$

$$= \frac{86}{120} = \frac{43}{60}$$

To say that we have a way that works of course is not to say that we have the best way. Sometimes bd is not the smallest number that can be used as a common denominator (not the **least common denominator**). For instance, in the case (illustrated in the margin) in which the denominators are 12 and 10, the least common multiple of the numbers is not 120 but 60. We need only multiply 12 by 5 and 10 by 6.

$$\frac{5}{12} + \frac{3}{10} = \frac{5 \cdot 5}{12 \cdot 5} + \frac{3 \cdot 6}{10 \cdot 6} = \frac{25 + 18}{60} = \frac{43}{60}$$

Subtraction can again be thought of as the addition of a negative number. Thus,

$$\frac{4}{5} - \frac{2}{3} = \frac{4}{5} + \left(-\frac{2}{3}\right) = \frac{4}{5} + \frac{-2}{3} = \frac{12}{15} + \frac{-10}{15} = \frac{2}{15}$$

Equivalently, we may subtract fractions with the same denominator by subtracting the numerator of the second from that of the first and assigning the common denominator.

$$\frac{4}{5} - \frac{2}{3} = \frac{12}{15} - \frac{10}{15} = \frac{12 - 10}{15} = \frac{2}{15}$$

MULTIPLICATION

Multiplying two fractions is a straightforward procedure. Simply multiply the numerators and multiply the denominators. Thus,

$$\frac{a}{b} \cdot \frac{c}{d} = \frac{ac}{bd}$$

$$\frac{2}{3} \cdot \frac{5}{4} = \frac{10}{12} = \frac{5}{6}$$

In particular,

$$a \cdot \frac{c}{d} = \frac{a}{1} \cdot \frac{c}{d} = \frac{ac}{d}$$

Note this last product. A common mistake is to multiply $a \cdot (c/d)$ by multiplying both c and d by a. The fact that it is common does not change the fact that it is a mistake. The correct result is

$$a \cdot \frac{c}{d} = \frac{ac}{d}$$

$$3 \cdot \frac{5}{12} = \frac{15}{12} = \frac{5}{4}$$

$$2\frac{3}{4} = 2 + \frac{3}{4} = \frac{8}{4} + \frac{3}{4} = \frac{11}{4}$$

$$2 \cdot \frac{3}{4} = \frac{6}{4} = \frac{3}{2}$$

There is an unfortunate notation that can be misleading. We refer to the practice of writing $4\frac{1}{3}$ to mean $4 + \frac{1}{3}$. If one means to write $4 \cdot \frac{1}{3}$ but omits the dot, the difficulty is obvious. We can only offer this counsel. When writing, don't omit dots that you mean to be there; when reading, assume that the writer has included all the dots intended, and that $2\frac{3}{4}$ therefore means $\frac{11}{4}$.

Multiplication can be simplified if we correctly use opportunities to cancel common factors.

$$\frac{56}{27} \cdot \frac{9}{35} = \frac{\cancel{7} \cdot 8 \cdot \cancel{9}}{\cancel{9} \cdot 3 \cdot \cancel{7} \cdot 5} = \frac{8}{15}$$

This cancellation of common factors is allowed because $ka/kb = a/b$. Note that there is no similar cancellation property for common terms.

$$\frac{5 + \cancel{3}}{7 + \cancel{3}} = \frac{5}{7} \quad \text{Wrong!} \qquad \frac{5 + 3}{7 + 3} = \frac{8}{10} = \frac{4 \cdot \cancel{2}}{5 \cdot \cancel{2}} = \frac{4}{5} \quad \text{Right}$$

DIVISION

Division of fractions is easily accomplished by remembering that the numerator and the denominator of a fraction can both be multiplied by the same number without changing the value of the fraction. If we choose this multiplier to be the least common multiple of the two individual denominators, we will have made a wise choice. Here is an example.

$$\frac{\frac{5}{6}}{\frac{7}{9}} = \frac{\frac{5}{6} \cdot 18}{\frac{7}{9} \cdot 18} = \frac{\frac{5 \cdot 3 \cdot \cancel{6}}{\cancel{6}}}{\frac{7 \cdot 2 \cdot \cancel{9}}{\cancel{9}}} = \frac{5 \cdot 3}{7 \cdot 2} = \frac{15}{14}$$

You may have learned a rule, "invert and multiply" which accomplishes the same thing *if* you remember which fraction to invert. The fraction in the denominator is the correct choice.

$$\frac{\frac{5}{6}}{\frac{7}{9}} = \frac{5}{6} \cdot \frac{9}{7} = \frac{5 \cdot \cancel{3} \cdot 3}{\cancel{3} \cdot 2 \cdot 7} = \frac{15}{14}$$

To handle complex fractions (double deck fractions), we suggest that you use the first method: multiply top and bottom by the least common multiple of all individual denominators.

$$\frac{\frac{2}{3} + \frac{7}{12}}{\frac{9}{40} - \frac{1}{9}} = \frac{\left(\frac{2}{3} + \frac{7}{2 \cdot 2 \cdot 3}\right) 3 \cdot 3 \cdot 2 \cdot 2 \cdot 2 \cdot 5}{\left(\frac{9}{2 \cdot 2 \cdot 2 \cdot 5} - \frac{1}{3 \cdot 3}\right) 3 \cdot 3 \cdot 2 \cdot 2 \cdot 2 \cdot 5} = \frac{240 + 210}{81 - 40} = \frac{450}{41}$$

Finally, we mention that all the familiar laws (commutative, associative, distributive) hold for fractions as well as for integers. Notice how these laws are used in the following calculation.

$$\frac{2}{3}\left(\frac{3}{4} - \frac{3}{10}\right) + 4\left(\frac{3}{2} \cdot \frac{7}{5}\right) - \frac{\frac{5}{6}}{\frac{5}{3}}$$

$$= \frac{2}{3} \cdot \frac{3}{4} - \frac{2}{3} \cdot \frac{3}{10} + \left(4 \cdot \frac{3}{2}\right) \cdot \frac{7}{5} - \frac{\frac{5}{6} \cdot 6}{\frac{5}{3} \cdot 6}$$

$$= \frac{1}{2} - \frac{1}{5} + 6 \cdot \frac{7}{5} - \frac{5}{10}$$

$$= \frac{1}{2} - \frac{1}{5} + \frac{42}{5} - \frac{1}{2} = \frac{41}{5}$$

BACK TO THE POOL PROBLEM

We had previously come to the conclusion that in one hour, the main faucet, the garden hose, and the fire hose working together would fill $(\frac{1}{12} + \frac{1}{30} + \frac{1}{5})$ of the pool. Addition gives

$$\frac{1}{12} + \frac{1}{30} + \frac{1}{5} = \frac{5}{60} + \frac{2}{60} + \frac{12}{60} = \frac{19}{60}$$

In one hour, $\frac{19}{60}$ (that is, nineteen sixtieths) of the pool will be filled.

What should be done with this information? Recall our various strategies. Let's try a problem with simpler numbers. What if one third of the pool could be filled in an hour? Then it would take three hours to fill the entire pool. How did we get it? We looked for a multiple of $\frac{1}{3}$ that would give 1. What number multiplies $\frac{19}{60}$ to give 1? The answer is $\frac{60}{19} = 3\frac{3}{19}$. It will take a little over three hours to fill the pool. To be more specific, it will take

$$\frac{3}{19} \cdot 60 = \frac{180}{19} \approx 10$$

minutes more than three hours. Homer's friends can drop in by 8:10 P.M.

SUMMARY

A rational number, more commonly called a fraction, is any number that can be expressed as a quotient of two integers. Such numbers obey all the rules familiar to us from our work with integers. In addition, we have the following useful facts.

Multiplication: $\quad a \cdot \dfrac{c}{d} = \dfrac{ac}{d}$

$$\frac{a}{b} \cdot \frac{c}{d} = \frac{ac}{bd}$$

$$\frac{a}{b} = \frac{ak}{bk}$$

Division: $\dfrac{a/b}{c/d} = \dfrac{ad}{bc}$

Addition and Subtraction: Begin by getting a common denominator

Signs: $\dfrac{-a}{-b} = -\dfrac{a}{-b} = -\dfrac{-a}{b} = \dfrac{a}{b}$

PROBLEM SET 2.3

Calculators can only handle fractions by changing them to decimals, the subject of the next section. Thus in doing this problem set, your calculator won't help you much.

1. Reduce each of the following fractions by canceling common factors between numerator and denominator.
   (a) $\frac{6}{8}$ (b) $\frac{9}{12}$ (c) $\frac{24}{27}$ (d) $\frac{16}{36}$
   (e) $\frac{60}{45}$ (f) $-\frac{63}{81}$ (g) $\frac{81}{108}$ (h) $\frac{63}{33}$

2. Find the reduced form of each fraction.
   (a) $\frac{9}{27}$ (b) $\frac{18}{12}$ (c) $\frac{24}{36}$ (d) $\frac{17}{37}$
   (e) $\frac{56}{72}$ (f) $-\frac{39}{26}$ (g) $\frac{256}{272}$ (h) $-\frac{49}{42}$

3. Perform the indicated operations and reduce.
   (a) $\frac{3}{4} \cdot \frac{5}{4}$ (b) $\frac{3}{4} + \frac{5}{4}$ (c) $\frac{3/5}{4/4}$
   (d) $3 \cdot \frac{16}{27}$ (e) $3 + \frac{16}{27}$ (f) $\frac{3}{4} \cdot \frac{6}{15} \cdot \frac{5}{2}$
   (g) $\frac{3/9}{4/16}$ (h) $\frac{9}{11} \cdot \frac{33}{5} \cdot \frac{15}{18}$ (i) $\frac{3}{5} + \frac{5}{4}$
   (j) $\frac{8}{9} - \frac{5}{12}$ (k) $\frac{4}{5} - \frac{3}{20} + \frac{3}{10}$ (l) $\frac{11}{24} + \frac{2}{3} - \frac{5}{12}$
   (m) $\dfrac{\frac{2}{3} + \frac{1}{5}}{\frac{5}{3} - \frac{1}{5}}$ (n) $\dfrac{\frac{8}{9} - \frac{2}{27}}{\frac{8}{9} + \frac{2}{27}}$

   (o) $1 - \dfrac{2}{2 + \frac{3}{4}}$ (p) $-\frac{2}{3}(\frac{5}{4} - \frac{1}{12})$
   (q) $\frac{1}{3}[\frac{1}{2}(\frac{1}{4} - \frac{1}{3}) + \frac{1}{6}]$

4. Perform the indicated operations and reduce.
   (a) $\frac{9}{15} \cdot 5$ (b) $3 \cdot \frac{5}{6} \cdot \frac{1}{6}$
   (c) $3 + \frac{5}{3}$ (d) $3 + \frac{5}{3} + \frac{1}{6}$
   (e) $\frac{5}{6} - \frac{11}{12}$ (f) $\frac{5}{4} \cdot \frac{8}{25} \cdot \frac{5}{6}$
   (g) $\frac{11}{15} + \frac{3}{4} - \frac{5}{6}$ (h) $\frac{9}{24}/\frac{15}{12}$
   (i) $\dfrac{\frac{3}{4}}{\frac{5}{5}} \cdot \frac{7}{5}$ (j) $\frac{4}{5} - \frac{3}{20} + \frac{3}{10}$
   (k) $\frac{5}{12} - \frac{5}{27} + \frac{3}{4}$ (l) $6/\frac{9}{7}$
   (m) $\dfrac{\frac{2}{3} + \frac{3}{4}}{\frac{2}{3} - \frac{3}{4}}$ (n) $\dfrac{\frac{3}{50} - \frac{1}{2} + \frac{4}{5}}{\frac{4}{25} + \frac{7}{10}}$
   (o) $2 + \dfrac{\frac{3}{4} + \frac{5}{12}}{\frac{2}{3}}$ (p) $\frac{3}{2}(\frac{8}{9} - \frac{5}{6})$
   (q) $-\frac{1}{3}[\frac{2}{5} - \frac{1}{2}(\frac{1}{3} - \frac{1}{5})]$

5. John can mow the lawn in 4 hours, Jill can do it in 6 hours. How long will it take them working together (each with a lawn mower)?

6. Henry estimates that he can paint the garage in 10 hours, James thinks he could do it in 8 hours. If both estimates are correct, how long would it take working together?

7. On July 2, 1980, Gulf Oil was selling at $41\frac{1}{8}$, Sperry at $48\frac{1}{4}$, and Pondorosa at $11\frac{3}{8}$, all given in dollars per share. If Betty bought 100 shares of each, what did it cost her (neglecting broker fees)?

8. Evelyn owns 56 shares of ABC. Yesterday it fell from $12\frac{7}{8}$ to $11\frac{3}{4}$. How much did she lose?

9. Wayne owns 72 shares of XYZ. Yesterday it rose from $23\frac{3}{8}$ to $25\frac{1}{4}$. How much did he gain?

10. A Right Whale has an average length of about 12 meters. An encyclopedia showed a picture of one using a scale of $\frac{1}{300}$. How long was the picture in centimeters?

11. An owner of a five-room house had the following expenses during 1 year: electricity $240, heat $520, water $130, repairs $310, and fire insurance $185. He uses one room as an office and is therefore entitled to declare $\frac{1}{5}$ of the house expenses as a business deduction on his income tax form. What does this amount to?

12. Sharon lived in New York for 5 months last year and is now working on her New York state income tax form. During the year she has driven her car 17,750 miles. To compute the gasoline tax that she has paid in New York, she needs to estimate the miles driven in New York. What estimate should she use?

13. I measure the width of each of 15 boards I have purchased to be $7\frac{5}{16}$ inches. If I lay the boards side by side, how wide a space will they cover assuming there are no gaps between them?

14. A picture $9\frac{1}{2}$ by 11 inches is reduced to one-third its original size (that is, each dimension is reduced to one-third its original size). What is the area of the reduced picture? How does this area compare with that of the original?

15. Find a fraction between $\frac{13}{40}$ and $\frac{1}{3}$. Hint: the average would be such a number.

Your textbook sold in 1980 for $16.95 in the bookstore. The publisher, Harper & Row, nets about 80 percent of that amount. The co-authors, Wayne Roberts and Dale Varberg, share in a 50-50 split of royalties which are 15 percent of the publisher's net. Dale Varberg pays 39 percent of his income in taxes (federal and state). How much did Varberg realize for himself when a book was sold in 1980?

2.4 / Decimals and Percents

The person most interested in the answer to the question posed above is Dale Varberg. However, the question is worthy of general discussion and is answered in this section not only because students often wonder how much authors receive in royalties but also because it is typical of one kind of problem that most people encounter repeatedly in life. Words like percent, net, 50-50 split, and income tax are in the vocabulary of most Americans. The key word is percent, and it is intimately related to the notion of decimal notation. That is where we begin.

The word "decimal" comes from the Latin word *decimus* which means tenth. Because our numeration system is based on 10, it is often called the decimal system. In this system, there are just 10 symbols, called digits (0, 1, 2, 3, 4, 5, 6, 7, 8, 9). Any number can be written using these 10 digits; this is possible because a digit has different values depending on the place it occupies. The following display illustrates this concept for the number 5034.278.

thousands	hundreds	tens	units		tenths	hundredths	thousandths
5	0	3	4	.	2	7	8

CONVERTING A DECIMAL TO A FRACTION

If you have learned to read 0.278 as "two hundred seventy-eight thousandths," you will not be surprised when somebody writes

$$0.278 = \frac{278}{1000}$$

Fundamentally, this is correct, because, by definition,

$$0.278 = 2\left(\frac{1}{10}\right) + 7\left(\frac{1}{100}\right) + 8\left(\frac{1}{1000}\right)$$

By the properties of fractions (Section 2.3), we can rewrite this as

$$\frac{2}{10} + \frac{7}{100} + \frac{8}{1000} = \frac{2}{10}\cdot\frac{100}{100} + \frac{7}{100}\cdot\frac{10}{10} + \frac{8}{1000}$$

$$= \frac{200}{1000} + \frac{70}{1000} + \frac{8}{1000} = \frac{278}{1000}$$

What about 5034.278? Can it be written as a fraction? The answer is yes.

$$5034.278 = 5034 + 0.278$$

$$= \frac{5034}{1} + \frac{278}{1000}$$

$$= \frac{5,034,000}{1000} + \frac{278}{1000}$$

$$= \frac{5,034,278}{1000}$$

CONVERTING A FRACTION TO A DECIMAL

When a fraction has a power of 10 as its denominator, it is easy to change it to a decimal. For example,

$$\frac{13}{100} = 0.13 \qquad \frac{27}{10} = \frac{20}{10} + \frac{7}{10} = 2.7$$

In most situations, we are forced to use long division to accomplish the conversion. The process for $\frac{3}{8}$ and $\frac{11}{64}$ is

```
      0.375              0.171875
  8) 3.000         64) 11.000000
     2 4                6 4
     ───                ───
      60                 460
      56                 448
     ───                 ───
      40                 120
      40                  64
     ───                 ───
                         560
                         512
                         ───
                         480
                         448
                         ───
                         320
                         320
                         ───
```

$$\frac{3}{8} = 0.375 \qquad \frac{11}{64} = 0.171875$$

```
    0.8333          0.6666
6) 5.0000       3) 2.0000
   4 8             1 8
   ───             ───
    20              20
    18              18
    ──              ──
    20              20
    18              18
    ──              ──
    20              20
    18              18
    ──              ──
    20              20
    18              18
    ──              ──
     2               2
```

In these two examples, the division process eventually terminates; sometimes it doesn't, as we illustrate with $\frac{5}{6}$ in the margin. It is clear that we will continue to get 3's indefinitely. In this case we decide what kind of decimal-place accuracy we desire and round off. If we want $\frac{5}{6}$ written as a decimal rounded off to four decimal places, we write

$$\tfrac{5}{6} = 0.8333$$

However,

$$\tfrac{2}{3} = 0.6667$$

Note that we round up if the remainder is half or more of the divisor.

All of these conversions can, of course, be done on a calculator; and most calculators will even round off according to the rule just stated.

OPERATIONS ON DECIMALS

We must know how to add, subtract, multiply, and divide decimals. Little needs to be said about addition and subtraction other than: (1) Be sure to line up the decimal points; and (2) supply as many zeros as you need. Here are two examples:

$$1.301 + 0.2543 + 36.2 \qquad 24.321 - 11.9234$$

```
    1.3010              24.3210
     .2543              11.9234
   36.2000              ───────
   ───────              12.3976
   37.7553
```

The key to multiplication of two decimals is to write one under the other, multiply as you would whole numbers, and then add the number of decimal places in the two factors to obtain the position of the decimal point in the answer.

```
    7.14   ⟵ Two decimal places
    0.043  ⟵ Three decimal places
    ─────
    2142              ↓
    2856
  ───────
  0.30702  ⟵ Five decimal places
```

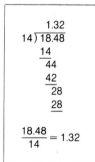

```
       1.32
   14) 18.48
       14
       ──
        44
        42
        ──
         28
         28
```

$$\frac{18.48}{14} = 1.32$$

Division is the hardest of the four operations to describe. If the divisor is a whole number, we set up as for long division and put the decimal point of the answer directly above that of the dividend. Thus $18.48/14 = 1.32$, as the marginal box shows.

If the divisor is itself a decimal, we change the division problem to an equivalent one in which the divisor is a whole number.

The process is illustrated in the margin for

$$\frac{2.8098}{0.18} = \frac{2.8098(100)}{0.18(100)} = \frac{280.98}{18}$$

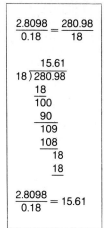

CALCULATORS AND DECIMALS

We warned you in the last section that a calculator would not be too helpful in working problems where the answer is to be expressed as a quotient of two integers, such as $\frac{3}{7}$. That is because the calculator will convert $\frac{3}{7}$ to its decimal equivalent and report an answer of .42857143. When we wish to do calculations in decimal notation, however, a calculator can be a great help. The last example above, for instance, is easily calculated by entering

$$2.8098 \boxed{\div} .18 \boxed{=}$$

The answer of 15.61 will be displayed immediately.

A calculator's real value becomes obvious in problems that involve several operations. It is important, however (as we have already emphasized), that you pay attention to the order in which your calculator performs operations. On the TI 30, we calculate

$$\frac{3.14 + 5.76}{4.85 - 1.73} \cdot \frac{57.13}{9.74}$$

by entering

$$3.14 \boxed{+} 5.76 \boxed{=} \boxed{\div} \boxed{(} 4.85 \boxed{-} 1.73 \boxed{)} \boxed{\times} 57.13 \boxed{\div} 9.74 \boxed{=}$$

to get 16.731724.

Is this answer correct? Did we press a wrong key? We could check by doing the problem over, preferably choosing some (correct) variation of the options available for entering the problem. We could, for example, enter

$$\boxed{(} 3.14 \boxed{+} 5.76 \boxed{)} \boxed{\times} 57.13 \boxed{\div} \boxed{(} 4.85 \boxed{-} 1.73 \boxed{)} \boxed{\div} 9.74 \boxed{=}$$

The answer should be the same.

A better idea than doing the problem twice is to begin the problem, not with your fingers, but with your head. Make a quick estimate first; then do the actual calculation (Guess, then demonstrate!). The idea here is that if you do make a mistake in entering something, particularly if you misplace a decimal point or mix up the order of operations, your answer will be thrown off far enough from the estimate so as to alert you to the fact that an error has been made.

For the calculation above, you might have estimated

$$\frac{3.14 + 5.76}{4.85 - 1.73} \cdot \frac{57.13}{9.74} \approx \frac{9}{3} \cdot \frac{60}{10} = 18$$

an approximation compatible with the calculated result.

PERCENTS

Percent means "hundredths." Thus 31 percent (written 31%) means thirty-one hundredths. To write a percent as a fraction, drop the percent sign and supply a denominator of 100. To change a percent to a decimal, drop the percent sign and move the decimal point two places left.

$$31\% = \frac{31}{100} = 0.31$$

$$12.5\% = \frac{12.5}{100} = \frac{125}{1000} = 0.125$$

Suppose that a store announces a sale in which every item is to be sold at a 33 percent discount. On a dress marked $85.79, you should then expect a discount of

$$(85.79)(.33) = \$28.31$$

The price to be paid is

$$\$85.79 - \$28.31 = \$57.48$$

A more direct way to find the price would have been to observe that the price is $100\% - 33\% = 67\%$ of the marked price; that is, multiply $(85.79)(0.67)$.

Now consider the problem posed in the opening box of this section. To solve it, we simply multiply

$$(16.95) \times (.80) \times (.15) \times (.50) \times (.61)$$

$$\left(\begin{array}{c}\text{Retail}\\\text{price}\end{array}\right) \quad \left(\begin{array}{c}\text{To get}\\\text{net price}\end{array}\right) \quad \left(\begin{array}{c}15\%\\\text{royalty}\end{array}\right) \quad \left(\begin{array}{c}\text{Varberg's}\\\text{``50''}\end{array}\right) \quad \left(\begin{array}{c}\text{After}\\\text{taxes}\end{array}\right)$$

Our calculator gives 0.62. Varberg will receive 62 cents on each book sold after taxes.

SUMMARY

The decimal system, so named because it is based on ten, relies on assigning different values to a digit according to its location with respect to a point (naturally called the decimal point). Computations using decimals must be made using some scheme (usually called an algorithm) that keeps digits in their correct relative place. Any number representable as a fraction can also be represented by a decimal. The conversion is accomplished by long division, and often requires some "rounding off." This rounding off, together with varying capacity for storing strings of digits, can account for differences when comparing the results from two different calculators.

Percent means hundredths. In calculations, 14 percent is always written, therefore, as $\frac{14}{100}$ or 0.14.

PROBLEM SET 2.4

1. Convert to fractions.
   - (a) 0.3
   - (b) 0.45
   - (c) 0.689
   - (d) 1.35
   - (e) 12.12
   - (f) 39%

2. Convert to fractions.
   - (a) 0.19
   - (b) 0.341
   - (c) 134.2
   - (d) 12%
   - (e) 12.3%
   - (f) 7.93%

Problems 3–6 can be done either by hand or on a calculator. Of course, the latter is easier but we think you should do some of them the hard way to make sure you remember how hand calculations are done.

3. Convert each of the following to decimals, rounding off your answer to four decimal places.
   - (a) $\frac{6}{25}$
   - (b) $\frac{5}{7}$
   - (c) $\frac{5}{9}$
   - (d) $\frac{131}{63}$
   - (e) $1 + \frac{5}{12}$

4. Convert to decimals, rounded off to five decimal places.
   - (a) $\frac{5}{13}$
   - (b) $\frac{11}{32}$
   - (c) $\frac{8}{7}$
   - (d) $\frac{237}{53}$
   - (e) $1 + \frac{006}{12}$

5. Perform the indicated operations, rounding off your answer to three decimal places.
   - (a) $6.5 + 31.92 - 4.013$
   - (b) $(2.96)(3.8)$
   - (c) $(0.0056)(386.45)$
   - (d) $(2.432)(6.93 - 4.21)$
   - (e) $38.6/70$
   - (f) $0.09267/2.4$

6. Perform the indicated operations, rounding off your answer to four decimal places.
   - (a) $57.12 - 1.03 - 5.0012$
   - (b) $(4.6)(3.7491)$
   - (c) $(0.0591)(35.72)$
   - (d) $(2.51)(5.341 - 8.235)$
   - (e) $166.4/4000$
   - (f) $3.51/0.065$

7. Change each of the following percents to decimals.
   - (a) 36%
   - (b) $8\frac{1}{2}\%$
   - (c) 9.2%
   - (d) 13.41%
   - (e) 151%
   - (f) $7\frac{3}{4}\%$

8. Change each decimal to a percent.
   - (a) 0.32
   - (b) 0.02
   - (c) 0.0234
   - (d) 0.1417
   - (e) 2.53
   - (f) 1.361

$\boxed{c}$ 9. Do the following calculations, rounded off to four decimal places, on a calculator. To avoid errors, begin by making some rough approximations combined with mental arithmetic to get an estimate of the answer.
   - (a) $\dfrac{5.791 - 3.462}{32.576}$
   - (b) $5.791 - \dfrac{3.462}{32.576}$
   - (c) $\dfrac{5.791}{32.576} - 3.462$
   - (d) $\dfrac{5.791}{32.576 - 3.462}$
   - (e) $3.4\left(\dfrac{5}{13} - \dfrac{6}{7}\right)$
   - (f) $\dfrac{3.4}{13}\left(2.1 + \dfrac{9}{13}\right)$

c 10. Follow the instructions of Problem 9.

(a) $\dfrac{32.3}{14.5 - 7.1}$

(b) $\dfrac{32.3}{14.5} - 7.1$

(c) $\dfrac{32.3}{14.5} - \dfrac{32.3}{7.1}$

(d) $32.3\left(\dfrac{1}{14.5} - \dfrac{1}{7.1}\right)$

(e) $\dfrac{32.3}{(14.5)(7.1)}$

(f) $\dfrac{32.3}{14.5}\Big/ 7.1$

11. A certain worker gets \$9.16 per hour with time and one-half for any time over 40 hours per week. What are his total wages in a week in which he worked 49 hours?

c 12. The worker of Problem 11 has deductions as follows: FICA 6.13%, federal withholding tax 18.2%, state withholding tax 8.1%, and medical insurance \$11.32. What is his take home pay for the week?

13. Cromwell's Department Store is going out of business with every item to be discounted 35%. Amy bought shoes marked \$36.75 and a dress marked \$118.92. What was her total outlay?

14. A furniture store of dubious reputation has announced a big sale in which every item will be sold at a discount of 20%. However, the night before the sale the manager replaced every sales tag with one marked up 20% over the original price. What will an item originally marked at \$100 sell for during the big sale?

c 15. Kay Trig and Joy Math have written a precalculus book which sells for \$18.79 at the bookstore. The publisher nets 77% of this and Kay and Joy have agreed on a 65-35 split of the 15% royalty which is figured on the publisher's net. How much does Kay receive on every book that is sold?

c 16. Suppose that food costs are inflating at 10.2% per year. How much will the groceries that cost \$98.17 today cost two years from now?

c 17. In Nomansland, which has a graduated income tax (as does the United States), a person pays 12% tax on income up to \$9000, 15% on that portion of income between \$9000 and \$16,000, and 18% on all income over \$16,000. How much tax will Naomi pay on income of \$19,241?

c 18. In a certain country, the tax rate is the same as the number of thousands of dollars of income. (If the income is \$6000, the tax rate is 6%.) What is the optimal income, assuming all incomes are in whole thousands? Hint: Experiment.

c 19. The swimming pool problem stated in the opening box of Section 2.3 is easy to do on a calculator especially if it has a $\boxed{1/x}$ key. To solve that problem press

$$12\ \boxed{1/x}\ \boxed{+}\ 30\ \boxed{1/x}\ \boxed{+}\ 5\ \boxed{1/x}\ \boxed{=}\ \boxed{1/x}$$

The answer 3.158 is in hours. Figure out why this works and then solve Problems 5 and 6 of Section 2.3 this way.

c 20 The world's fastest human can run 100 yards in 9.1 seconds. How many miles per hour is that?

c 21. A flea weighing 0.00001 of an ounce can jump $7\frac{3}{4}$ inches. If a man could jump in proportion to his weight as well as a flea, how far could a 180 pound man jump?

[c] 22. If a certain car loses 14% of its value each year, what will it be worth after three years if it costs $8650 when new?

[c] 23. In selling a house, the seller often has to pay "points" on a house that is financed through FHA. Typically this is 4 percentage points for each 1% that conventional bank loan rates exceed FHA rates. How much will "points" cost the seller on a loan of $61,000 if conventional loan rates are 10.9% and FHA loan rates are 9.75%? (Of course, the seller usually passes this cost on to the buyer by increasing the price of the house.)

[c] 24. A businessman depreciates each car in his fleet 25% each year (that is, he considers its value at the end of a year to be 75% of what it was at the beginning of the year.) He sells a car during the year in which its value falls below $1000. During what year will he sell a car that he bought for $8500?

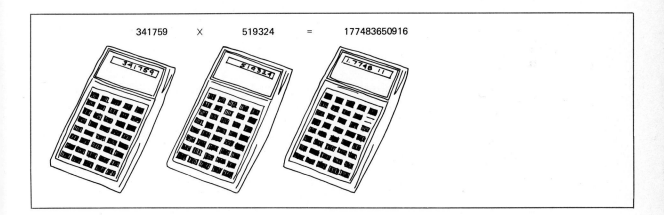

341759 X 519324 = 177483650916

2.5 / Exponents and Scientific Notation

When two large numbers are multiplied on a calculator, the answer has too many digits to fit into the display. Faced with this problem, many simple calculators give up and print E or Error. But a more sophisticated model will display the answer in a different notation (scientific notation) as indicated in the example above. Before trying to explain it, we need to review an old idea, the use of exponents to indicate the power of a number.

EXPONENTS

To abbreviate
$$3 \cdot 3 \cdot 3 \cdot 3 \cdot 3 \cdot 3 \cdot 3 \cdot 3 \cdot 3 \cdot 3 \cdot 3 \cdot 3 \cdot 3 \cdot 3 \cdot 3 \cdot 3 \cdot 3$$

most ordinary people and all mathematicians write 3^{18}. The number 18 is called an **exponent;** it tells you how many 3's to multiply together. The number 3^{18} is called a power of 3 and is read "three to the eighteenth."

The idea generalizes. If n is a positive integer and b is any number,

$$b^n = \underbrace{b \cdot b \cdot b \cdot \cdots \cdot b}_{n \text{ factors}}$$

Thus, $b^3 = b \cdot b \cdot b$ and $b^6 = b \cdot b \cdot b \cdot b \cdot b \cdot b$. How do we write the product of 1000 b's? (*HONEY* is not the answer we have in mind.) The product of 1000 b's is written as b^{1000}.

Notice that

$$b^3 \cdot b^6 = \underbrace{(b \cdot b \cdot b)}_{3} \underbrace{(b \cdot b \cdot b \cdot b \cdot b \cdot b)}_{6}$$

$$= \underbrace{b \cdot b \cdot b \cdot b \cdot b \cdot b \cdot b \cdot b \cdot b}_{9}$$

$$= b^9 = b^{3+6}$$

To multiply one power of b by another power of b, you add the exponents; that is,

$$b^m \cdot b^n = b^{m+n}$$

Similarly,

$$(b^6)^2 = b^6 \cdot b^6 = b^{6+6} = b^{6 \cdot 2}$$

To raise a power to another power, you multiply the exponents; that is,

$$(b^m)^n = b^{m \cdot n}$$

For example $(2^3)^4 = 2^{12}$ and $(10^3 \, 10^2)^6 = (10^5)^6 = 10^{30}$

Often we need to simplify quotients like

$$\frac{3^5}{3^5} \qquad \frac{2^{11}}{2^4} \qquad \frac{6^3}{6^8}$$

The first is simple enough; it equals 1.

Moreover

$$\frac{2^{11}}{2^4} = \frac{2^7 \cdot 2^4}{2^4} = 2^7 = 2^{11-4}$$

and

$$\frac{6^3}{6^8} = \frac{6^3}{6^3 \cdot 6^5} = \frac{1}{6^5} = \frac{1}{6^{8-3}}$$

These examples suggest the following three-pronged rule.

$$\frac{b^m}{b^n} = \begin{cases} 1 & \text{if} \quad m = n \\ b^{m-n} & \text{if} \quad m > n \\ \dfrac{1}{b^{n-m}} & \text{if} \quad n > m \end{cases}$$

This rule is correct but awkward; that's why we haven't put a box around it.

ZERO AND NEGATIVE EXPONENTS

So far we haven't used symbols like 3^0 and 3^{-4}. If they are to be admitted to the family of powers, we insist that they obey the house rules (the two rules that are boxed in). Now by the first of these rules

$$3^0 \cdot 3^7 = 3^{0+7} = 3^7$$

which implies that $3^0 = 1$. This suggests that we define

$$b^0 = 1$$

Here b can be any number except 0 (we leave 0^0 undefined).

To make sense of 3^{-4}, we apply the first rule again. According to it,

$$3^{-4} \cdot 3^4 = 3^{-4+4} = 3^0 = 1$$

This means that 3^{-4} is the reciprocal of 3^4; that is, $3^{-4} = 1/3^4$. It seems that we must, for $b \neq 0$, define

$$b^{-n} = \frac{1}{b^n}$$

Having done this, the complicated quotient rule simplifies. For example,

$$\frac{b^4}{b^4} = 1 = b^0 = b^{4-4}$$

and

$$\frac{b^3}{b^8} = \frac{b^3}{b^3 \cdot b^5} = \frac{1}{b^5} = b^{-5} = b^{3-8}$$

In general,

$$\frac{b^m}{b^n} = b^{m-n} \qquad b \neq 0$$

Let's summarize what we have done. We have given meaning to the symbol b^n where n is any integer. Moreover, we have done it in such a way that all the familiar rules for exponents hold. There are actually five such rules (three that we have discussed and two others). We put them all together for handy reference.

Rules for Exponents

If m and n are any integers,

1. $b^m \cdot b^n = b^{m+n}$
2. $(b^m)^n = b^{m \cdot n}$
3. $\dfrac{b^m}{b^n} = b^{m-n}, \qquad b \neq 0$
4. $(ab)^n = a^n b^n$
5. $\left(\dfrac{a}{b}\right)^n = \dfrac{a^n}{b^n}, \qquad b \neq 0$

Here are three examples that are worth studying:

$$(10^3 \cdot 10^{-5} \cdot 10^6)^2 = (10^4)^2 = 10^8$$

$$\frac{10^4 \cdot 10^{-2}}{10^5} = \frac{10^2}{10^5} = 10^{-3}$$

$$\frac{3.24 \times 10^7}{1.08 \times 10^3} = \frac{3.24}{1.08} \times \frac{10^7}{10^3} = 3 \times 10^4$$

Exponential expressions can get quite complicated. One helpful thing is this: *You can always move a factor from denominator to numerator or vice versa by changing the sign of its exponent.* Thus

$$\frac{2^3 4^{-2} 5^6}{2^{-2} 4^3 5^2} = \frac{2^3 2^2 5^6 5^{-2}}{4^3 4^2} = \frac{2^5 5^4}{4^5} = \left(\frac{2}{4}\right)^5 5^4 = \left(\frac{1}{2}\right)^5 5^4 = \frac{625}{32}$$

SCIENTIFIC NOTATION

One of the conveniences of our decimal system of writing numbers is this. To multiply a number by 10, move its decimal point one place to the right; to divide a number by 10, move its decimal point one place to the left. Thus,

$$323 = 3.23 \times 100 = 3.23 \times 10^2$$
$$32.3 = 3.23 \times 10 = 3.23 \times 10^1$$
$$3.23 = 3.23 \times 1 = 3.23 \times 10^0$$
$$0.323 = 3.23/10 = 3.23 \times 10^{-1}$$
$$0.0323 = 3.23/100 = 3.23 \times 10^{-2}$$

The numbers in the right-hand column are said to be in scientific notation. To be precise, we say that a positive member is expressed in **scientific notation** when it is written in the form

$$c \times 10^n$$

where n is an integer and $1 \leq c < 10$.

Here is an alternate way of describing scientific notation. Consider standard position for the decimal point of a number to be right after the first nonzero digit, as in 4.2715. Then take any number with the same sequence of digits, for example, 4271.5. Count the number of places that its decimal point lies to the right or left of standard position. That number (positive if to the right, negative if to the left) is the exponent on 10 in the scientific notation for the number. Thus,

$$4271.5 = 4.2715 \times 10^3$$

3 places

$$0.00042715 = 4.2715 \times 10^{-4}$$

4 places

Now return to the picture that opened this section. We wanted to calculate

$$341759 \times 519324$$

Our calculator (a TI30) responded with

$$1.7748 \quad 11$$

This is its way of indicating

$$1.7748 \times 10^{11}$$

which in familiar notation is

$$177480000000$$

The exact answer (done by hand) is

$$177483650916$$

Our calculator did the best it could, rounding off the answer to five digits.

Two observations are in order. First, all but the simplest calculators (which show no answer at all) are going to show in scientific notation answers that require too many digits for the display in standard notation. Second, they are going to round off the answer to some prescribed number of digits (this number varies with calculator models).

Here are two further examples which will clarify what we have said and also serve to explain the use of negative exponents. To calculate 0.000349/695000, enter

$$.000349 \boxed{\div} \ 695000 \boxed{=}$$

The answer

$$\boxed{5.0216 \quad -10}$$

will appear in the display. To enter a number, say, 4.13×10^4, in scientific notation press 4.13 $\boxed{\text{EE}}$ 4. The $\boxed{\text{EE}}$ key tells the calculator that you are going to *enter exponent*. Thus to calculate

$$\frac{4.13 \times 10^4}{2.51 \times 10^{-5}}$$

press

$$4.13 \ \boxed{\text{EE}} \ 4 \ \boxed{\div} \ 2.51 \ \boxed{\text{EE}} \ 5 \ \boxed{\text{\textbackslash/}} \ \boxed{=}$$

and you will obtain the answer

$$\boxed{1.6454 \quad 09}$$

which stands for

$$1.6454 \times 10^9$$

THE $\boxed{y^x}$ KEY

This is the key for raising numbers to powers. To calculate 3^7, simply press 3 $\boxed{y^x}$ 7 $\boxed{=}$ and the answer 2187 will appear in the display. To calculate $(1.03)^{-9}$, press

$$1.03 \ \boxed{y^x} \ 9 \ \boxed{\text{\textbackslash/}} \ \boxed{=}$$

and you will get 0.766417. About all you need to remember is that your calculator will insist that y be positive. It doesn't know how to handle $(-3)^5$; if you try to calculate this power, it will simply say E or Error.

In an expression that involves arithmetic operations as well as raising to powers, the latter are done first. For example,

$$2^3 + 3 \cdot 2^4 = 8 + 3 \cdot 16 = 8 + 48 = 56$$

Similarly, to calculate this expression on a calculator, press

$$2 \ \boxed{y^x} \ 3 \ \boxed{+} \ 3 \ \boxed{\times} \ 2 \ \boxed{y^x} \ 4 \ \boxed{=}$$

Your calculator will automatically do the operations in the right order.

SUMMARY

The notion of a positive integer exponent is expanded to allow 0 and negative exponents by using the definitions

$$a^0 = 1 \qquad a^{-n} = \frac{1}{a^n}$$

This enables us to state succinctly the five rules for exponents displayed in this section. It also paves the way for scientific notation, a device that enables us to write any positive number as the product of a number between 1 and 10, and a power of 10. Calculators use scientific notation when the number otherwise requires for display more digits than the calculator can display.

PROBLEM SET 2.5

1. Evaluate, writing the answer without exponents.
   (a) 2^4 (b) $(-2)^4$ (c) -2^4
   (d) 2^{-4} (e) $(-2)^{-4}$ (f) $(\frac{1}{2})^4$
   (g) $(\frac{1}{2})^{-4}$ (h) $2^4 2^3$ (i) $(2^4)^2$
   (j) $\dfrac{2^4 2^{-2}}{2}$ (k) $(2^5 2^{-3})^2$ (l) $\dfrac{2^{-2} 2^6}{2^5}$
   (m) $2^3(2^2 - 2^4)$ (n) $2^0 + 3 \cdot 2^2$ (o) $(2^2 + 2^3)^2$

2. Evaluate, writing your answer without exponents.
   (a) 3^2 (b) 3^{-2} (c) $(-3)^2$
   (d) -3^2 (e) $(\frac{1}{3})^2$ (f) $(\frac{1}{3})^{-2}$
   (g) $3^2 3^3$ (h) $(3^2)^2$ (i) $\dfrac{3^2 3^4}{3^3}$
   (j) $\dfrac{3^{-2} 3^{-5}}{3^{-4}}$ (k) $(3^{-2} 3^4)^2$ (l) $(3^0)^3$
   (m) $3^2(3^3 - 3^2)$ (n) $3 + 2 \cdot 3^2$ (o) $(3^2 + 3)^{-1}$

3. Simplify
   (a) $\dfrac{2^3 3^2 4^{-2}}{2^4 3}$ (b) $\dfrac{2^6 3^3 4^2}{2^8 3^2 4}$

4. Simplify
   (a) $(\frac{2}{3})^2(\frac{27}{8})$ (b) $\frac{2}{3}(\frac{3}{4})^2$

[c] 5. Use your calculator to evaluate.
   (a) 2^{12} (b) 3^9
   (c) $5 \cdot 2^{13}$ (d) $4 \cdot 3^{11}$
   (e) $5 + 3 \cdot 2^{11}$ (f) $2 \cdot 3^5 + 4 \cdot 3^6$
   (g) $120(1.03)^{12}$ (h) $475(1.02)^{20}$
   (i) $(1.03)^{-12}$ (j) $450(1.02)^{-18}$

[c] 6. Evaluate
   (a) 4^9 (b) 9^4
   (c) $3 \cdot 4^7$ (d) $5 \cdot 9^{-5}$
   (e) $5 + 2 \cdot 6^5$ (f) $2 \cdot 5^4 - 3 \cdot 5^3$
   (g) $432(1.04)^9$ (h) $372(1.03)^{-6}$
   (i) $36 + 54(1.05)^4$ (j) $472[1 - (1.03)^{-10}]$

7. Write each of the following in scientific notation.
   (a) 4325   (b) 5,120,000
   (c) 0.00513   (d) 0.2341

8. Write each of the following in ordinary notation.
   (a) 5.134×10^4 (b) 4.13×10^{-2}
   (c) 6.12×10^6 (d) 3.134×10^{-4}

$\boxed{c}$ 9. Perform the indicated computation on your calculator but then write the answer it gives in ordinary notation.
   (a) $(532161)(743526)$
   (b) $(4.31 \times 10^5)(2.51 \times 10^2)$

$\boxed{c}$ 10. Calculate
$$\frac{5.145 \times 10^4}{3.234 \times 10^{-6}}$$

$\boxed{c}$ 11. The formula for the volume of a cube of side x is $V = x^3$. Find the volume of a cube of side 32.5 inches.

$\boxed{c}$ 12. The area of a circle is given by $A = \pi r^2$ where r is the radius and $\pi = 3.1415927$. Find the area of a circle of radius 4.235 centimeters. Note: Some calculators have a key for π.

$\boxed{c}$ 13. The volume of a sphere is given by $V = \frac{4}{3}\pi r^3$ where r is its radius (see Problem 12 for π). Find the volume of a sphere of radius 10.37 inches.

$\boxed{c}$ 14. Find the volume of the earth assuming it is a sphere of radius 4000 miles (see Problem 13).

$\boxed{c}$ 15. With the passing of each year, a certain car loses 19 percent of its value. If it cost $5600 new, what will it be worth at the end of 10 years? Hint: At the end of 1 year, it is worth $5600(0.81)$, at the end of 2 years $5600(0.81)(0.81) = 5600(0.81)^2$.

$\boxed{c}$ 16. A heavy drill press loses 11 percent of its value at the beginning of a year by the end of the year. If it costs $12,750 new, what will it be worth at the end of 7 years? (See Problem 15.)

$\boxed{c}$ 17. After making $9000 in its first year, a certain company has doubled its profits in each succeeding year. What did it make during the second year? The third year? The twelfth year?

$\boxed{c}$ 18. If inflation runs at 10 percent per year each year from now on, what will a refrigerator that costs $380 today cost 8 years from now? Hint: One year from now it will cost $380 + (0.10)380$ or $380(1.10)$; 2 years from now it will cost $380(1.10)(1.10) = 380(1.10)^2$.

$\boxed{c}$ 19. If $500 is put in the bank today at 10 percent interest compounded annually, it will be worth $500 + (0.10)500 = 500(1.10)$ at the end of a year, $500(1.10)^2$ at the end of 2 years, and so on. Find out how much it will be worth at the end of 12 years. Note: Compound interest will be discussed thoroughly in Section 4.5.

$\boxed{c}$ 20. If $480 is put in the bank at 8 percent interest compounded annually, what will it be worth at the end of 9 years? (See Problem 19).

Chapter 3
Algebraic Methods

"Algebra is a merry science," Uncle Jakob would say. "We go hunting for a little animal whose name we don't know, so we call it x. *When we bag our game we pounce on it and give it its right name."*
A. EINSTEIN

HOW MANY GEESE?
"We aren't 100! If you take twice our
number and add half our number, and
add a quarter of our number, and
finally add you, the result is 100."

3.1 / Solving Equations

Do you remember the story of the geese in Section 1.2? Here again
are the words of the leader of the flock: "We aren't 100! If you take
twice our number and half our number, and add a quarter of our
number, and finally add you, the result is 100, but . . . well, you figure it out." The stork was smart. He drew some lines on the
ground, asked some clever questions, and soon discovered that 36
was the answer. Let's take a different tack.

What is it that we want to find? Why, the number of geese in the
flock. Let's give this number a name. (Now, we certainly can't call
it 36 or 27 or 19. That would be to pretend we already knew.) We
choose a name that for the moment has a value unknown to us or
anybody else. We could use ____ like the stork, or ☐ like some
fifth-grade textbooks but an old algebra tradition suggests that we
use the letter x. *Let x stand for the number of geese in the flock.*

Next we must find a relationship that x satisfies. And we must
state this relationship in an algebraic way. Read the goose's statement again, note how the stork translated it, and finally see how it
can be stated algebraically.

English Language	*Stork Picture*	*Algebra*
If you take twice our number	———	$2x$
and		+
half our number	——	$\frac{1}{2}x$
and add		+
a quarter of our number	—	$\frac{1}{4}x$
and finally add		+
you		1
the result is 100		= 100

In algebra, we make a practice of writing our results horizontally in what is called an *equation:*

$$2x + \tfrac{1}{2}x + \tfrac{1}{4}x + 1 = 100$$

Finally, we solve the equation.

Given equation:	$2x + \frac{1}{2}x + \frac{1}{4}x + 1 = 100$
Multiply by 4:	$8x + 2x + x + 4 = 400$
Combine like terms:	$11x + 4 = 400$
Add -4:	$11x = 396$
Multiply by 1/11:	$x = 36$

We conclude that there are 36 geese in the flock. That is, of course, the answer the stork got by a different method back in Section 1.2.

MORE ABOUT GEESE

The stork thought geese were stupid. Not so, said Old Macdonald, who had a farm. He claimed that geese have the finest mathematical and artistic tastes of all the animals. Consequently, he made all his geese pens in the shape of golden rectangles. These most perfect of all rectangles are discussed in Section 4.6 but for now all we need to know is that in a golden rectangle, the length is 1.62 times the width.

Here is Macdonald's problem. He has exactly 120 feet of fence with which he plans to build two identical geese pens. To maximize the use of his fence, he will abut the two pens along a common long side. What are the dimensions of the rectangles?

Following advice learned well in Chapter 1, we draw the picture shown in the margin, *letting x be the width of each pen.* Then $1.62x$ is their corresponding common length. There are four sides of size x and three of size $1.62x$. Their sum must be 120, that is,

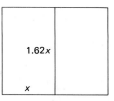

$$4x + 3(1.62x) = 120$$
$$4x + 4.86x = 120$$

Combine like terms:	$8.86x = 120$
Divide by 8.86:	$x = \dfrac{120}{8.86} \approx 13.54$

This means that the width of each pen is 13.54 feet; the length is $(1.62)(13.54) \approx 21.93$ feet.

MATHEMATICS OR X-ITUS

For many students, x marks the spot where arithmetic leaves off and real mathematics begins. Its appearance no doubt disappoints some of our readers who think it means a return to what they found to be both boring and difficult in high school. Others are delighted, feeling we have finally gotten to something significant.

Both groups may be helped by this advice. Don't confuse mathematics with *x*-itus. The aimless manipulation of *x*'s is not mathematics any more than a random collection of words is an essay. Mathematics is the process of reasoning from a clearly stated question or problem to an irrefutable answer or solution. For us, the use of a letter, usually *x*, to stand for an unknown number is a device, a strategy in solving problems. The correct manipulation of *x*'s, as the correct spelling of words, is not the chief end; it is a means to something more important.

HOW TO SOLVE AN EQUATION

Getting the right equation for a word problem is the topic of the next section. But having gotten the equation, we still must solve it. The two geese examples illustrate the fundamental rules on which equation solving is based.

Rule 1. You may add (or subtract) the same quantity on both sides of an equation.

Rule 2. You may multiply (or divide) both sides of an equation by the same quantity, provided it isn't zero.

With these rules at our disposal, we suggest a step-by-step procedure for solving an equation. But don't follow it slavishly; sometimes, you will see a shortcut; occasionally, you will want to do the steps in a different order.

Step 1. Remove any parentheses that appear in the equation. Often this will require use of the distributive law: $a(b + c) = ab + ac$.

Step 2. Multiply both sides by a number chosen to clear of fractions or decimals.

Step 3. Add terms as necessary to both sides until all terms involving *x* are on one side and the remaining terms on the other side. This puts the equation in the form $ax = b$.

Step 4. Multiply both sides by $1/a$ (or divide both sides by a), thus determining *x*.

Step 5. Check your answer by substituting it in the original equation.

Now refer to our solution of the two geese problems to see how we used these steps in solving their equations. Then consider the following additional examples.

$$\begin{aligned}
\text{Original equation:} \quad & 3(x - 2) - 5.9 = 1.9x - 4.2 \\
\text{Remove parentheses:} \quad & 3x - 6 - 5.9 = 1.9x - 4.2 \\
& 3x - 11.9 = 1.9x - 4.2
\end{aligned}$$

Multiply by 10:	$10(3x - 11.9) = 10(1.9x - 4.2)$
	$30x - 119 = 19x - 42$
Add 119:	$30x = 19x + 77$
Add $-19x$:	$11x = 77$
Multiply by 1/11	$x = 7$
Check:	$3(7 - 2) - 5.9 \overset{?}{=} 1.9(7) - 4.2$
	$9.1 = 9.1$

Original equation:	$\frac{1}{4}x - 1 = \frac{2}{3}x - \frac{1}{6}$
Multiply by 12:	$12(\frac{1}{4}x - 1) = 12(\frac{2}{3}x - \frac{1}{6})$
	$3x - 12 = 8x - 2$
Add $-3x$	$-12 = 5x - 2$
Add 2:	$-10 = 5x$
Divide by 5:	$-2 = x$
Check:	$\frac{1}{4}(-2) - 1 \overset{?}{=} \frac{2}{3}(-2) - \frac{1}{6}$
	$-\frac{3}{2} = -\frac{9}{6}$

SUMMARY

Many word problems are best solved by the methods of algebra. This involves (in the language of Chapter 1) transforming the problem to one of solving an equation. For the latter, there is a clear procedure to follow. We have illustrated it in four examples.

PROBLEM SET 3.1

1. Solve the following equations.
   (a) $2x + 9 = 17$
   (b) $3x + 8 = 20 + x$
   (c) $2x + 9 = 4 + 5x$
   (d) $3x + 4 = 9 + 5x$
   (e) $2x + 3 + x = 5x - 9$
   (f) $2(x + 1) = 4(x - 1) - 2$
   (g) $-3(x + 2) + x = 3x + 1$
2. Solve.
   (a) $3x + 2 = 20$
   (b) $5x + 6 = 24 + 3x$
   (c) $11x + 11 = 5 + 5x$
   (d) $4x + 6 = 12 + x$
   (e) $x + 4 + 5x = 12 - x + 7$
   (f) $4(x - 2) = 5(x + 1) + 3$
   (g) $-2(x - 3) = 2x - 1$
3. Solve. Hint: First clear of fractions and decimals.
   (a) $0.8x + 11.7 = 2.5x + 4.9$
   (b) $0.3x - 2.1 = 2.4 - 1.2x$
   (c) $0.41x - 0.32x + 1.42 = 0.39x$
   (d) $\frac{2}{3}x + \frac{1}{6} = x - \frac{5}{6}$

(e) $\frac{3}{4}x - 1 = \frac{2}{3}x$

(f) $\frac{3}{5}(x-2) = \frac{7}{10}$

4. Solve.

(a) $\frac{1}{3}x - 4 = \frac{3}{5}x - \frac{8}{5}$

(b) $\frac{1}{2}x + \frac{2}{3} = x + \frac{1}{6}$

(c) $\frac{2}{7}(x + 2) = \frac{3}{14}x - 1$

(d) $0.3x - 5.1 + 2.3x = 1.9x - 0.2$

(e) $0.4(x - 3) = 0.8(x + 2)$

(f) $0.31x + 0.29 = 0.3x - 1.9$

5. Write an algebraic expression using x that represents each English phrase. For example, a number plus one-half its square could be represented by $x + \frac{1}{2}x^2$.

(a) One-half the sum of a number and its square.

(b) The sum of 3 and twice a number.

(c) The sum of three consecutive even numbers if the smallest is x.

(d) The perimeter (distance around) a square of side x.

(e) The area (length times width) of a rectangle of width x if the length is 3 less than twice the width.

(f) The volume of a box (length times width times height) with square base of x meters and height 3 more than one-half the width.

(g) The value in cents of the coins in Ellen's purse if it contains x nickels, three times as many dimes as nickels, 4 fewer pennies than dimes, and no other coins.

6. Follow the directions of Problem 5.

(a) One-third the product of a number and four times its square.

(b) Fifty-two percent of the amount by which a number exceeds 10.

(c) The total cost of the tickets for x children, three more adults than children, if adult tickets are $3.50 each and child's tickets are $1.25 each.

(d) The number of miles traveled if I drove 55 miles per hour for x hours.

(e) The time it took to drive 400 miles if Susan drove at x miles per hour.

(f) The cost of gas at $1.55 per gallon for a 600 mile trip if my car gets x miles per gallon.

(g) Amy's weekly pay if she worked 40 hours at $6.80 per hour and x additional hours at time and one-half.

7. The sum of three consecutive positive integers is 636.

(a) Let x be the smallest of the three. What are the next two integers?

(b) What expression represents the sum of the three integers written in part (a)? Set this expression equal to 636.

(c) Find the three integers.

8. The sum of three consecutive even numbers is 192. Find them, following the procedure outlined in Problem 7.

9. The sum of three consecutive integers is 63 more than the smallest one.

(a) Let s represent the smallest integer. What expressions represent the next two integers?

(b) Write an equation, using the fact that the sum of the three integers represented in part (a), decreased by 63, must equal s.

(c) Find the three integers.

10. A number plus a third of that number and a fifth of that number is equal to 46.

(a) Let n be that number. What expressions represent a third and a fifth of that number?

(b) Translate the statement of the problem into a mathematical equation.

(c) Find the number.

11. A rectangle is 4 feet longer than twice its width; its perimeter is 200 feet.

(a) Let w equal the width. What is the length?

(b) Using the expressions from part (a), what is the perimeter?

(c) Find the dimensions of the rectangle.

12. The width of a rectangle is one-third its length and its perimeter is 300 centimeters. Find its width, following the procedure outlined in Problem 11.

13. Farmer Brown has 208 meters of fence with which he plans to fence three identical adjoining pens as shown in the margin. If the length of each pen is 2 feet more than its width, what are the dimensions of each pen? Hint: Let x be the width of one of the pens. In terms of x, label each side in the diagram. Then what?

14. The side of one square is 8 feet less than twice the side of another square and the sum of their perimeters is 112 feet. Find the sides of both squares.

[c] 15. The radius of one circle is 8 meters longer than that of another and the sum of their perimeters is 120 meters. Find both radii. Hint: The formula for the circumference of a circle is $C = 2\pi r$. Use $\pi = 3.14159$.

[c] 16. A running track has the shape of a square with a semicircle at each end as shown in the margin. If it is a 300 yard track, what is the width of the square?

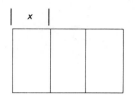

René Descartes
1596–1650

Rules for the Direction of the Mind

1. Reduce any kind of problem to a mathematical problem.
2. Reduce any kind of mathematical problem to a problem of algebra.
3. Reduce any problem of algebra to the solution of a single equation.

3.2 / Problems That Lead to One Equation

Descartes has left a permanent mark on history. He was a philosopher, physicist, and mathematician. His rules for direction of the mind may be overstated, and they cannot always be followed. But they do suggest the direction we want to pursue in this section.

GETTING THE EQUATION

For most people, the really challenging part of solving a word problem is finding the right equation. Here we are greatly helped if we remember the strategies we learned in Chapter 1. It is still important to *clarify the question*, to make sure that we understand what is said and what is asked. A picture or a chart will help us *organize the given information*. To avoid foolish mistakes, it is wise to make a preliminary *guess* at the answer, but we must not be satisfied until we *demonstrate* the correctness of our final answer. To translate a problem from one stated in English to one represented by an algebraic equation is to *transform the problem*. These are general guidelines; but we can be more specific.

Step 1.   *Read the problem quickly to get an idea what it's about and then very carefully to understand every word.*

Step 2.   *Clearly specify what unknown number x is to represent.* Usually, this will be the quantity you are asked to find. But if this is the length of a rectangle and you know its width is half as much, it may be better to call the width x and the length $2x$, thereby avoiding fractions.

I hope that I shall shock a few people in asserting that the most important single task of mathematical instruction in the secondary schools is to teach the setting up of equations to solve word problems.
G. Polya

Step 3. *Draw a picture if possible.* On it, mark the given information and the unknown number *x*. List any formulas that might be helpful (for example, $C = 2\pi r$ or $A = \pi r^2$ if it involves a circle).

Step 4. *Express the conditions of the problem in an equation involving x.* Somewhere in the statement of the problem (perhaps in the picture), two quantities are seen to be equal. If at least one of them involves *x*, this gives the equation.

Step 5. *Solve the equation.* Section 3.1 suggested ways to do this.

Step 6. *Check your answer.* Do it two ways. First, ask yourself whether your answer makes common sense (an age for a grandmother of either 20 or 120 should be questioned). Then check to see that your answer fits the conditions of the problem.

Now it's time to illustrate these principles.

AN AGE OLD PROBLEM

When a father was 42, his son was 8. Now he is three times as old as his son. How old is the son?

We begin by specifying the unknown. *Let x denote the son's age right now.* Next, we draw a picture that describes the problem.

Be Definite
Do not say x = age.
Do not say x = son's age.
Do say x = son's age now.

Now read the first sentence of the problem (and look at the picture). It says that the father was (and therefore always will be) 34 years older than his son. So his age now is $x + 34$. And the second sentence says (father's age now) is (three times son's age now)

$$x + 34 = 3x$$

Solving this equation is easy.

Add $-x$: $34 = 2x$
Multiply by $\frac{1}{2}$: $17 = x$

Thus, the son is 17 and the father is $17 + 34 = 51$.

Do these answers make sense? Yes. If we had gotten 170 and 204, we would have immediately gone back to look for a mistake. In any case, it is easy to check our answers. We note that 51 is three times 17, and that, under these circumstances, when the son was 8, the father was 42.

A MIXTURE PROBLEM

How many liters of pure sulfuric acid should be added to 5 liters of 15 percent sulfuric acid to obtain a solution that is 50 percent sulfuric acid?

We'll begin with a picture that displays all the given information in a visual way. Note that we have *let x represent the number of liters of pure acid to be added.* But how do we get an equation? We look at the amount of *pure* acid in the three containers in the figure.

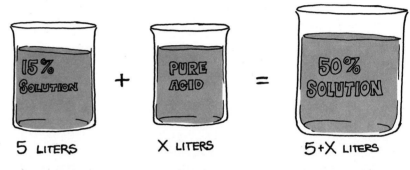

$$\begin{pmatrix}\text{The pure acid} \\ \text{we start with}\end{pmatrix} \text{plus} \begin{pmatrix}\text{the pure acid} \\ \text{we add}\end{pmatrix} \text{is} \begin{pmatrix}\text{the pure acid} \\ \text{we end with}\end{pmatrix}$$
$$(0.15)(5) \quad + \quad x \quad = \quad (0.50)(5+x)$$

To solve the equation, we might use the following steps.

Remove parentheses:	$0.75 + x = 2.50 + 0.50x$
Multiply by 100:	$75 + 100x = 250 + 50x$
Add -75:	$100x = 175 + 50x$
Add $-50x$:	$50x = 175$
Divide by 50:	$x = \dfrac{175}{50} = 3.5$

The answer of 3.5 liters of pure acid to be added seems reasonable. You can check that it is exactly right.

A DISTANCE-RATE-TIME PROBLEM

Karen Milbank has rented a motorboat for 6 hours from a river resort. She was told that the boat will travel 5 miles per hour upstream and 10 miles per hour downstream. How far upstream can she go and still return the boat to the resort within the allotted 6-hour time period?

We can be sure that the formula $D = RT$ is going to be important; it plays a crucial role in all distance-rate-time problems. This formula says that if I travel at 30 miles per hour for 6 hours, I will cover a distance of $D = RT = (30 \times 6) = 180$ miles. On the other hand, it also says that if I travel 100 miles at 20 miles per hour, it will take me $T = D/R = 100/20 = 5$ hours.

Now Karen wants to know how far upstream she dares to go. *Let us call that distance x miles.* The picture in the margin helps us think about the problem but here a chart that summarizes the given information is even more valuable.

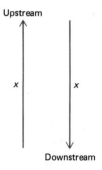

Upstream

Downstream

	Upstream	*Downstream*
Distance (miles)	x	x
Rate (miles per hour)	5	10
Time (hours)	$x/5$	$x/10$

There is a piece of information that we have not used and it is the key to the problem. The total time allowed is 6 hours. That is the sum of the time going upstream and returning downstream.

$$\frac{x}{5} + \frac{x}{10} = 6$$

After multiplying both sides by 10, we have

$$2x + x = 60$$
$$3x = 60$$
$$x = 20$$

A 20 mile trip upstream sounds reasonable. To check that it is correct, note that Karen will need 4 hours to go 20 miles upstream and 2 hours to return. That is a total of 6 hours.

A GAS MILEAGE PROBLEM

Rodney Roller bought a compact car that goes 15 miles farther on a gallon of gas than did his old desert yacht. He finds that he can now get to Grandma Uphill's house on 4 gallons of gas whereas it used to require 9 gallons. How far does she live from Rodney?

We will surely need the fact that

$$\text{miles per gallon} = \frac{\text{distance traveled}}{\text{gallons used}}$$

or equivalently

$$\text{distance traveled} = (\text{gallons used})(\text{miles per gallon})$$

Since the problem asks how far Grandma Uphill lives from Rodney,

Grandma's house

x

Rodney's house

it seems sensible to let x represent that distance. Thus,

$$x = \text{(gallons used)} \cdot \text{(miles per gallon)}$$

Depending on whether we are referring to the old car or the new car this gives

$$x = 9 \cdot \text{(miles per gallon with old car)}$$
$$x = 4 \cdot \text{(miles per gallon with new car)}$$

This almost asks that we equate the two expressions for x. When we do, we get

$$9(\text{miles per gallon with old car}) = 4(\text{miles per gallon with new car})$$

It may seem unfortunate that our unknown x has now disappeared but this apparent misfortune forces us to realize there are other unknowns in this problem. We don't know how many miles per gallon either car gets, though we do know how they are related.

This turns out to be one of those problems which yields to the strategy of asking a related question: Can we find the miles per gallon the old car got? *Let m be that number.* Then, $m + 15$ is the number of miles per gallon with the new car. And our equation becomes

$$9m = 4(m + 15)$$

This equation is easily solved for m.

$$9m = 4m + 60$$
$$5m = 60$$
$$m = 12$$

If m is 12, then $x = 9(12) = 108$. It is 108 miles to Grandma Uphill's house. You can check that this answer is correct.

SUMMARY

Word problems come in many forms and therefore demand considerable creativity in their solution. That's why we cannot give a simple procedure that will always work. However, the six-step plan we gave should help with most problems. Two of those steps are worth emphasizing again.

Let x represent the unknown of the problem (or perhaps a related unknown). Be very specific about what x represents; write it down. Remember that it must be a number.

Draw a picture. A good picture will guide your thinking and often suggest important relationships.

PROBLEM SET 3.2

1. If 6 times a number is increased by 10 the result is 8 times the number, less 12.
   (a) Let the number be *x*. What expression represents 6 times the number, increased by 10? Eight times the number, less 12?
   (b) Translate the problem into a mathematical equation.
   (c) Find the number.

2. One number is 5 more than twice another number. Their sum is 80. Find the numbers. Does the procedure outlined in Problem 1 help?

3. A student has scores of 61 and 77 on the first two exams. What must her score on the third exam be if her three-exam average is to be 75? Hint: The average of three numbers is their sum divided by 3.

4. If the student in Problem 3 wants an average of 80, what must her third score be?

5. A father is three times as old as his son, but 15 years from now he will be only twice as old as his son.
   (a) Let *x* designate the son's age now. What is the father's age now?
   (b) What will be the age of the son in 15 years? The father? Express the fact that the father is (in 15 years) twice as old as the son.
   (c) How old is the son now?

6. John is four times as old as Amy is now, but in 16 years he'll only be twice as old. How old is John now? Follow the procedure outlined in Problem 5.

7. Rachel is 3 years younger than Joel. Forty-one years ago, she was two-thirds his age. How old is Rachel?

8. Susan's grandmother is 11 times as old as Susan, but in 45 years Susan will be half as old as her grandmother. What are their ages?

9. If a basketball weighs 9.5 ounces plus half of its own weight, how much does it weigh?

10. John now weighs one-third what his brother Robert does. If both put on 30 pounds, John's weight will be one-half of his brother's. Find their weights.

11. A tank contains 10 gallons of 5 percent salt brine. An 8 percent salt brine is to be obtained by evaporating some of the water in the original mixture.
    (a) Note that the original mixture contains 10(.05) gallons of salt.
    (b) Let *x* represent the number of gallons of water to be evaporated. What expression represents the number of gallons of salt in the tank after evaporation?
    (c) How does the salt in the tank before evaporation compare with the salt in the tank after evaporation?
    (d) How much water should be evaporated?

12. The cooling system in Rodney Roller's car has a capacity of 14 quarts. At the present time, 25 percent of the coolant is antifreeze. To protect his car for a Minnesota winter, he ought to have a 50 percent mixture.

(a) Suppose he drains off x quarts of his present mixture. Before adding any pure antifreeze, how much antifreeze will he have in his tank?

(b) What algebraic expression will represent the quarts of antifreeze in the tank after refilling?

(c) How many quarts of antifreeze does he want in his tank?

(d) How many quarts should he drain off?

13. How many gallons of a 60 percent solution of nitric acid should be added to 10 gallons of a 30 percent solution to obtain a 50 percent solution of the acid?

14. How much commercial bleach, which contains 0.51 pound of sodium hypochlorite per gallon, should be added to 1 gallon of water to obtain a solution that contains 0.06 pound of sodium hypochlorite per gallon?

15. The Joneses plan to put in a concrete drive from their garage to the street. The drive is to be 36 feet long, and they have been advised to make it 4 inches thick. Since there is an extra delivery charge for less than 4 cubic yards of ready-mixed concrete, the Joneses decide to make the drive just wide enough to use 4 cubic yards. How wide should they make it? Hint: Put everything in the same units. Volume is length times width times height. And there are 27 cubic feet in a cubic yard.

16. Not content merely to keep up with the Joneses, the Smiths decide to put in a concrete drive 48 feet long and 6 inches thick. How wide should it be to use up all the concrete in a 9 cubic yard load?

17. When you rent a boat to fish by trolling in a river, you are told that it will take you upstream at 3 miles per hour and bring you back at 7 miles per hour. You set out at 6:00 A.M. going upstream, having rented the boat until noon.

(a) Let t represent the time spent going upstream. How much time will be spent going downstream?

(b) Express in algebraic form the fact that the distance traveled upstream equals the distance traveled downstream.

(c) At what time should you turn around?

18. Going at a certain speed it took Mary 8 hours to get to Center City. If she could have driven 10 miles an hour faster, it would only have taken $6\frac{2}{3}$ hours. How far did she drive?

19. Trains traveling on parallel tracks start at noon from cities 800 miles apart. One train travels at 40 miles per hour, the other at 75 miles per hour. At what time will they meet?

20. A group of students plans to charter a flight to take them to a ski resort. They are told that they will have to pay so much each if 75 students sign, but can save $65 each if they can find another 25 who want to go. What is the price if only 75 go? Assume the total price stays the same.

21. Susan Sharp, manager of a shoe department, must sell a certain pair of shoes for $18.60 to make her usual profit. She knows the store is planning a storewide sale in which everything will be sold at a 20 percent discount. How should she price the shoes to maintain her profit?

22. While filling out your income tax, you suddenly realize that you can deduct the sales tax on a boat you bought last summer. You find the canceled check which is made out for $2588.60. You remember that this included $7.50 for a license. If the sales tax in your state is 6 percent, how much tax did you pay?

23. John drives 20,000 miles each year and gasoline costs $1.80 per gallon. His present car gets 20 miles per gallon. If he bought the new model of the same car, he figures he would save $360 per year in gasoline costs. How many miles per gallon does the new model get?

24. In Problem 23, you saw that increasing John's gas mileage from 20 to 25 miles per gallon saved him $360 in gasoline expense in 1 year. Would increasing his mileage to 30 miles per gallon save an additional $360? First guess and then make sure of your answer.

25.* A classic problem is the courier problem. If a column of men 3 miles long is marching at 5 miles per hour, how long will it take a courier on a motorcycle traveling at 25 miles per hour to deliver a message from the end of the column to the front and then return?

26.* How long after 4:00 P.M. will the minute hand on a clock overtake the hour hand?

27.* What is the first time after twelve o'clock noon when the hands of the clock will be together again?

28.* At what time between four and five o'clock do the hands form a straight line?

29. Jack can do a certain job in 4 days, and Jill can do it in 3 days. How long will it take them working together?

30.* A can do a piece of work in one-third the time B can; B can do it in three-fourths the time C can; all together they can do it in 12 days. How long would it take A to do it alone?

31.* The combined age of a ship and its boiler is 48 years. The ship is twice as old as the boiler was when the ship was half as old as the boiler will be when the boiler is three times as old as the ship was when the ship was three times as old as the boiler. How old is the ship?

The Professor's Question
In my backyard is an assortment of boys and dogs. Counting heads, I get 21; counting legs, I get 70. How many boys and how many dogs are in my backyard?

3.3 / Problems That Lead to Two Equations

A variant of the problem posed above is the starting point for an amusing but brilliant essay on setting up equations by the master problem poser and problem solver, George Polya [*Mathematical Discovery,* vol. 1 (New York: Wiley, 1962), Chapter 2]. Polya suggests three different ways the professor might have mused about his problem.

TRIAL AND ERROR

Let's see. The answers must be whole numbers (fractional boys and dogs don't exist either in math books or backyards). There are 21 animals in my backyard. They can't all be boys, for then I would be able to count only 42 legs. Nor can they all be dogs, for then I would get 84 legs. Surely the answers are somewhere between these two extremes. I'll try 11 boys and 10 dogs. That gives 62 legs —too few. I need fewer boys and more dogs. I'll try 10 boys, 8 boys, and 7 boys. There it is! Seven boys and 14 dogs have 70 legs.

BRILLIANT IDEA

I think I have a better idea. I'll close my eyes and imagine that I see the boys and dogs engaged in a weird new game. Every boy is hopping along on one leg, and each dog is standing on its two hind

Boys	Dogs	Legs
21	0	42
0	21	84
11	10	62
10	11	64
8	13	68
7	14	70

A BRILLIANT IDEA

legs. Now I can count only 35 legs touching ground. And the number 35—I can think of it as counting each boy once and each dog twice. If I subtract the total number of animals, namely, 21, I'll have the number of dogs. There it is! There have to be 35 − 21 = 14 dogs; and that leaves 7 boys.

ALGEBRA

Trial and error is time-consuming and inefficient, especially in problems with many possibilities. And I can't rely on a brilliant idea coming along for every problem. I would like a systematic method that depends on neither hard work nor sudden visions. Such is the method of algebra. I must translate the problem into symbols and set up equations.

English	*Algebra*
In my backyard are a certain number of boys	x
and	
a certain number of dogs	y
I count 21 heads	$x + y = 21$
and 70 legs	$2x + 4y = 70$

Now I have two unknowns, x and y, but I also have two equations relating them. I want to find the values for x and y which satisfy both equations at the same time. This can be done by a method called *elimination of one unknown.* But to do it, I'll need another rule (in addition to the two in Section 3.1).

> *Rule 3.* You may add one equation to another (or subtract one equation from another).

This is how I use it:

Multiply the first equation by -2: $\qquad\qquad -2x - 2y = -42$
Write down the second equation: $\qquad\qquad\quad\;\, 2x + 4y = \;\;70$
Add the two equations: $\qquad\qquad\qquad\qquad\quad\;\;\, 2y = \;\;28$

Multiply by $\frac{1}{2}$: $\qquad\qquad\qquad\qquad\qquad\qquad\qquad\; y = \;\;14$

Substitute $y = 14$ in one of the original equations: $\quad x + 14 = \;\;21$

Add -14 $\qquad\qquad\qquad\qquad\qquad\qquad\qquad\quad\;\;\, x = \quad 7$

The key idea is this. Multiplying the first equation by -2 makes the coefficients of x the same, but of opposite sign, in the two equations. Addition of the two equations eliminates x, leaving one equation in a single unknown, y. The resulting equation can be solved by the methods of Section 3.1.

ANOTHER EXAMPLE

Homer has 50 coins in his pocket, all in nickels and dimes, and altogether he has $3.50. How many nickels and how many dimes has he?

Here is a simple way to set up this problem. *Let x be the number of nickels and y the number of dimes.*

English		*Algebra*
$\begin{pmatrix}\text{The number}\\\text{of nickels}\end{pmatrix}$ plus $\begin{pmatrix}\text{the number}\\\text{of dimes}\end{pmatrix}$ is 50		$x + y = 50$
$\begin{pmatrix}\text{The value of}\\\text{the nickels}\\\text{in cents}\end{pmatrix}$ plus $\begin{pmatrix}\text{the value of}\\\text{the dimes}\\\text{in cents}\end{pmatrix}$ is 350		$5x + 10y = 350$

And here is the way to solve the two equations:

Multiply the first equation by -5: $\qquad\quad -5x - \;\;5y = -250$
Write down the second equation: $\qquad\qquad\;\; 5x + 10y = \;\;\;350$
Add the two equations: $\qquad\qquad\qquad\qquad\qquad\;\; 5y = \;\;\;100$

Multiply by $\frac{1}{5}$: $\qquad\qquad\qquad\qquad\qquad\qquad\qquad\;\; y = \quad\;\; 20$

Substitute $y = 20$ into one of the
original equations: $\qquad\qquad\qquad\qquad x + 20 = \quad\;\; 50$

Add -20: $\qquad\qquad\qquad\qquad\qquad\qquad\qquad\;\;\; x = \quad\;\; 30$

Thus, Homer has 30 nickels and 20 dimes. That is certainly 50 coins. The nickels are worth $1.50, the dimes, $2.00, a total of $3.50.

A TWO-UNKNOWN MIXTURE PROBLEM

Suppose that you are applying for a job at Harold's Nut Shop. To test your ability, Harold points to two bins of nuts, walnuts costing $1.30 per pound and cashews costing $2.30 per pound. He tells you to prepare 25 pounds of a nut mix to be sold at $1.56 per pound. How many pounds of each kind should you use?

On an empty paper sack, you should write something like the following.

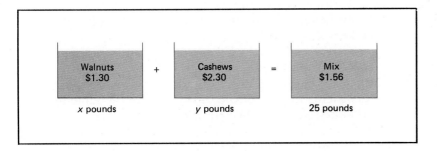

Setting it up:

x = number of pounds of walnuts
y = number of pounds of cashews
$$1.30x + 2.30y = (1.56)(25) \quad \text{(looking at values)}$$
$$x + y = 25 \quad \text{(looking at pounds)}$$

To solve, multiply the second equation by -1.30 and then add the two equations.

$$1.30x + 2.30y = 39.00$$
$$-1.30x - 1.30y = -32.50$$
$$y = 6.50$$

Now substitute $y = 6.50$ into the second of the original equations.

$$x + 6.50 = 25$$
$$x = 18.50$$

The conclusion is that you should mix 18.5 pounds of walnuts and 6.5 pounds of cashews. You might use your calculator to check on these answers. If it has algebraic logic, press

$$1.30 \boxed{\times} 18.5 \boxed{+} 2.30 \boxed{\times} 6.5 \boxed{=} \boxed{\div} 25 \boxed{=}$$

You will get $1.56 as desired.

SUMMARY

Some word problems involve several unknowns. In such cases, it is often best to label each of the unknowns with a different letter. Then try to find as many equations as there are unknowns. Finally, solve the equations simultaneously. Here are some guidelines for the case of two unknowns.

Let x represent one of the unknown quantities and y the other, being very specific about both. Draw a picture or make a chart.

Translate the conditions of the problem into equations. Two equations must be found.

Multiply the equations by appropriate numbers so that the coefficients for one of the unknowns are the same but of opposite sign in the two equations.

Add the two equations, thus eliminating an unknown.

Solve the resulting single equation for its unknown.

Substitute the value obtained into one of the original equations and solve for the other unknown.

Check your answers.

PROBLEM SET 3.3

1. Solve the following pairs of equations for x and y. Hint: In part (d), multiply the first equation by 3 and the second by 5; then add. Use a similar method in (e) and (f).

(a) $2x + 4y = 14$
$4x + y = 14$

(b) $3x + y = 17$
$2x + y = 12$

(c) $3x + 2y = 1$
$2x - 2y = 14$

(d) $3x + 5y = 2$
$2x - 3y = -5$

(e) $2y + 5x = 10$
$3y + 2x = 26$

(f) $3x + 8y = 0$
$7y + 6x = 5$

2. Solve for x and y.

(a) $3x + 4y = 7$
$x + 2y = 2$

(b) $2x + y = 14$
$8x + y = 26$

(c) $2x - 7y = 3$
$x + 7y = 12$

(d) $3x + 4y = 17$
$4x + 3y = 18$

(e) $x + y = 9$
$\frac{2}{3}x + y = 8$

(f) $3x + 5y = 0$
$5x + 7y = 2$

3. Solve for x and y. Hint: First clear of decimals and fractions.

(a) $3.5x + 2.5y = 7.5$
$2x + 5y = 9$

(b) $\frac{1}{4}x + \frac{1}{3}y = \frac{5}{6}$
$\frac{2}{5}x - \frac{4}{5}y = 1$

4. Solve for x and y.

(a) $2.1x - 1.2y = -2.7$
$0.7x + 2.4y = 5.1$

(b) $\frac{3}{4}x - \frac{5}{8}y = \frac{1}{2}$
$\frac{2}{3}x + \frac{5}{6}y = \frac{1}{3}$

5. The sum of two numbers is 17. If you take two-thirds of the first and subtract the second, you will get 8.

(a) If x and y are the two numbers, what is $x + y$?

(b) What algebraic expression describes two-thirds of the first, subtract the second?

(c) Find the two numbers.

6. Two numbers are so related that their sum is 35, and three times the one number is twice the other number. Find the numbers.

7. A boy has some quarters and nickels with a total value of $6. There are three times as many nickels as quarters.

(a) Let n be the number of nickels, q the number of quarters. Measured in cents, what is the value of his nickels? Of his quarters? The sum of these values must be 600.

(b) Express algebraically that there are three times as many nickels as quarters.

(c) How many coins of each kind does he have?

8. The price of admission tickets to a certain theater was $2.00 for adults and $1.50 for children. If 410 tickets were sold and the total receipts were $765, how many adults and how many children attended the theater?

9. A grocer has some Aromatic coffee worth $3.84 per pound and some Caffineo coffee worth only $2.64 per pound. She wishes to mix them so as to get 10 pounds worth $3.18 per pound.

(a) Let x be the number of pounds of Caffineo, y the number of pounds of Aromatic to be used in the mix. What is $x + y$?

(b) Measured in cents, what is the value of the Caffineo put into the mix? The value of the Aromatic in the mix? The value of the mix?

(c) How much of each coffee should she use?

10. A solution that is 40 percent alcohol is to be mixed with one that is 90 percent alcohol to obtain 100 liters of 60 percent alcohol solution. How much of each should be used?

11. If the length of a rectangle is decreased by 10 meters and the width is increased by 8 meters, the resulting figure will be a square whose area is the same as that of the rectangle.

   (a) If the length and the width of the given rectangle are l and w, respectively, what expressions describe the length and width of the new rectangle?
   (b) What algebraic equation states that the new rectangle is a square?
   (c) What algebraic equation states that the rectangles have equal area?
   (d) What are the dimensions of the given rectangle?

12. Professor Witquick, who had visited on his way to the office with a man digging a ditch, also stopped on the way home. Noting that the man's head was now below ground level, Witquick quipped, "Well, I see you've gotten a head in your work." Appreciating the pun, the worker replied that the top of his head was now as far below ground level as it had been above ground that morning and that his feet were now twice as far underground. "Ah," said Witquick, "but you'll have to tell me your height if you expect me to figure the depth of the ditch." "I'm 5 feet 9 inches," replied the worker. How deep is the ditch? Hint: Draw two pictures, one for the morning, one for now. Let x be the height of the worker's head above ground in the morning and y the depth of his feet below ground at that time. Then $2y$ is the depth of the ditch now.

13. A rowing crew that has been practicing on a river would like to know its rate in still water. The crew was able to row 16 miles downstream in 1 hour but took 2 hours to row back.
   (a) If x is the rate at which the crew rows in still water and y is the rate at which the stream carries the boat, what is the rate going downstream? Upstream?
   (b) Use $D = RT$ to get an equation describing the trip downstream. Do the same for the trip upstream.
   (c) Find the rate at which the crew rows in still water.

14. A fisherman motored upstream at a steady pace for 4 hours and then turned around and went back in 2 hours. The next day he needed 6 hours to travel 48 miles upstream. Assuming that the boat was going at full throttle both days and that the rate of the current remained constant, find out how fast the boat can go in still water.

15. Susan Sharp paid $2400 for some dresses and coats—$20 for each dress and $50 for each coat. She sold the dresses at a 20 percent markup and the coats at a 50 percent markup and made in all a profit of $900. How many dresses and how many coats did she buy?

16* Workers in a certain factory are classified in two groups, depending upon the skill required for their jobs. Group A workers are paid $4.00 per hour; group B workers are paid $2.50. In negotiations for a new contract, the union demands that the hourly wages for group B workers be brought up to two-thirds of those for group A workers. The company has 55 employees in group A and 40 in group B, all of whom work a 40-hour week. If the company is prepared to increase its weekly payroll by $2880, what hourly rates should it propose for each class of workers?

17* A woman leaves her fortune to her children. The first child receives \$1000 plus one-tenth of what is left; the second receives \$2000 plus one-tenth of what is left; the third receives \$3000 plus one-tenth of what is left; and so on. It turns out that each of the children receives the same amount. How many children are there, and how much does each receive?

18* Three mutually tangent circles have centers A, B, and C and radii a, b, and c, respectively. The lengths of segments AB, BC, and CA are 17, 23, and 12, respectively. Find the lengths of the radii.

Part II
Finding Order

ORDER

People seem to appreciate patterns, whether they are found in the mosaics of a great artist, the web of a common spider, or the molecules of a salt crystal. This appreciation extends to number patterns; most folks are at least momentarily intrigued when it is pointed out to them that

$$1^3 + 2^3 = (1 + 2)^2$$
$$1^3 + 2^3 + 3^3 = (1 + 2 + 3)^2$$
$$\vdots \qquad \qquad \vdots$$

There are people of course who resent too much organization or structure in their lives, who seemingly thrive on chaos and disharmony. And there are philosophers who believe that our world is ultimately chaotic, that it has no rhyme or reason. Consciously or unconsciously, most scientists reject this view. They can hardly do otherwise. The aim of science is to explain the universe — to explain it in terms of principles, rules, and laws. Asked how he arrived at the theory of relativity, Einstein replied that he had discovered it because he was so "firmly convinced of the harmony of the universe." In writing (with Infeld) a popular book on science, he said that he had tried "to give some idea of the eternal struggle of the inventive human mind for a fuller understanding of the laws governing physical phenomena."

Mathematicians are firmly committed to the same struggle. The prince of mathematicians, Carl Gauss, took as his motto, "Thou, nature, art my goddess; to thy laws my services are bound." Yet there is a difference between mathematics and science. The mathematician's laboratory is not so much the physical world of birds, bees, and molecules as it is the world of numbers, geometric shapes, and algebraic formulas. And here we find structure, laws, and patterns which may surpass in beauty those of the physical world itself.

But patterns, whether in the score of a symphony or in the arrangement of numbers in a sequence, are more easily recognized when we are trained to look for them. And so in Part II, we take as our theme the task of discovering patterns, of looking for order where at first none seems to exist. In our search for ordered phenomena, we move across many branches of mathematics — sequences, counting, probability, statistics, networks. We even introduce our readers to the magical world of modern computers, partly to show that they are just electronic devices designed to carry out the instructions of a well-ordered human mind.

Problem solving continues to be important. Almost every section begins with a problem; it inevitably ends with a host of others. There will be abundant opportunity to practice the strategies we learned in Part I. But now we have a loftier goal. We do not simply solve a single problem; we look at all its relatives. We try to discern a principle that underlies a whole class of problems. In short, we search for order in all that we do.

Leonhard Euler (1707–1783)

Born in Switzerland to a minister who wanted him to study theology, Leonhard Euler instead took up mathematics, completing his formal study at age 15. He was soon publishing papers and at age 19 won a prize from the French Académie of Sciences for his work on the masting of ships. Thus was launched the career of the most prolific writer on mathematics who ever lived. His collected works, when completed, will fill 74 volumes.

Euler made contributions to all fields of pure and applied mathematics. His interests spanned algebra, geometry, number theory, calculus, physics, astronomy, music, and such practical subjects as ship design, cartography, insurance, and canal building. His prodigious memory allowed him to write over 400 articles after he was totally blind. By the time of his death he had been accorded universal respect, and it was said that all the mathematicians of Europe were his pupils. In his breadth of interests and in his uncanny ability to find pattern and order in every subject, Leonhard Euler best represents the spirit of Part II.

Chapter 4
Numerical Patterns

"The northern ocean is beautiful," said the Orc, "and beautiful the delicate intricacy of the snowflake before it melts and perishes, but such beauties are as nothing to him who delights in numbers, spurning alike the wild irrationality of life and the baffling complexity of nature's laws."
J. L. SYNGE

4.1 / Number Sequences

Most students first meet number sequences on intelligence tests. Given a few numbers arranged according to some pattern, young scholars are encouraged to exhibit their mental powers by filling in a few more numbers—say two more, as in the problems we pose here.

a. 2, 4, 6, 8, □, □, . . .
b. 2, 4, 8, 16, □, □, . . .
c. 1, 8, 27, 64, □, □, . . .
d. 2, 5, 10, 17, □, □, . . .
e. 1, $\frac{1}{2}$, $\frac{1}{3}$, $\frac{1}{4}$, □, □, . . .

Strictly speaking there is no definite answer to these problems. Different people, looking at the same arrangement, see different patterns; in fact, the same person looking at the same arrangement may see different patterns at different times. The Rorschach Test in psychology is given with the idea that people reveal something about their mental outlook in telling what they see when looking at an inkblot. Number sequences are used in intelligence tests with the idea that people will reveal something about their mental ability in telling what they see when looking at number patterns. It is assumed in such tests that people will look for as simple a pattern

Keep It Simple
The principle of looking for a
simple pattern that explains a
given phenomenon is a guiding
principle in science.

"Our experience justifies us in
believing that nature is the
realization of the simplest
conceivable mathematical ideas."
A. Einstein

as possible and (as opposed to their grandmother's phone number) a pattern that others are likely to recognize.

Consider, for example, sequence d:

$$2, 5, 10, 17, \square, \square, \ldots$$

The chances are that most people, noting the pattern of adding successively larger odd integers (i.e., 3, 5, and 7) will conclude that the numbers in the boxes should be 26 and 37. But a more imaginative soul may observe that 3, 5, 7 are the first three odd primes and conclude that the answers are

$$17 + 11 = 28$$
$$28 + 13 = 41$$

The principle of preferring the simple solution suggests that we use the first answers; but if the person giving the second answers is as forceful as he is imaginative, it will be hard to convince him that he is wrong—because he isn't.

We mention in passing that the sequence

$$1, 1, 2, 3, 5, 8, \square, \square, \ldots$$

which was the subject of our initial cartoon, was first studied by Leonardo Fibonacci in 1202. The reader who has decided that the next two numbers are 13 and 21 may already be caught in the web surrounding this intriguing sequence. Some enthusiasts have made a life-long hobby of studying this sequence; they have formed the Fibonacci Society, and they regularly publish the *Fibonacci Quarterly*. In Section 4.6 we shall see some of the reasons for their fascination.

SOME DEFINITIONS

We have used the word "sequence" in the familiar sense to describe a series of things following one another in a definite order. Thus a **number sequence** is just an ordered arrangement of numbers. Relying on the common habit of reading from left to right, we have written in example a:

$$2, 4, 6, 8, \ldots$$

to indicate that 2 comes first, 4 comes second, 6 comes third, etc. The dots mean that the numbers in the sequence continue indefinitely; there is a hundredth term, a millionth term, etc.

A general sequence can be written:

$$a_1, a_2, a_3, a_4, \ldots$$

with a_1 designating the first term, a_2 the second, etc. Note that the subscript indicates the position of the term within the sequence. In the example above,

$$a_1 = 2 \qquad a_2 = 4 \qquad a_3 = 6 \qquad a_4 = 8 \qquad \ldots$$

Now we ask a question. What is a_{21}? The keen observer, noting that in this sequence you can get the value of a term by doubling its subscript, answers 42.

Mathematicians love formulas. They provide not only their means of livelihood but also a way of packing huge amounts of information into short sentences. Let n be a counting number. What is the value of a_n? That is, can we give an algebraic formula for the nth term of our sequence? For the example discussed above, we saw that

$$a_n = 2n$$

There, in that short line, is all the information you need to construct the sequence. No longer need we speak of the sequence 2, 4, 6, 8, ... We can speak of the sequence $a_n = 2n$, and we will have no more trouble writing down the 99th or 121st term than in writing the 21st.

$$a_{99} = 2 \cdot 99 = 198 \qquad a_{121} = 2 \cdot 121 = 242$$

We call a formula like this an **explicit formula.** It tells us how a_n is related to n, that is, how the value of the term a_n is related to its subscript.

There is another way of describing the pattern we observe in a sequence. We can observe how a term relates to previous terms. In our example, we always add 2 to a term to find the next one. A concise way of saying this is

$$a_n = a_{n-1} + 2$$

For example, the fourth term equals the fourth minus one term (or third term) plus 2. Now, knowing that $a_1 = 2$, we can find all the other terms provided we list them in order. Thus

$$a_2 = a_1 + 2 = 2 + 2 = 4$$
$$a_3 = a_2 + 2 = 4 + 2 = 6$$
$$a_4 = a_3 + 2 = 6 + 2 = 8$$

A formula that tells how the general term is related to the previous term (or terms) is called a **recursion formula.**

In Table 4-1 below we illustrate these two methods of expression for the five examples introduced at the beginning of the section. The reader should verify his or her understanding by checking them carefully.

TABLE 4-1		
Sequence	**Recursion Formula**	**Explicit Formula**
a. 2, 4, 6, 8, . . .	$a_n = a_{n-1} + 2$	$a_n = 2n$
b. 2, 4, 8, 16, . . .	$b_n = 2b_{n-1}$	$b_n = 2^n$
c. 1, 8, 27, 64, . . .	$c_n = ?$	$c_n = n^3$
d. 2, 5, 10, 17, . . .	$d_n = d_{n-1} + (2n - 1)$	$d_n = n^2 + 1$
e. 1, $\frac{1}{2}$, $\frac{1}{3}$, $\frac{1}{4}$, . . .	$e_n = ?$	$e_n = 1/n$

We have left two question marks where the formulas are not obvious. Sometimes it seems impossible to find either a recursion formula or an explicit formula (e.g., no one knows a formula for the nth prime). But the challenge is always there; find a formula. Sometimes, as in many computer-related problems, recursion formulas are important. In other cases, explicit formulas are the most useful; they allow us to answer almost any question concerning terms of the sequences right away. What is the fourteenth term in sequence d? $d_{14} = 14^2 + 1 = 197$. What is the sum of the fifth and sixth terms in sequence c?

$$c_5 + c_6 = 5^3 + 6^3 = 125 + 216 = 341$$

TRIANGULAR AND SQUARE NUMBERS

The Greeks were fascinated by numbers that arise in a geometric way. Consider the following diagrams.

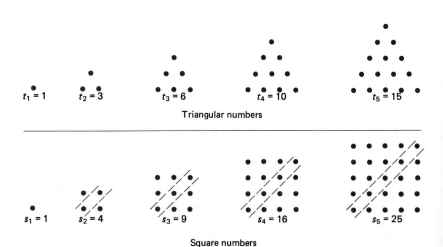

Triangular numbers

Square numbers

We seek explicit formulas for both t_n and s_n. The latter is easy; $s_n = n^2$. The former requires some ingenuity. Let us look, for example, at the s_4 diagram. Above and below the indicated diagonal we have triangles, each consisting of six dots. There are four dots on the diagonal itself. Thus, $s_4 = 2 \cdot 6 + 4 = 2t_3 + 4$. A similar look at s_2, s_3, s_5, etc., leads to

$$s_2 = 2t_1 + 2$$
$$s_3 = 2t_2 + 3$$
$$s_4 = 2t_3 + 4$$
$$s_5 = 2t_4 + 5$$

and, in general,

$$s_{n+1} = 2t_n + (n + 1)$$

But

$$s_{n+1} = (n + 1)^2 = n^2 + 2n + 1$$

and so

$$n^2 + 2n + 1 = 2t_n + n + 1$$
$$n^2 + n = 2t_n$$

$$\boxed{\frac{n(n + 1)}{2} = t_n}$$

Now we can find the 100th triangular number. It is just

$$t_{100} = 100(101)/2 = 5050$$

Here is a problem that challenged the great mathematician Leonhard Euler. Are there any square triangular numbers, that is, numbers that are both square and triangular? A tedious check (yes, even mathematicians find calculating tedious) reveals the four numbers 1, 36, 1225, and 41,616. Are there more? The answer is an emphatic yes; Euler found that there are infinitely many and, as early as 1730, he gave formulas for them. Here are his results, one an explicit formula and the other a recursion formula:

$$u_n = \frac{(17 + 12\sqrt{2})^n + (17 - 12\sqrt{2})^n - 2}{32}$$
$$u_n = 34u_{n-1} - u_{n-2} + 2$$

We expect that most readers will, if they pause to think about it, wonder how Euler discovered these formulas. But once they have been given, it is easy to check that they work, at least for small values of n.

SUMMARY

A number sequence a_1, a_2, a_3, . . . is an ordered arrangement of numbers. We find it convenient if the sequence can be described by a formula which can take one of two forms. An explicit formula (e.g., $a_n = 3n$) relates the value of the nth term a_n directly to n. A recursion formula (e.g., $a_n = a_{n-1} + 3$) relates the nth term to the previous term or terms. Explicit formulas are especially useful, an important one being that for the nth triangular number, namely, $t_n = n(n+1)/2$.

PROBLEM SET 4.1

1. Try to find a pattern and use it to fill in the boxes.
   (a) 3, 6, 9, 12, □, □, . . .
   (b) 21, 18, 15, 12, □, □, . . .
   (c) 1, $\frac{1}{2}$, $\frac{1}{4}$, $\frac{1}{8}$, □, □, . . .
2. Fill in the boxes.
   (a) 1, 3, 5, 7, □, □, . . .
   (b) 2, 8, 32, 128, □, □, . . .
   (c) 2, $\frac{2}{3}$, $\frac{2}{9}$, $\frac{2}{27}$, □, □, . . .
3. In each case, an explicit formula is given. Find the indicated terms.
   (a) $a_n = 3n + 2$; $a_6 = $ □, $a_{10} = $ □
   (b) $a_n = 1/2n$; $a_9 = $ □, $a_{20} = $ □
   (c) $a_n = (\frac{1}{2})^n$; $a_3 = $ □, $a_6 = $ □
   (d) $a_n = 15 - 2n$; $a_6 = $ □, $a_{12} = $ □
4. Find the indicated terms.
   (a) $a_n = 2n - 9$; $a_3 = $ □, $a_{12} = $ □
   (b) $a_n = (-1)^n(2n)$; $a_2 = $ □, $a_5 = $ □
   (c) $a_n = (\frac{2}{3})^n$; $a_4 = $ □, $a_7 = $ □
5. Give an explicit formula for each sequence in Problem 1. Hint: Try listing the terms of the sequence in a row, each term directly under a general term, as illustrated in the margin for the sequence 1, 4, 9, 16, . . . Then try to relate the value of the term to the subscript above it.

   $a_1\ a_2\ a_3\ a_4 \cdots$
   $1^2\ 2^2\ 3^2\ 4^2 \cdots$
   $a_n = n^2$

6. Give an explicit formula for each sequence in Problem 2.
7. In each case, an initial term (or terms) and a recursion formula are given. Find a_5. Hint: First find a_2, a_3, a_4.
   (a) $a_1 = 1$; $a_n = 3a_{n-1}$
   (b) $a_1 = 2$; $a_n = a_{n-1} + 3$
   (c) $a_1 = 4$; $a_n = \frac{1}{2}a_{n-1}$
   (d) $a_1 = 1$, $a_2 = 2$; $a_n = 2a_{n-1} - a_{n-2}$
8. Find a_6 for each of the following.
   (a) $a_1 = 1$; $a_n = 2a_{n-1} + 1$
   (b) $a_1 = 4$; $a_n = 3a_{n-1}$
   (c) $a_1 = 6$; $a_n = \frac{1}{3}a_{n-1}$
   (d) $a_1 = 2$, $a_2 = 3$; $a_n = a_{n-1} + a_{n-2}$
9. Give a recursion formula for each of the sequences in Problem 1.
10. Give a recursion formula for each of the sequences in Problem 2.
11. Find a pattern in each of the following sequences and fill in the boxes.
    (a) 2, 6, 18, 54, 162, □, □, . . .
    (b) 11, 9, □, 5, 3, □, . . .
    (c) $\frac{1}{2}$, $\frac{2}{3}$, $\frac{3}{4}$, $\frac{4}{5}$, $\frac{5}{6}$, □, □, . . .

(d) $1, \frac{1}{8}, \frac{1}{27}, \frac{1}{64}, \frac{1}{125}, \Box, \Box, \ldots$

(e) $6, 3, \Box, \frac{3}{4}, \frac{3}{8}, \Box, \ldots$

(f) $2, 2, 4, 6, 10, \Box, \Box, \ldots$

(g) $2, 2, 4, 8, 14, 26, \Box, \Box, \ldots$

(h) $2, 3, 5, 7, 11, 13, 17, \Box, \Box, \ldots$

(i) $1, 3, 6, 10, 15, 21, \Box, \Box, \ldots$

12. In each case in Problem 11, try to find a recursion formula or an explicit formula (or both, if you can) for the sequence.

13. Let $A_n = a_1 + a_2 + a_3 + \cdots + a_n$. Find A_6 for each of the following. Hint: Begin by finding $a_1, a_2, a_3, a_4, a_5, a_6$.

   (a) $a_n = 2^n$

   (b) $a_n = n + 2$

   (c) $a_n = (-1)^n$

   (d) $a_n = n^2 - n$

   (e) $a_1 = 8, a_n = \frac{1}{2}a_{n-1}$

   (f) $a_1 = 1, a_2 = 2, a_n = a_{n-1} - a_{n-2}$

14. Let $a_n = (\frac{1}{2})^n$ and $A_n = a_1 + a_2 + \cdots + a_n$.

   (a) Find a_1, a_2, a_3, a_4, a_5.

   (b) Find A_1, A_2, A_3, A_4, A_5.

   (c) Guess at an explicit formula for A_n.

15. Let a_n be the nth digit in the decimal expansion of $\frac{1}{7}$. Using long division, obtain $a_1, a_2, \ldots, a_{14}$. Do you observe a pattern? Find a_{53}. Do the same for the decimal expansion of $\frac{5}{13}$.

16. Find $1 + 2 + 3 + \cdots + 1000$. Hint: It's a triangular number.

17* Consider the sequence $1, 1 + 3, 1 + 3 + 5, \ldots$. If we call the nth term a_n, guess at an explicit formula for a_n. Now look at the diagrams for the square numbers. Do you see anything?

18* The numbers $1, 5, 12, \ldots$ are called pentagonal numbers (see diagrams at the right). Call the resulting sequence p_n. Find p_4, p_5, and p_6. Can you guess at an explicit formula for p_n. If so, what is p_{100}?

19* If $a_1 = 3, a_2 = 5$, and $a_n = 3a_{n-1} - 2a_{n-2}$, find an explicit formula for a_n.

20*
$$\begin{array}{rcl} 1 & = & 1 \\ 2 + 3 + 4 & = & 1 + 8 \\ 5 + 6 + 7 + 8 + 9 & = & 8 + 27 \\ 10 + 11 + 12 + 13 + 14 + 15 + 16 & = & 27 + 64 \end{array}$$

Guess the general law suggested by these examples and try to write it using good mathematical notation.

c 21. The recursion formula for the square triangular numbers is $u_n = 34u_{n-1} - u_{n-2} + 2$. Use this formula and the fact that $u_1 = 1$ and $u_2 = 36$ to calculate u_3, u_4, u_5, and u_6.

c 22. The explicit formula for the square triangular numbers is given near the end of this section. Use it to calculate u_3, u_4, u_5, and u_6. Compare your answers to those obtained in Problem 21. Can you account for any discrepancies?

c 23. There is a well-known recursion formula for finding $\sqrt{A}$. Let $a_1 = a/2$ and $a_n = \frac{1}{2}(a_{n-1} + A/a_{n-1})$. For $A = 20$, find a_2, a_3, a_4, and a_5. Compare a_5 with the result obtained when you calculate $\sqrt{20}$ using the square root key. Now follow the same procedure for $A = 312$, but also calculate a_6 and a_7.

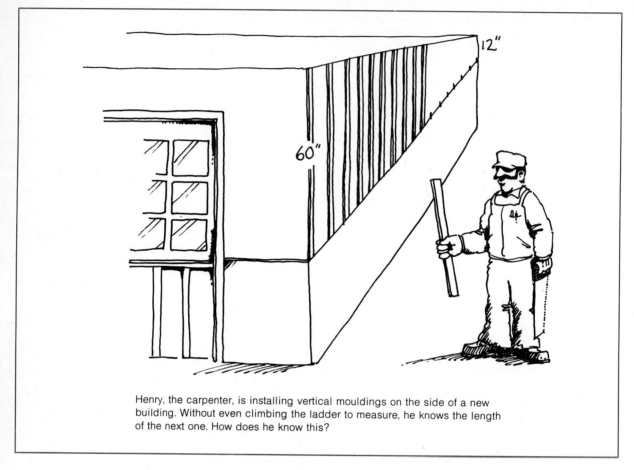

Henry, the carpenter, is installing vertical mouldings on the side of a new building. Without even climbing the ladder to measure, he knows the length of the next one. How does he know this?

4.2 / Arithmetic Sequences

Actually, Henry doesn't need much math to figure out his problem. He simply observes four things: The first moulding is 60 inches long; the last moulding is 12 inches long; there are 17 mouldings altogether; and each moulding is a fixed amount (say t inches) shorter than its neighbor to the left. From here on, it's downhill all the way. For to go from the first to the seventeenth moulding requires a total decrease of $16t$ inches. Thus $16t = 60 - 12 = 48$, or $t = 3$ inches. And so the next moulding, which happens to be the eleventh, has the length $60 - (10)(3) = 30$ inches.

Now consider the following number sequences. When you see the pattern, fill in the boxes.

a. 2, 5, 8, 11, □, □, . . .
b. 8, 8.4, 8.8, 9.2, □, □, . . .
c. 7, 3, −1, −5, □, □, . . .

"Did you ever notice that remarkable coincidence?" F. Scott Fitzgerald wrote in 1928 to British writer Shane Leslie. "Bernard Shaw is 61 years old, H, G. Wells is 51, G. K. Chesterton is 41, you're 31 and I'm 21—all the great authors of the world in arithmetic progression."

What feature do these sequences have in common? Just this. A term is obtained from its predecessor by the addition of a fixed number (3, 0.4, and −4, respectively). Such a sequence is called an **arithmetic sequence.** It has the general form $a_1, a_2, a_3, \ldots$ with the recursion formula

$$a_n = a_{n-1} + d$$

where d is the fixed number, called the **common difference,** that we must add to go from one term to the next.

Of course we also want an explicit formula. The diagram at the right should help you discover it. Note that the number of d's to be added is 1 less than the subscript of the term; that is

$$\boxed{a_n = a_1 + (n-1)d}$$

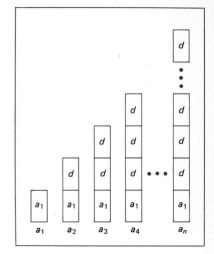

With this formula we can answer all kinds of questions. What is the 101st term in sequence a? In sequence b?

$$a_{101} = 2 + (100)(3) = 302$$
$$b_{101} = 8 + 100\,(0.4) = 48$$

Can you find the 59th term in sequence c?

SUMS OF ARITHMETIC SEQUENCES

Here is an old problem. Calculate the sum

$$1 + 2 + 3 + \cdots + 98 + 99 + 100$$

Without testifying to its accuracy, we pass along an anecdote about it taken from E. T. Bell's book, *Men of Mathematics* (New York: Simon and Schuster, 1937). Carl Gauss was 10 years old when he was admitted to his first arithmetic class. His teacher, one Buttner by name, was in the habit of obtaining an hour or so of quiet for himself by requiring the class to do long addition problems. One day he asked his pupils to add all the numbers from 1 to 100. Hardly had he made the assignment when young Carl flung his slate on the teacher's desk with the correct answer on it. This is how he thought about the problem:

$$1 + 2 + \cdots + 49 + 50 + 51 + 52 + \cdots + 99 + 100$$

There are 50 pairs, each totaling 101. Therefore the answer is $(50)(101) = 5050$. Bell observes with admirable restraint, "[This trick] is very ordinary once it is known, but for a boy of ten to find it instantaneously by himself is not so ordinary."

The same trick that Gauss used works, with slight modification,

Carl F. Gauss (1777–1855)
Carl Gauss was born to humble parents in Brunswick, Germany. An infant prodigy who discovered a mistake in his father's accounts at age 3, he became the greatest mathematician of the nineteenth century. He gave the first proof of the Fundamental Theorem of Algebra and contributed to all branches of mathematics as well as physics. He and Weber invented the telegraph.

for any arithmetic sequence. Let $a_1, a_2, a_3, \ldots$ be such a sequence and consider

$$A_n = a_1 + a_2 + a_3 + \cdots + a_{n-2} + a_{n-1} + a_n$$

Because the sequence is arithmetic, each of the indicated pairs has the same sum and there are $n/2$ of them. Answer: $A_n = (n/2)(a_1 + a_n)$. But we cheated, even though our answer is right. We implicitly assumed that n was even, since we divided the terms of A_n into pairs. To get around this difficulty, write A_n twice, once forward and once backward, and add the results.

$$
\begin{array}{llll}
A_n = a_1 & + a_2 & + \cdots + a_{n-1} & + a_n \\
A_n = a_n & + a_{n-1} & + \cdots + a_2 & + a_1 \\
\hline
2A_n = (a_1 + a_n) & + (a_2 + a_{n-1}) & + \cdots + (a_{n-1} + a_2) & + (a_n + a_1)
\end{array}
$$

Each group on the right has the same value, and there are n such groups. Thus

$$2A_n = n(a_1 + a_n)$$

$$\boxed{A_n = \frac{n}{2}(a_1 + a_n)}$$

Consider again the garage moulding problem introduced at the beginning of the section. Here is a second question about it. How much moulding material is needed to make all 17 mouldings?

$$A_{17} = \tfrac{17}{2}(a_1 + a_{17}) = \tfrac{17}{2}(60 + 12) = 612$$

It will take 612 inches of moulding material.

SOME APPLICATIONS

Imagine a mammoth pile of logs stacked in the manner shown at the left. Let us suppose that there are 113 logs in the bottom row. How many logs are there in the whole pile? Note that this amounts to adding

$$1 + 2 + 3 + \cdots + 113$$

The formula above gives us

$$A_{113} = \tfrac{113}{2}(1 + 113) = \tfrac{113}{2}(114) = 6441$$

Consider next a salary problem. Suppose that you have been offered a starting annual salary of $18,000 with annual increments of $1500 for each of the succeeding 39 years of your career. How much will you make during the last year of work and how much during your whole 40-year career?

The sequence to be considered is

$$18,000, \ 19,500, \ 21,000, \ \ldots$$

For it we must calculate a_{40} and A_{40}

$$a_{40} = a_1 + (40 - 1)d = 18,000 + 39(1500) = \$76,500$$
$$A_{40} = \tfrac{40}{2}(a_1 + a_{40}) = 20(18,000 + 76,500) = \$1,890,000$$

Here is a trickier problem. Slugger Brown must make a difficult decision. Both the Podunk Possums and the Tooterville Toads have offered her a 5-year baseball contract.

> Podunk contract: Starting salary of $40,000 per year with annual increases of $8000 each.
> Tooterville contract: Starting salary of $40,000 per year with semiannual increases of $2000 each.

Brown likes the Tooterville manager best but considers Podunk's offer to be superior financially. Should she opt for the manager or the money? The answer is—she should do some mathematics. Actually, Tooterville is offering more money and the best manager.

To see why, draw two 5-year time lines indicating the salary obtained during each period (salary indicated in thousands).

Podunk   40   48   72

Tooterville   20   22   24   26   38

Both sequences are arithmetic. Thus last-period salaries and 5-year totals can be obtained from the two shaded formulas. Note that we must use $n = 5$ in the case of annual raises and $n = 10$ in the case of semiannual raises.

$$\text{Podunk total:} \quad \tfrac{5}{2}(40,000 + 72,000) = \$280,000$$
$$\text{Tooterville total:} \quad \tfrac{10}{2}(20,000 + 38,000) = \$290,000$$

SUMMARY

A sequence that increases (or decreases) by always adding the same number d is called an arithmetic sequence. For such a sequence, there are two important formulas, one giving the nth term a_n and the other giving the sum A_n of the first n terms:

$$a_n = a_1 + (n - 1)d$$
$$A_n = a_1 + a_2 + \cdots + a_n = \frac{n}{2}(a_1 + a_n)$$

PROBLEM SET 4.2

1. Fill in the boxes.
   (a) 1, 3, 5, 7, □, □, . . .
   (b) 5, 8, 11, 14, □, □, . . .
   (c) 100, 96, 92, 88, □, □, . . .
2. Fill in the boxes.
   (a) 17, 15, 13, □, □, . . .
   (b) 2.0, 2.5, 3.0, □, □, . . .
   (c) 100, 150, 200, □, □, . . .
3. Determine the common difference d and the fortieth term for each sequence in Problem 1. Hint: To find the fortieth term use $a_n = a_1 + (n - 1)d$ with $n = 40$.

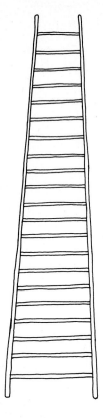

4. Determine d and the thirtieth term for each sequence in Problem 2.
5. Find A_{40}, B_{40}, and C_{40} for the sequences in Problem 1. Hint: To find A_{40}, recall that you found a_{40} in Problem 3, and then use $A_n = (n/2)(a_1 + a_n)$ with $n = 40$.
6. Find A_{30}, B_{30}, and C_{30} for the sequences in Problem 2.
7. Calculate:
   (a) $2 + 4 + 6 + \cdots + 100$
   (b) $1 + 3 + 5 + \cdots + 99$
   (c) $3 + 6 + 9 + \cdots + 99$
8. Calculate:
   (a) $1 + 3 + 5 + \cdots + 199$
   (b) $5 + 10 + 15 + \cdots + 600$
   (c) $8 + 12 + 16 + \cdots + 64$
9. A ball falls 16 feet during the first second, 48 feet during the second, 80 feet during the third, etc. How far does it fall during the tenth second? What is the total distance it falls in 10 seconds?
10. How many integers are there between 100 and 300 that are multiples of 7? What is the sum of these integers?
11. The bottom rung of a tapered ladder is 60 centimeters long; the top one is 40 centimeters long. If there are 21 rungs, how much rung material was needed to make the ladder?
12. The terms 2, a, b, c, 3.2, . . . form an arithmetic sequence. Determine a, b, and c.
13. A Christmas Club collects 7 cents the first week, 14 cents the second, 21 cents the third, etc., for 50 weeks. How much will it be worth at that time?
14. A contractor who fails to complete a building in a specified amount of time must pay a penalty of $100 for each of the first 8 days of extra time required. After that the penalty is increased by $10 per day. If he goes 30 days over, what is his total penalty?
15. At a club meeting with 200 people present, everyone shook hands with every other person exactly once. How many handshakes were there? Suggestion: Number the people from 1 to 200. Let person 1 shake hands with each of the 199 other people. Now let person 2 shake hands—with how many people? etc.
16. A clock strikes once at one o'clock, twice at two o'clock, etc. How many times does it strike between 8:15 A.M. on Monday morning and the same time on the next day?
17. A pile of logs has 120 logs in the first layer, 119 in the second layer, 118 in the third, etc. How many logs are there in the pile if there are 57 layers?
18.* Consider the diagrams of the first three triangular, square, and pentagonal numbers shown at the left. Note the appearance of arithmetic sequences and use this fact to find explicit formulas for t_n, s_n, and p_n.
19.* Consider the list

$$
\begin{array}{rl}
1 & = 1 \\
2 + 3 + 4 & = 1 + 8 \\
5 + 6 + 7 + 8 + 9 & = 8 + 27 \\
10 + 11 + 12 + 13 + 14 + 15 + 16 & = 27 + 64
\end{array}
$$

What does the nth row look like? Consider adding all these equali-

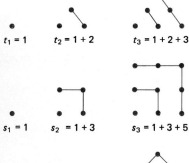

$t_1 = 1$ $t_2 = 1 + 2$ $t_3 = 1 + 2 + 3$

$s_1 = 1$ $s_2 = 1 + 3$ $s_3 = 1 + 3 + 5$

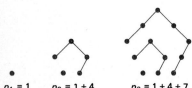

$p_1 = 1$ $p_2 = 1 + 4$ $p_3 = 1 + 4 + 7$

ties down through the nth row. The left sum is $1 + 2 + 3 + \cdots + n^2$. What is the right sum? Can you use this to obtain an explicit formula for $1^3 + 2^3 + 3^3 + \cdots + n^3$?

20.* Consider the list

$$
\begin{array}{rl}
1 & = 1^3 \\
3 + 5 & = 2^3 \\
7 + 9 + 11 & = 3^3 \\
13 + 15 + 17 + 19 & = 4^3
\end{array}
$$

What does the nth row look like? Now add all these equalities through the nth row. Simplify. Compare with Problem 19.

With the advent of paper money, Silas, who used to like to run his fingers through his gold coins, now likes to contemplate his fortune piled up in crisp, new $1 bills. Poor Silas started with just $1, but he has found a way to double his money each year. How high a pile does he have after 40 years? (Assume a $1 bill is about 0.01 inch thick.)

4.3 / Geometric Sequences

Poor Silas is going to need a telescope. In fact, his stack of dollar bills reaches almost to the moon. Let's see why.

If we list the height of the stack of bills at the beginning and then at the end of the first year, end of the second year, etc., we obtain the sequence

$$0.01,\ 2(0.01),\ 2^2(0.01),\ 2^3(0.01),\ \ldots$$

Our job is to calculate $2^{40}(0.01)$. Now of course we could do this calculation exactly. But it makes little sense to do so. Clearly we are only interested in a "ballpark" answer. To that end we use these approximations: 1 foot is about 10 inches, 1 mile is about 5000 feet, and 2^{10} is about 1000 (since $2^{10} = 1024$). Thus

$$
\begin{aligned}
2^{40}(0.01) &= 2^{10}2^{10}2^{10}2^{10}\,(0.01) \\
&\approx (1000)(1000)(1000)(1000)(0.01) \\
&= 10{,}000{,}000{,}000 \text{ inches} \\
&\approx 1{,}000{,}000{,}000 \text{ feet} \\
&\approx 200{,}000 \text{ miles}
\end{aligned}
$$

The moon is about 240,000 miles away. Certainly by year 41, Silas' pile of bills will extend beyond the moon.

We have just considered a *doubling sequence*. Try to discover a similar feature in the following sequences and then fill in the boxes.

    a. 2, 6, 18, 54, □, □, . . .
    b. 16, 8, 4, 2, □, □, . . .
    c. 1, −2, 4, −8, □, □, . . .

Each of these is a **geometric sequence.** Any term in such a sequence is obtained from its predecessor by multiplication by a fixed number $(3, \tfrac{1}{2}, -2,$ respectively, in the three examples). It has the general form $a_1, a_2, a_3, \ldots$ with the recursion formula

$$a_n = ra_{n-1}$$

The fixed number r is called the **common ratio.**

We seek an explicit formula for a_n. Note that

$$
\begin{aligned}
a_1 &= a_1 \\
a_2 &= ra_1 \\
a_3 &= ra_2 = r(ra_1) = r^2a_1 \\
a_4 &= ra_3 = r(r^2a_1) = r^3a_1 \\
&\ \ \vdots
\end{aligned}
$$

$$a_n = r^{n-1}a_1$$

It is easy to put this formula to use. For the three sequences above, we have

$$
\begin{aligned}
a_7 &= 3^6(2) = 729(2) = 1458 \\
b_{11} &= (\tfrac{1}{2})^{10}(16) = \frac{1}{2^{10}} \cdot 2^4 = \frac{1}{2^6} = \frac{1}{64} \\
c_8 &= (-2)^7(1) = -128
\end{aligned}
$$

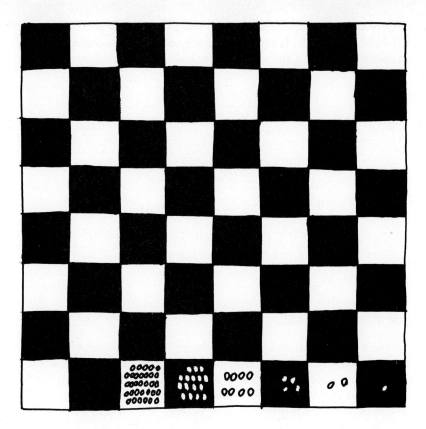

SUMS OF GEOMETRIC SEQUENCES

There is a legend associated with geometric sequences and chess-boards. When Sessa invented the game of chess, the king of Persia was so pleased that he promised to fulfill any request—to the half of his kingdom. Sessa, with an air of modesty, asked for one grain of wheat for the first square of the chessboard, two for the second square, four for the third, etc. The king was amused at this odd request but immediately commanded one of his servants to get a sack of wheat and begin counting it out. It was soon apparent that the king could not fulfill his promise. It would have taken the world's total production of wheat for several centuries. And to count the required number of grains at the rate of five per second would have taken 100 billion years.

What is the mathematics behind Sessa's request? He asked for

$$1 + 2 + 2^2 + 2^3 + \cdots + 2^{63}$$

grains of wheat, the sum of a geometric sequence with a common ratio of 2.

Let's look at the general problem. Consider $A_n = a_1 + a_2 + \cdots + a_n$, where $a_1, a_2, \ldots$ is a geometric sequence with a common ratio of r. Multiply A_n by r and subtract the product from A_n as indicated:

$$A_n = a_1 + ra_1 + r^2a_1 + \cdots + r^{n-1}a_1$$
$$\underline{rA_n = \qquad ra_1 + r^2a_1 + \cdots + r^{n-1}a_1 + r^na_1}$$
$$A_n - rA_n = a_1 + 0 \quad + 0 \quad + \cdots + 0 \qquad - r^na_1$$
$$A_n(1-r) = a_1(1-r^n)$$

$$\boxed{A_n = \frac{a_1(1-r^n)}{1-r} = \frac{a_1(r^n-1)}{r-1}}$$

Thus Sessa's request was for

$$A_{64} = \frac{1(2^{64}-1)}{2-1} = 2^{64} - 1 \approx 2^{64} \approx 2^4(1000)^6$$
$$= 16,000,000,000,000,000,000$$

grains of wheat.

FOR THE MUSICALLY INCLINED

For the Greeks, music and mathematics were inseparably linked. Plato and Aristotle taught that one must study the *quadrivium,* consisting of arithmetic, geometry, astronomy, and music in order to understand the laws of the universe. Mathematics and music still appear close to each other in most college catalogs, but this is due more to administrative respect for the alphabet than educational philosophy. Discussing geometric sequences affords us the opportunity to demonstrate a genuine interconnection.

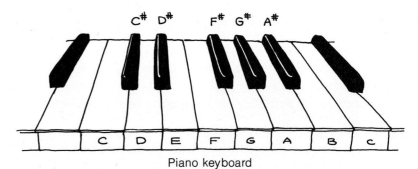

Piano keyboard

In the equally tempered scale, to which all keyed instruments have been tuned since the days of Bach (1685–1750), the intervals between the 12 halftones are said to be equal. What this really means is that the frequencies of vibration for these 12 tones form a geometric sequence. If middle C has frequency f (f is usually chosen to be about 262), the other frequencies are as follows.

C	C♯	D	D♯	E	F	F♯	G	G♯	A	A♯	B	C̄
f	rf	r^2f	r^3f	r^4f	r^5f	r^6f	r^7f	r^8f	r^9f	$r^{10}f$	$r^{11}f$	$r^{12}f$

Moreover, the frequency of high C, labeled C̄, is just twice that of middle C, so that $r^{12} = 2$, or

$$r = \sqrt[12]{2} = 1.05946 \ldots$$

You don't have to be a music major to know that when two piano keys are struck together you may or may not hear a sound that you like. What is it that makes some concords (intervals) pleasing while others are positively jarring? It turns out that the concords most pleasant to the human ear are those with frequencies that form small-integer ratios (e.g., $\frac{2}{1}$, $\frac{3}{2}$, $\frac{4}{3}$, $\frac{5}{4}$). Here are some pleasant concords familiar to all music students.

Interval Name	Interval	Ratio of Frequencies
Octave	C to C̄	$(1.05946)^{12} = 2 = \frac{2}{1}$
Perfect fifth	C to G	$(1.05946)^7 = 1.49831 \approx \frac{3}{2}$
Perfect fourth	C to F	$(1.05946)^5 = 1.33484 \approx \frac{4}{3}$
Major third	C to E	$(1.05946)^4 = 1.25992 \approx \frac{5}{4}$

Actually none of the ratios, other than that of the octave, is exactly a small-integer ratio. For example, the perfect fifth has a ratio of 1.49831 which is a bit off from $\frac{3}{2} = 1.50000$. It is these slight discrepancies that make it difficult for a violinist to play with a piano, since the violinist is naturally inclined to play a perfect fifth so that the ratio involved is exactly $\frac{3}{2}$.

Many alternatives to the equally tempered scale have been proposed, but none considered satisfactory. One that is well-known is the so-called diatonic scale which has the following frequencies (for the white keys):

C	D	E	F	G	A	B	C̄
256	288	320	$341\frac{1}{3}$	384	$426\frac{2}{3}$	480	512

Note that the interval C to G has the ratio

$$\frac{384}{256} = \frac{3(128)}{2(128)} = \frac{3}{2}$$

while the interval C to F has the ratio

$$\frac{341\frac{1}{3}}{256} = \frac{\frac{1024}{3}}{256} = \frac{1024}{3(256)} = \frac{4(256)}{3(256)} = \frac{4}{3}$$

This looks promising. However, equal tempering has been sacrificed (i.e., the interval between successive notes is not constant), and this means that transpositions between different keys will be troublesome. The reader who is interested in other possible scales can consult H. Steinhaus, *Mathematical Snapshots* (New York: Oxford University Press, 1960), pp. 40–43.

SUMMARY

A geometric sequence is built by multiplying any given term by a fixed number r to find the next term. Two important formulas for geometric sequences are

$$a_n = r^{n-1}a_1$$

$$A_n = a_1 + a_2 + \cdots + a_n = \frac{a_1(1 - r^n)}{1 - r} = \frac{a_1(r^n - 1)}{r - 1}$$

PROBLEM SET 4.3

1. Fill in the boxes.
   (a) 2, 8, 32, 128, □, □, . . .
   (b) 27, 9, 3, 1, □, □, . . .
   (c) 3, −3, 3, −3, □, □, . . .
2. Fill in the boxes.
   (a) 3, 6, 12, □, □, . . .
   (b) 4, 2, 1, □, □, . . .
   (c) 7, 0, 0, □, □, . . .
3. Write an explicit formula for the fortieth term in each sequence in Problem 1; that is, write formulas for a_{40}, b_{40}, and c_{40}.
4. Write an explicit formula for each of the sequences in Problem 2.
5. Use a formula we have derived to find the sum of the first five terms in each sequence in Problem 1.
6. Find the sum of the first seven terms in each sequence in Problem 2.
7. A certain culture of bacteria doubles every 24 hours. If there are 1000 bacteria initially, how many will there be after 10 full days?
8. "As I was going to St. Ives,
   I met a man with seven wives.
   Every wife had seven sacks,
   Every sack had seven cats,
   Every cat had seven kits.
   Kits, cats, sacks and wives,
   How many were going to St. Ives?"
9. In Problem 8, how many were coming from St. Ives?
10. Take a large sheet of paper and consider folding it in half 40 times (actually this is physically impossible, but we can still consider it mentally). If the paper is 0.01 inch thick, approximately how high will the resulting stack of paper be?

11. Suppose that the sheet of paper in Problem 10 winds up having an area of 1 square inch after 40 folds. What was its area originally? Approximate.

12. A tropical water lily grows so rapidly that each day it covers a surface double in size the area it covered the day before. At the end of 30 days it completely covers a lake. How long would it take two of these lilies to cover the same lake?

[c] 13. Let $a_n = (1.023)^n$. Find a_{40} and A_{40}.

[c] 14. Let $b_n = (.921)^n$. Find b_{30} and B_{30}.

[c] 15. A home is increasing in value by 10% each year (that is, its value is 10% more at the end of each year than it was at the beginning of the year). How much will a home worth $75,000 today be worth at the end of 10 years?

[c] 16. Suppose that inflation is running at 12% per year so that an item costing $1.00 today will cost $1.12 one year from now. Thus in constant value dollars, $1.00 will be worth $1.00/1.12 = \$.89286$ one year from now. How much will today's dollar be worth 20 years from now?

[c] 17. If your boss offered you the choice of $10,000.00 per month or $0.01 for the first working day, $0.02 for the second, $0.04 for the third, and so on for the 20 working days, which would you choose? Why?

[c] 18.* G. P. Jetty (worth $2 billion) has hired a secretary for $0.01 on January 1, $0.02 on January 2, $0.04 on January 3, etc. About when will he go broke?

[c] 19.* One-third of the air in a container is removed with each stroke of a vacuum pump. What fraction of the original amount of air remains in the container after five strokes? How many strokes are necessary to remove 99% of the air?

20.* A ball is dropped from a height of 10 feet. At each bounce it rises to two-thirds of its previous height. How far will it have traveled altogether (up and down) by the time it hits the floor for the 1000th time? An answer correct to two decimal places is good enough. Note: $(\frac{2}{3})^{1000}$ is negligibly small.

21.* Recall the formula for the sum of a geometric sequence?

$$A_n = \frac{a_1(1 - r^n)}{1 - r} = \frac{a_1}{1 - r} - \frac{a_1 r^n}{1 - r}$$

If r is between 0 and 1, then as n gets larger and larger, A_n gets closer and closer to $a_1/(1 - r)$. Why? To what number does A_n get closer and closer in each of the following?

(a) $A_n = \dfrac{3}{10} + \dfrac{3}{100} + \cdots + \dfrac{3}{10^n}$

(b) $A_n = \dfrac{7}{10} + \dfrac{7}{100} + \cdots + \dfrac{7}{10^n}$

(c) $A_n = \dfrac{9}{10} + \dfrac{9}{100} + \cdots + \dfrac{9}{10^n}$

Do your answers seem reasonable?

22.* The ideas in Problem 21 lead to a great deal of advanced mathematics. If a_1, a_2, . . . is a geometric sequence with a ratio r between 0 and 1, we say that its total sum is $a_1/(1 - r)$. Use this fact to calculate the values of the following infinite repeating decimals.

(a) 0.3333 . . .   Note: 0.3333 . . . $= \frac{3}{10} + \frac{3}{100} + \frac{3}{1000} + \cdots$.

(b) 0.1111 . . .

(c) 0.5555 . . .

(d) 0.3232323232 . . .

(e) 0.134134134 . . .

Can you demonstrate that any infinite repeating decimal is a rational number (a ratio of two integers)?

23.* Suppose the government spends an extra billion dollars without raising taxes. What is the total increase in spending due to this action? Assume that each business and individual saves 20% of its income and spends the rest, so that of the initial $1 billion spent by the government 80% is respent by individuals and businesses, and then 80% of that, etc. In economics, this is called the multiplier effect.

24.* Why are transpositions from one key to another so simple with the equally tempered scale, but troublesome with the diatonic scale?

One hears young people asking for a cause. The cause is here. It is the biggest single cause in history: simply because history has never presented us with such a danger . . . . No more people than the earth can take. That is the cause.
C. P. Snow (1969)

4.4 / Population Growth

C. P. Snow is only one of a long list of doomsayers that stretches way back to Thomas Malthus in 1798. None of them has proved very reliable in predicting population growth. But all agree on the trend (up!) and the end result (catastrophe!) However, neither mathematical predictions nor vivid graphs have so far affected very much people's zeal for multiplying.

Most people tend to think of growth as a linear process. A quantity is growing **linearly** if in each time period (of specified duration) it increases by a fixed amount. If town A starts with a population of 100 and adds 25 people each year, its population is growing linearly. However, a quantity is growing **exponentially** if during each time period it increases by a fixed multiple of the amount present at the beginning of that period. If town B starts with a population of 100 and has at the end of 1 year $100(1.25) = 125$ people, at the end of 2 years $125(1.25) \approx 156$ people, etc., its population is growing exponentially.

A graphical comparison will illustrate the difference between these two types of growth in a dramatic way. While both towns grow the same amount during the first year, the exponentially growing town actually grows more than three times as much over a 12-year period. As a matter of fact, in 12 years a town growing exponentially at only 10% per year will overtake a town growing linearly at the rate of 25%. An exponentially growing curve will always eventually conquer a linearly growing curve.

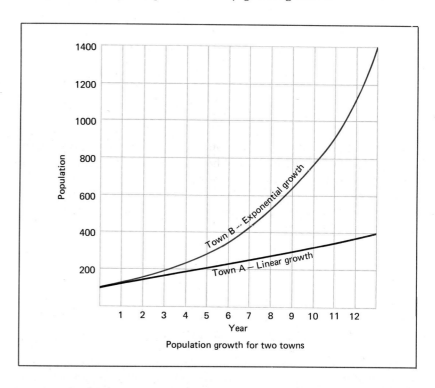

Population growth for two towns

We should note an assumption we have made in connection with population growth. Both in our discussion and in drawing nice, smooth curves, we have implicitly assumed that population grows continuously with time. Of course, it really doesn't; it grows at distinct times and by unit amounts. But here, as throughout all of science, we are attracted by simplicity and elegance. We are willing to sacrifice exactness in order to obtain a model we can work with easily. A model that allowed for discontinuous jumps at random times would be intolerably complicated and certainly not appropriate for our present purposes.

RELATION TO EARLIER WORK

We have already met the ideas discussed above. If we look at the population of town A at yearly intervals, we obtain the sequence

$$100, 125, 150, 175, \ldots$$

which is an arithmetic sequence. The corresponding sequence for town B is

$$100, 100(1.25), 100(1.25)^2, 100(1.25)^3, \ldots$$

which is a geometric sequence. Linear corresponds to arithmetic, exponential to geometric. The words "linear" and "exponential" are generally used to describe continuous processes; the adjectives "arithmetic" and "geometric" are applied to sequences.

APPLICATION TO WORLD POPULATION

It seems reasonable to suppose that the number of births (as well as deaths) during a year represents a fixed percentage of the people living at the beginning of the year. Thus on purely theoretical grounds we are led to exponential growth as an appropriate model for world population. The historical evidence as shown in the accompanying graph gives support to this thesis, at least for the last century.

> I said that population, when unchecked, increased in a geometrical ratio, and subsistence for man in an arithmetical ratio.
> *Thomas Malthus, 1798*

Consider now the following problem based on actual figures (rounded off for ease of calculation).

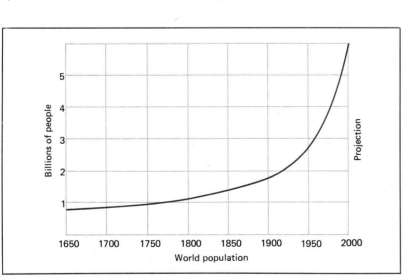

World population

TABLE 4-2	
n	$(1.02)^n$
5	1.104
10	1.219
15	1.346
20	1.486
25	1.641
30	1.811
35	2.000
40	2.208
45	2.438
50	2.692
55	2.972
60	3.281
65	3.623
70	4.000
75	4.416
80	4.875
85	5.383
90	5.943
95	6.562
100	7.245
105	7.998
110	8.831
115	9.750

Assumptions. (1) World population is increasing at a rate of 2% per year. (2) It takes 1 acre to provide food for one person. (3) The world has 8 billion acres of arable land. (4) World population was 4 billion in 1975.

Question. When will the world reach the maximum population it can support?

Solution. We assume that the population is growing exponentially at 2% per year. Thus, in 1975, 1976, 1977, . . . , it had a population (in billions) of

$$4, 4(1.02), 4(1.02)^2, \ldots$$

Since the world can support a population of only 8 billion people, our problem is to determine n so that

$$4(1.02)^n = 8$$
$$(1.02)^n = 2$$

In Table 4-2, we find that $(1.02)^{35} = 2$, so n is approximately 35. In $1975 + 35 = 2010$, the world will reach maximum population. Of course, long before that mass starvation is likely to occur as a result of inequities in food distribution. In Problem Set 4.4, we explore the effect of changes in assumptions. The conclusions are alarming under any assumptions that are realistic. Unchecked population growth is not a good idea; but then neither are most of the checks (war, famine, and plagues) used so far.

DOUBLING TIMES AND HALF-LIVES

Perhaps you have already noticed a characteristic feature of exponential growth. If an exponentially growing body doubles in size in n years, it will double again in the next n years. We call n its **doubling time.** If the growth rate for world population is 2% per year, the doubling time is 35 years. Thus a population of 4 billion in 1975 will grow to 8 billion by 2010, 16 billion by 2045, 32 billion by 2080, etc. This again emphasizes the spectacular nature of exponential growth.

But not everything grows; some things decline or decay exponentially, in which case we talk not about doubling time, but rather halving time — or half-life. The **half-life** of a given substance is the time it will take for half of what exists to disappear. Thus one-half of a substance will disappear during its first half-life, one-half of the remainder during its second half-life (leaving one-fourth of the original), etc. Notable examples are radioactive elements, all of which are believed to decay exponentially. A given

Population growth must inevitably level off. The only question is; Will it be because of high death rates or low birth rates?

amount of radium, for example, decays so that after 1690 years one-half of it remains, at the end of $2(1690) = 3380$ years only one-fourth of it remains, etc. We say that radium has a half-life of 1690 years. This law of decay is used as the basis for dating old objects. If an object contains radium and lead (radium changes to lead when it decays) in the proportion 1:3, it is believed that an amount of pure radium has decayed to one-fourth of its original amount. The object is two half-lives (3380 years) old.

We should emphasize that radioactive dating methods are based on assumptions (e.g., that decay is exactly exponential over long periods of time). Recent findings suggest that the problem of dating old objects is not as simple as the above description may imply.

SUMMARY

Many natural processes exhibit exponential growth (or decay). This means that the amounts of material present at the ends of consecutive time periods form a geometric sequence. There is evidence that the world's population (at least in the 1900–1975 period) is growing exponentially at about 2% per year. It is a matter of simple mathematical calculation to determine the dire consequences if this pattern continues.

PROBLEM SET 4.4

1. Fill in the boxes.
   (a) 3, 5, 7, 9, □, □, . . .
   (b) 200, 210, 220, 230, □, □, . . .
   (c) 2, 3, 4.5, 6.75, □, □, . . .
   (d) 2, 4, 8, 16, □, □, . . .
2. Fill in the boxes.
   (a) 4000, 4500, 5000, □□, □□, . . .
   (b) 4000, 4500, 5062.5, □□, □□, . . .
   (c) 4000, 3500, 3000, □□, □□, . . .
3. If the sequences in Problem 1 represent populations at the ends of years, which indicate linear growth and which indicate exponential growth?
4. In Problem 2, which suggest linear growth and which suggest exponential growth (or decay)?
5. A population is growing linearly at 10% per year. If it has 1000 members today, how many will it have after 5 years?
6. A population is growing exponentially at 10% per year. If it has 1000 members today, how many will it have after 5 years? Approximate.

7. A bacterial culture is growing exponentially. If it now contains 1000 bacteria and doubles its number every 3 hours, how many bacteria will there be after 6 hours? Nine hours? Thirty hours?

8. A manufacturing company adds 100 new workers each year. Is it growing exponentially? Linearly? If it has 10,000 workers today, how many will it have after 10 years? Twenty years? How long will it take to double in size? Quadruple in size?

9. Does the human body grow exponentially or linearly or neither over its lifetime? Explain.

10. A certain city grew so that its population was 1000, 1100, 1300, 1600, and 2000 in 1930, 1940, 1950, 1960, and 1970, respectively. Is it growing linearly or exponentially or neither? Do you see any pattern in its growth?

11. The radioactive element polonium has a half-life of 140 days. If the Chemtom Company has 200 grams today, approximately how much will it have after 3 years and 25 days?

12. Carbon 14 decays exponentially with a half-life of 5730 years. The concentration of carbon 14 in a piece of wood of unknown age is one-fourth of that in a modern piece of wood. How old do you estimate the old piece of wood to be?

13. The number of bacteria in a culture grew (exponentially) from 100 to 800 in 24 hours. How many bacteria were there after 8 hours?

14. Consider the problem on world population in this section. Suppose that through the use of fertilizers, insecticides, etc., food production can be increased so that $\frac{1}{4}$ acre will support one person. When will the world reach maximum population?

15. As in Problem 14, suppose $\frac{1}{4}$ acre can support a person and suppose further that 9 billion acres can be cultivated. When will the world reach maximum population?

16. Suppose we reduce the growth rate of the world's population to 1% per year, that $\frac{1}{4}$ acre can support a person, that there are 12 billion acres of arable land. When will the world reach maximum population? Note: $(1.01)^{70} = 2$.

[c] 17. The population of a certain country is 14,600,000 today and it is growing exponentially at 1.5% per year. What will its population be at the end of 20 years? Note: You will need a y^x key to do this and the next several problems.

[c] 18. How long will it take the country of Problem 17 to double in population? Hint: You must solve the equation $(1.015)^n = 2$. Experiment using the y^x key.

[c] 19. Town A is growing linearly at 5% per year while town B is growing exponentially at 1.5% per year. Both have a population of 1000 today.
    (a) Calculate the populations of both towns after 40, 60, 80, and 100 years.
    (b) About when will the population of town B overtake that of town A?

20.* Earlier we quoted Thomas Malthus. In modern language, he said that the world population increases exponentially and food production linearly. Why do these assumptions also lead to catastrophe?

Note that Malthus' assumption allows food production to grow indefinitely.

21* Assuming a population growth rate of 2% per year continues indefinitely, how long will it be until all the people of the world are packed together like sardines? Note: There are about 1,500,000 billion square feet of land area on the surface of the earth. Sardine-style packing for humans is about one person per square foot. As in the text, assume world population was 4 billion in 1975.

Suggested Reading

The Limits of Growth, Universe, D. H. Meadows, et al., New York, 1972.

Seeing the fragile constitution of their newborn babe, Methuselah's parents opened for him a savings account of $100 at 4% interest compounded annually. What was this account worth at the time of Methuselah's death?

4.5 / Compound Interest

We suppose that some modern readers will have to be told that Methuselah lived to the ripe age of 969 years. While we lament this lack of biblical knowledge, we are also distressed that so few people understand any more about compound interest than that it is something to be paid, and paid, and paid. It is the latter subject that we now investigate.

Actually the principle is very simple, and we already have all the necessary mathematical machinery. What we need to learn is the vocabulary. Suppose that M were to invest $1.00 (we call this the **principal**) in a savings account at 4% **simple interest.** What would happen? At the end of 1 year, the bank would add 4% of $1.00 or $0.04 to the account, at the end of a second year $0.04 more, etc. M's worth now and at the ends of successive years forms the arithmetic sequence

$$1.00, \ 1.04, \ 1.08, \ 1.12, \ 1.16, \ . \ . \ .$$

After 25 years, M will be worth

$$1.00 + 25(0.04) = \$2.00$$

But now suppose M invests his principal of $1.00 at 4% **compound interest,** with interest being compounded annually. What will happen? At the end of 1 year, the bank will add $0.04 to the account, just as before. But during the second year, the resulting $1.04 will draw interest. Thus the bank will add $(0.04)(1.04)$ to the account, bringing the total to

$$1.04 + (0.04)(1.04) = (1 + 0.04)(1.04) = (1.04)^2$$

at the end of 2 years. The process will continue, yielding the geometric sequence

$$1.00, \ 1.04, \ (1.04)^2, \ (1.04)^3, \ (1.04)^4, \ \ldots$$

After 25 years, M will be worth $(1.04)^{25}$. To calculate numbers like this, most bankers have a book of tables something like Table 4-3 (bankers carry these little books in their inside coat pockets—close to the heart). From the table, we learn that M's account will be worth $(1.04)^{25} = \$2.67$ after 25 years.

			TABLE 4-3		
			Compound Interest Table		
n	**$(1.01)^n$**	**$(1.02)^n$**	**$(1.04)^n$**	**$(1.08)^n$**	**$(1.12)^n$**
1	1.01000000	1.02000000	1.04000000	1.08000000	1.12000000
2	1.02010000	1.04040000	1.08160000	1.16640000	1.25440000
3	1.03030100	1.06120800	1.12486400	1.25971200	1.40492800
4	1.04060401	1.08243216	1.16985856	1.36048896	1.57351936
5	1.05101005	1.10408080	1.21665290	1.46932808	1.76234168
6	1.06152015	1.12616242	1.26531902	1.58687432	1.97382269
7	1.07213535	1.14868567	1.31593178	1.71382427	2.21068141
8	1.08285671	1.17165938	1.36856905	1.85093021	2.47596318
9	1.09368527	1.19509257	1.42331181	1.99900463	2.77307876
10	1.10462213	1.21899442	1.48024428	2.15892500	3.10584821
11	1.11566835	1.24337431	1.53945406	2.33163900	3.47854999
12	1.12682503	1.26824179	1.60103222	2.51817012	3.89597599
15	1.16096896	1.34586834	1.80094351	3.17216911	5.47356576
20	1.22019004	1.48594740	2.19112314	4.66095714	9.64629309
25	1.28243200	1.64060599	2.66583633	6.84847520	17.00006441
30	1.34784892	1.81136158	3.24339751	10.06265689	29.95992212
35	1.41660276	1.99988955	3.94608899	14.78534429	52.79961958
40	1.48886373	2.20803966	4.80102063	21.72452150	93.05097044
45	1.56481075	2.43785421	5.84117568	31.92044939	163.98760387
50	1.64463182	2.69158803	7.10668335	46.90161251	289.00218983
55	1.72852457	2.97173067	8.64636692	68.91385611	509.32060567
60	1.81669670	3.28103079	10.51962741	101.25706367	897.59693349
65	1.90936649	3.62252311	12.79873522	148.77984662	1581.87249060
70	2.00676337	3.99955822	15.57161835	218.60640590	2787.79982770
75	2.10912847	4.41583546	18.94525466	321.20452996	4913.05584077
80	2.21671522	4.87543916	23.04979907	471.95483426	8658.48310008
85	2.32978997	5.38287878	28.04360494	693.45648897	15259.20568055
90	2.44863267	5.94313313	34.11933334	1018.91508928	26891.93422336
95	2.57353755	6.56169920	41.51138594	1497.12054855	47392.77662369
100	2.70481383	7.24464612	50.50494818	2199.76125634	83522.26572652

Let's consider one more possibility. Suppose that M invests his $1.00 at 4% compound interest, with interest compounded quarterly. Interest rates are usually stated as annual rates; 4% interest compounded quarterly really means 1% per quarter. At the end of 3 months, the bank will add $(0.01)(1.00)$ or $0.01 to the account, and $1.01 will draw interest during the second quarter. The sequence generated by this process is

$$1.00, \ 1.01, \ (1.01)^2, \ (1.01)^3, \ \ldots$$

Consulting the table, we find that after 25 years (100 quarters) M's account will be worth $(1.01)^{100} = \$2.70$.

So far, the reader will have observed several things. Simple interest leads to an arithmetic sequence, and compound interest to a geometric sequence. Money placed at simple interest grows slowly; money placed at compound interest eventually grows very rapidly.

To make the latter fact more vivid, consider the problem of Methuselah's account. A principal of $100 invested at 4% interest compounded annually grows to $100(1.04)^{969}$ or about

$$\$3,201,148,300,000,000,000$$

after 969 years.

REGULAR SAVINGS

Suppose you decide to be a consistent saver. At the beginning of each year for the next 25 years, you plan to put $1000 in the bank. If the bank pays 8% interest compounded annually, how much will your account be worth at the end of 25 years?

Let's do the problem for $1 payments and then multiply the final result by 1000. In the accompanying diagram, we keep track of each $1 payment by means of time lines.

Yearly Saving of $1 at 8% Interest	
Payment Number	
1	$1 \quad 1.08 \quad (1.08)^2 \qquad\qquad (1.08)^{23} \ (1.08)^{24} \ (1.08)^{25}$
2	$1 \quad 1.08 \qquad\qquad (1.08)^{22} \ (1.08)^{23} \ (1.08)^{24}$
.	
.	
.	
.	
24	$1 \quad 1.08 \quad (1.08)^2$
25	$1 \quad 1.08$

The value of the account after 25 years is obtained by adding the entries in the last column.

$$1.08 + (1.08)^2 + \cdots + (1.08)^{24} + (1.08)^{25}$$

This is just a problem of summing a geometric sequence, and for this we developed a formula in Section 4.3:

$$A_{25} = \frac{a_1(1 - r^n)}{1 - r} = \frac{(1.08)[1 - (1.08)^{25}]}{1 - 1.08} = \frac{(1.08)(1 - 6.84847520)}{-0.08}$$
$$= \$78.954415$$

For payments of $1000, the result is

$$(1000)(78.954415) = \$78,954.42$$

To save $1000 per year is not a trivial goal, but with prospects like this, it is surprising that more people don't try.

HOUSE PAYMENTS

So far we have talked about saving. But most of us seem to be more concerned with borrowing. Suppose that you decide to purchase a house for $80,000 on which you put $20,000 down. The balance will be paid in equal installments at the end of each month for the next 25 years (300 months). If interest is at 9% compounded monthly, what will your payments be? Note that 9% annual interest corresponds to $i = \frac{3}{4}\% = 0.0075$ per month.

The first thing to realize is that $1.00 paid today and $1.00 paid 1 month from today have quite different values in the world of business. Why? Simply because $1.00 paid today could be put in the bank, draw interest for a month, and therefore be worth more than $1.00 a month from now. Put slightly differently, we ask what is today's value (the so-called present value) of $1.00 that is going to be paid a month from now? It is less than $1.00, but by how much? Recall that, if a principal of P is put in the bank today, it is worth $P(1 + i)$ dollars a month from now. Thus, if $1/(1+i)$ dollars is put in the bank today, it will be worth

$$\frac{1}{1+i}(1 + i) = \$1.00$$

a month from now. Therefore the present value of a payment of $1.00 a month from now is only $1/(1 + i) = \$0.9926$. Similarly, a payment of $1.00 two months from today is worth only $1/(1 + i)^2$ dollars today.

Payments of A dollars at the end of each of 300 months have the present values shown on the time line below.

> For which of you intending to build a tower, sitteth not down first, and counteth the cost, whether he have sufficient to finish it?
> *Luke 14:28*

$\dfrac{A}{1+i}$	$\dfrac{A}{(1+i)^2}$	$\dfrac{A}{(1+i)^3}$		$\dfrac{A}{(1+i)^{299}}$	$\dfrac{A}{(1+i)^{300}}$
1	2	3		299	300

We therefore want

$$60,000 = \frac{A}{1+i} + \frac{A}{(1+i)^2} + \frac{A}{(1+i)^3} + \cdots + \frac{A}{(1+i)^{299}} + \frac{A}{(1+i)^{300}}$$

The right side is again geometric; it can be summed by the formula in Section 4.3. We obtain

$$60,000 = \frac{\dfrac{A}{1+i}\left[1 - \left(\dfrac{1}{1+i}\right)^{300}\right]}{1 - \dfrac{1}{1+i}} = \frac{A\left[1 - \left(\dfrac{1}{1+i}\right)^{300}\right]}{i}$$

More generally, if a principal P is to be paid off in n periods with equal payments of size A assuming interest at i per period, then

$$P = \frac{A[1 - (1+i)^{-n}]}{i}$$

Solved for the payment A, it says that

$$A = \frac{iP}{1 - (1+i)^{-n}}$$

When we substitute $P = 60,000$, $i = 0.0075$, $n = 300$ and use a pocket calculator, we obtain $A = \$503.52$

To get the answer to a complicated problem is an achievement; to understand its implications is more significant. Spreading the payments over 25 years rather than paying $60,000 cash today has a whopping effect on what you actually pay. For in 25 years, you will make 300 payments amounting to $300(503.56) - \$151,056$. Behold how the trivial difference between $0.9926 today and $1.00 a month from now grows when compounded over 25 years.

Finally, we remark on the tremendous effect a reduction in the interest rate can make. At 5% interest rather than 9% ($i = 0.004167$ rather than $i = 0.0075$), the monthly payment would only be $350.82. And the total amount over 25 years would be $300(350.82) = \$105,246$.

SUMMARY

To make wise financial decisions, an investor or a borrower should understand compound interest. This is ultimately a matter of understanding geometric sequences. For if a dollar is invested at compound interest at the rate of i per period (for example, 8% interest compounded quarterly corresponds to $i = 0.02$), the value of that dollar now and at the end of succeeding periods is

$$1, \ 1 + i, \ (1+i)^2, \ (1+i)^3, \ (1+i)^4, \ \ldots$$

PROBLEM SET 4.5

For most of these problems you will need to refer to the table on page 121 or use a calculator with a y^x key.

1. Find:
   (a) $(1.02)^{50}$
   (b) $(1.12)^{15}$
   (c) $(1.08)^9$
   (d) $100\,(1.08)^{20}$

2. Find:
   (a) $(1.01)^{100}$
   (b) $(1.02)^{25}$
   (c) $(1.04)^{25}$
   (d) $1000\,(1.08)^9$

3. If you put \$100 in the bank today, how much will it be worth after 25 years
   (a) At 8% simple interest?
   (b) At 8% compound interest, compounded annually?
   (c) At 8% compound interest, compounded quarterly?

4. If you put \$1000 in the bank today, how much will it be worth after 10 years
   (a) At 12% simple interest?
   (b) At 12% compound interest, compounded annually?
   (c) At 12% compound interest, compounded every 2 months?

5. If you put \$100 in the bank today at 12% compound interest, compounded monthly, how much will it be worth after 100 months?

6. If you put \$50 in the bank today at 8% compound interest, compounded quarterly, how much will it be worth after $12\frac{1}{2}$ years?

7. How long (approximately) does it take money to double
   (a) At 12% simple interest?
   (b) At 12% compound interest, compounded annually?
   (c) At 12% compound interest, compounded every 2 months?
   (d) At 12% compound interest, compounded monthly?

8. How long (approximately) does it take money to double
   (a) At 8% simple interest?
   (b) At 8% compound interest, compounded annually?
   (c) At 8% compound interest, compounded quarterly?

9. What rate of simple interest is required to give the same yield as 8% interest, compounded quarterly, over a period of 10 years?

10. What rate of simple interest is equivalent to 12% interest, compounded monthly, over a period of 1 year?

11. Suppose a consumer buys a television set for \$200, using her charge account. If the store adds 1% of the unpaid balance to the bill each month it is not paid, and if she pays nothing until just after 12 charges have been made (that is, after 1 year),
    (a) How much does she then owe the firm?
    (b) How much interest (considered as simple interest) is she being charged?

12. Redo Problem 11, this time assuming there is no interest for the first month and only 11 charges are made.

c 13. Calculate:

   (a) $350\left(1+\dfrac{0.13}{12}\right)^{24}$
   (b) $475\left(1+\dfrac{0.11}{365}\right)^{365}$

14. What interest problems do the expressions in Problem 13 represent?

TABLE 4-4
INSTALLMENT BUYING – PRESENT VALUE OF AN ITEM AT $1.00 A MONTH

Number of Months	Interest		
	12%	15%	18%
1	0.99	0.99	0.99
2	1.97	1.96	1.96
3	2.94	2.93	2.91
4	3.90	3.88	3.85
5	4.85	4.82	4.78
6	5.80	5.75	5.70
7	6.73	6.66	6.60
8	7.65	7.57	7.49
9	8.57	8.46	8.36
10	9.47	9.35	9.22
11	10.37	10.22	10.07
12	11.26	11.08	10.91
13	12.13	11.93	11.73
14	13.00	12.77	12.54
15	13.87	13.60	13.34
16	14.72	14.42	14.13
17	15.56	15.23	14.91
18	16.40	16.03	15.67
19	17.23	16.82	16.43
20	18.05	17.60	17.17
21	18.86	18.37	17.90
22	19.66	19.13	18.62
23	20.46	19.88	19.33
24	21.24	20.62	20.03

c 15. If Horace puts $560 in the bank today at 11% interest, how much will it be worth at the end of 2 years if
   (a) Interest is compounded annually?
   (b) Interest is compounded monthly?
   (c) Interest is compounded daily?

c 16. Hugo invested $100 in a $2\frac{1}{2}$ year money market certificate that pays 12.5% interest compounded daily. How much will it be worth when it matures?

c 17. Mary invested $750 in a $2\frac{1}{2}$ year money market certificate that pays 14% interest compounded daily. How much will it be worth when it matures?

c 18.* If Susan Sharp saves $200 at the end of each month for the next 25 months and receives 12% compound interest, compounded monthly, how much will she be worth at the end of the 25 months? First write an expression for her worth and then evaluate it.

c 19.* In order to buy a new car, Rachel Rinker decides to put $100 in the bank every 3 months for the next 4 years. If she receives 8% compound interest, compounded quarterly, what will she have at the end of this 4-year period? Assume the first deposit is made at the end of 3 months.

c 20.* If Professor Witquick buys a house for $60,000 with $10,000 down and the rest to be paid in equal monthly payments over 30 years (360 payments), and if the interest is 9%, compounded monthly, what are his payments?

c 21.* John and Mary bought a house for $80,000, paying $10,000 down. The balance of $70,000 will be paid in equal installments at the end of each month for the next 30 years. What will these payments be if interest is 13% compounded monthly?

c 22.* David and Evelyn are financing a nortgage of $60,000 by making equal payments at the end of each month for 25 years. If interest is 11.5% compounded monthly, what will their payments be?

23.* Homer bought a watch costing $20.00 which he will pay for on the installment plan at $1.00 per month for 24 months. The salesperson pointed out that he will pay only $4.00 in interest, or $2.00 per year. On $20.00 this is 10% interest.
   (a) What is wrong with the salesperson's reasoning? Hint: Will Homer really use the full $20.00 for 2 years?
   (b) Table 4-4 shows the present value of an item at $1.00 per month. For example, an item paid for in 12 monthly installments at 15% interest should have a price tag of $11.08. What interest rate does Homer really pay?

24.* If an item is paid for in monthly installments of $1.00 for 20 months and has a price tag of $18.05, what is the actual interest rate (see Problem 23)?

25.* Rodney Roller bought a jalopy worth $160.30 but agreed to pay for it at $10.00 per month over 18 months. What interest rate did he pay? Hint: This is equivalent to buying an item worth $16.03 at $1.00 per month for 18 months (cf. Problem 23).

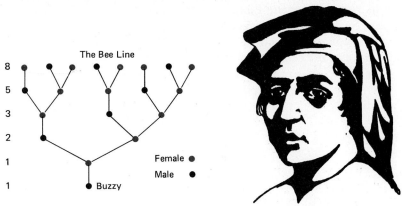

The Bee Line

8
5
3
2
1
1   ● Buzzy

Female ●
Male ●

**Leonardo Fibonacci
(ca. 1170–1250)**
Leonardo of Pisa was perhaps the greatest mathematician of the Middle Ages. After traveling in North Africa and the Orient, he wrote *Liber Abaci* which collected a host of arithmetic and algebraic facts and introduced the Hindu-Arabic numeration system to western Europe.

The Bee Line
A male bee has only a mother; a female has both a mother and a father. Thus Buzzy has one parent, two grandparents, three third-order parents, etc. How many tenth-order parents does he have?

4.6 / Fibonacci Sequences

Note the numbers ascending up the left-hand side of the diagram. They form the same sequence introduced at the beginning of Section 4.1, the Fibonacci Sequence. But Leonardo Fibonacci was not particularly interest in bees. He was more interested in rabbits. Here is his problem.

A pair of rabbits is mature enough to reproduce another pair after 2 months and will do so every month thereafter. If each new pair of rabbits has the same reproductive habits as its parents and none of them dies, how many pairs will there be at the end of 4 months? Thirty months? n months?

The Rabbit Habit

Number of months	Solid figures represent mature pairs	Number of pairs
1	ᙠᙠ	1
2	ᙠᙠ	1
3	ᙠᙠ ᙠᙠ	2
4	ᙠᙠ ᙠᙠ ᙠᙠ	3
5	ᙠᙠ ᙠᙠ ᙠᙠ ᙠᙠ ᙠᙠ	5
6	ᙠᙠ ᙠᙠ ᙠᙠ ᙠᙠ ᙠᙠ ᙠᙠ ᙠᙠ ᙠᙠ	8

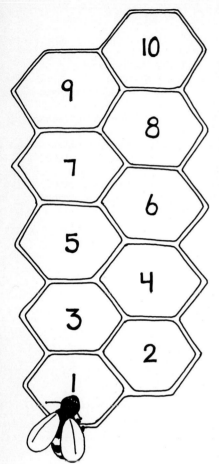

The results are shown in the table on page 127. Descending down its right side is the Fibonacci Sequence.

Suppose we are asked how many ways Buzzy can crawl over the hexagonal cells shown at the left. Naturally he can crawl only to an adjacent cell, and it is assumed that he insists that each move be a step up in the world. With these constraints, there is only one way he can reach cell 2, but there are two ways to reach cell 3 (directly from cell 1, or by way of cell 2). How many paths are there to cell 4? To cell 5? Make a chart showing the number of paths from cell 1 to cell n for $n = 2, 3, 4, 5, 6$. If you don't get stuck, you will see that once again Buzzy is entangling you in the Fibonacci web.

The Fibonacci Sequence has intrigued mathematicians for centuries, partly because it arises in unexpected places and partly because even an amateur can explore its remarkable properties. This we proceed to do now.

SOME FORMULAS

The crucial observation to make about the Fibonacci Sequence is that any term (after the first two) is the sum of the preceding two. In fact, we take this as our formal definition. The **Fibonacci Sequence** is the sequence $f_1, f_2, f_3, f_4, \ldots$ which has its first two terms f_1 and f_2 both equal to 1 and satisfies thereafter the recursion formula

$$f_n = f_{n-1} + f_{n-2}$$

With this formula we can gradually build up Table 4-5 shown on page 129. If Fibonacci were really correct in his assumptions about rabbit reproduction, one pair would result in 832,040 pairs after 30 months.

Note again the difficulty with a recursion formula. To use it, we must calculate all the terms preceding the one we want. It works, but it is time-consuming. Is there an explicit formula for f_n? Yes. It was apparently discovered by the Swiss mathematician Daniel Bernoulli in 1724.

$$f_n = \frac{1}{\sqrt{5}} \left[\left(\frac{1 + \sqrt{5}}{2} \right)^n - \left(\frac{1 - \sqrt{5}}{2} \right)^n \right]$$

It is too complicated to be of much use except for theoretical purposes, and we have no intention of trying to prove it. Perhaps you would like to check it for $n = 1, 2, 3$.

We return to the bee line problem to get at another idea. Suppose that Buzzy were interested in the total number of entries in his family tree back through nine generations. He would simply have to add

$$1 + 1 + 2 + 3 + 5 + 8 + 13 + 21 + 34$$

which turns out to be 88. More generally, set

$$F_n = f_1 + f_2 + f_3 + \cdots + f_n$$

and let's see if we can find another formula. Observe that

$$
\begin{aligned}
F_1 &= 1 & &= 2 - 1 = f_3 - 1 \\
F_2 &= 1 + 1 = 2 & &= 3 - 1 = f_4 - 1 \\
F_3 &= 1 + 1 + 2 = 4 & &= 5 - 1 = f_5 - 1 \\
F_4 &= 1 + 1 + 2 + 3 = 7 = 8 - 1 = f_6 - 1
\end{aligned}
$$

A pattern is emerging. It suggests the formula

$$F_n = f_{n+2} - 1$$

which is further confirmed by noting that

$$F_9 = 88 = 89 - 1 = f_{11} - 1$$

But of course we don't claim to have proved it, only to have accumulated some evidence (see Problem 19).

THE GOLDEN RATIO

Surely one of the most remarkable properties of the Fibonacci Sequence is that the ratio of two consecutive terms is alternately larger or smaller than an important number, the so-called **golden ratio,** a number of great interest to the Greeks and symbolized by one of their letters, namely, ϕ (phi). Its value is

$$\phi = \frac{1 + \sqrt{5}}{2} = 1.618033989 \ldots$$

As n gets larger and larger, the ratio of f_{n+1} to f_n gets closer and closer to this value. This fact can be demonstrated using the explicit formula for f_n that we gave earlier.

The golden ratio takes its name from its progenitor, the golden rectangle, which is constructed as follows. Beginning with a square $GBCD$ of side length 2, locate the midpoint P of GB as shown in the margin. Use center P and radius PC (which by the Pythagorean Theorem has length $\sqrt{5}$) to draw an arc, thus determining a point O on the extension of GB. Finally, locate L so that $GOLD$ forms a rectangle, a so-called **golden rectangle** in which the ratio of the length to the width is the aforementioned golden ratio ϕ. The golden rectangle enchanted the Greeks and appears over and over in ancient architecture and art, one prominent example relating

TABLE 4-5	
n	f_n
1	1
2	1
3	2
4	3
5	5
6	8
7	13
8	21
9	34
10	55
11	89
12	144
13	233
14	377
15	610
16	987
17	1597
$\vdots$	$\vdots$
30	832040

TABLE 4-6		
n	f_n	f_{n+1}/f_n
1	1	1
2	1	2
3	2	1.3
4	3	1.66
5	5	1.60
6	8	1.625
7	13	1.615
8	21	1.619
9	34	1.618
10	55	1.618

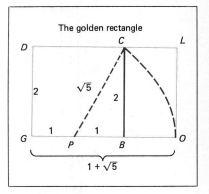

The golden rectangle

to the dimensions of the Parthenon. That the golden rectangle has genuine geometric significance can be seen by studying a modern geometry book, for example, H. S. M. Coxeter, *Introduction to Geometry* (New York: Wiley, 1962).

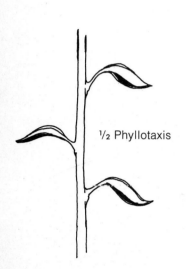

½ Phyllotaxis

The Parthenon at Athens

BOTANICAL APPLICATIONS

The Fibonacci numbers appear mysteriously in nature in a phenomenon called **phyllotaxis** (literally "leaf arrangement"). The leaves of many plants spiral around and up the branch or stem. On the branches of certain trees (e.g., elm and basswood), the leaves occur alternately on one side and then the other, that is, leaves occur at one-half turns around the branch. We call this 1/2 phyllotaxis. In other trees (e.g., beech and hazel), the leaves occur at one-third turns around a branch (1/3 phyllotaxis). Oak and apricot exhibit 2/5 phyllotaxis, poplar and pear 3/8, willow, and almond 5/13, etc. The remarkable fact is that nature displays a fascinating tendency to pick ratios that involve Fibonacci numbers.

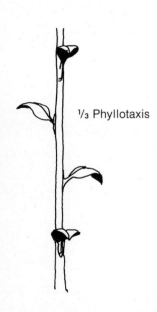

⅓ Phyllotaxis

   Another example can be found in the heads of certain flowers, especially the sunflower. The seeds are distributed over the head in spirals, one set unwinding clockwise and the other counterclockwise. If one counts the numbers of these two kinds of spirals, they are almost always Fibonacci numbers. Small heads may have 13:21 or 21:34 combinations, large heads 34:55, 55:89, or even 89:144 combinations.

SUMMARY

The Fibonacci Sequence has fascinated people in all walks of life since its discovery about 1200 A.D. It is determined by the recursion formula $f_n = f_{n-1} + f_{n-2}$, together with the condition that $f_1 = f_2 = 1$. Among its many occurrences are some, perhaps fanciful, relating to bees and rabbits; others, less artificial, are found in geometry and botany.

PROBLEM SET 4.6

1. Find f_{18}, f_{19}, f_{20} (Use Table 4-5 to save yourself some work).
2. Let $e_n = e_{n-1} + e_{n-2}$, but with $e_1 = 0$, $e_2 = 1$. Find $e_3, e_4, e_5, e_6, e_7, e_8, e_9$. How does this sequence compare with the Fibonacci Sequence?
3. Recall that $F_n = f_1 + f_2 + \cdots + f_n$. Thus $F_2 = f_1 + f_2 = 1 + 1 = 2$ and $F_3 = f_1 + f_2 + f_3 = 1 + 1 + 2 = 4$. Find $F_4, F_5, F_6, F_7, F_8, F_9$. Do your results agree with the formula $F_n = f_{n+2} - 1$?
4. Let $E_n = e_1 + e_2 + \cdots + e_n$ (see Problem 2). Find $E_4, E_5, E_6, E_7, E_8, E_9$. Can you guess a formula for E_n?

5. Fill in the boxes.
   (a) $f_1 = \square$
   (b) $f_1 + f_3 = \square$
   (c) $f_1 + f_3 + f_5 = \square$
   (d) $f_1 + f_3 + f_5 + f_7 = \square$
   (e) $f_1 + f_3 + \cdots + f_{2k-1} = \square$

6. Fill in the boxes.
   (a) $f_2 = \square$
   (b) $f_2 + f_4 = \square$
   (c) $f_2 + f_4 + f_6 = \square$
   (d) $f_2 + f_4 + f_6 + f_8 = \square$
   (e) $f_2 + f_4 + \cdots + f_{2k} = \square$

7. Look at the bee line again (at the beginning of Section 4.6).
   (a) Write the first few terms in the sequence that gives the number of male ancestors at each level above Buzzy.
   (b) Do the same for female ancestors. What do you notice?

8. The bee line problem is based on some big assumptions that were not explicitly stated. Can you identify some of them?

9. Fill in the boxes.
$$f_n: 1, 1, 2, 3, 5, 8, 13, 21, \ldots$$
$$(f_n)^2: 1, 1, 4, \square, \square, \square, \square, \square, \ldots$$
$$(f_n)^2 + (f_{n+1})^2: 2, 5, 13, \square, \square, \square, \square, \square, \ldots$$
$$f_{2n+1}: 2, 5, 13, \square, \square, \square, \square, \square, \ldots$$
   Guess at a relationship.

10. Fill in the boxes.
$$1^2 + 1^2 = 1 \cdot 2$$
$$1^2 + 1^2 + 2^2 = 2 \cdot 3$$
$$1^2 + 1^2 + 2^2 + 3^2 = \square \cdot \square$$
$$1^2 + 1^2 + 2^2 + 3^2 + 5^2 = \square \cdot \square$$
   Conjecture a formula for $f_1^2 + f_2^2 + \cdots + f_n^2$.

11. Fill in the boxes.
$$2 \cdot 1 - 1^2 = 1$$
$$3 \cdot 1 - 2^2 = -1$$
$$5 \cdot 2 - 3^2 = \square$$
$$8 \cdot 3 - 5^2 = \square$$
$$13 \cdot 5 - 8^2 = \square$$
   Conjecture a formula for $f_{n+1}f_{n-1} - (f_n)^2$.

12. The Lucas Sequence is determined by $g_1 = 1$, $g_2 = 3$, and $g_n = g_{n-1} + g_{n-2}$.
   (a) Write out the first 10 terms for g_n.
   (b) Fill in the boxes.
$$f_n: 1, 1, 2, 3, 5, 8, \ldots$$
$$g_n: 1, 3, 4, 7, 11, 18, \ldots$$
$$f_{n-1} + f_{n+1}: ?, 3, 4, \square, \square, \square, \ldots$$
$$f_{2n}: 1, 3, 8, \square, \square, \square, \ldots$$
$$f_n g_n: 1, 3, 8, \square, \square, \square, \ldots$$
   Conjecture two relationships.

13. Let $h_1 = a$, $h_2 = b$, and $h_n = h_{n-1} + h_{n-2}$. We call h_n a generalized Fibonacci Sequence.
   (a) Write out the first 10 terms of h_n.

(b) Show that $(h_1 + h_2 + \cdots + h_{10}) = 11 h_7$.

14. Each of the following is a generalized Fibonacci Sequence. Fill in the boxes.
    (a) 9, 10, $\square$, $\square$, . . .
    (b) 6, $\square$, 14, $\square$, . . .
    (c) 11, $\square$, $\square$, 43, . . .

15. A sequence whose terms are the sum of the preceding three is somewhat facetiously called a "Tribonacci" sequence.
    (a) If the first three terms are all 1, find the next five terms.
    (b) Write the recursion formula in mathematical symbols.

16.* Let $k_1 = 1$ and $k_n = 1 + 1/k_{n-1}$. Write out the first eight terms of k_n. Do you observe anything?

17.* Let $\phi = (1 + \sqrt{5})/2$ be the golden ratio. Show that $\phi = 1 + 1/\phi$ and that $\phi^2 - \phi - 1 = 0$.

18.* How many different strings of pluses and minuses are there of length n in which no two pluses are adjacent? Experiment. Note, for example, that $+-+$, $--+$, $+--$, $-+-$ and $---$ are the strings of length 3. What is the recursion formula?

19.* It is one thing to guess at a result and quite another to demonstrate rigorously that it is correct. Try to prove each of the facts below. One of your main tools will be Mathematical Induction. This is really a form of deduction and is one of the basic tools of mathematicians. Let P_n be a statement about the counting number n. If P_1 is true and if the truth of $P_1, P_2, \ldots, P_k$ implies the truth of P_{k+1}, then P_n is true for every counting number n.
    (a) $F_n = f_1 + f_2 + \cdots + f_n = f_{n+2} - 1$
    (b) $f_1 + f_3 + \cdots + f_{2n-1} = f_{2n}$ (Problem 5)
    (c) $f_2 + f_4 + \cdots + f_{2n} = f_{2n+1} - 1$ (Problem 6)
    (d) $f_{2n+1} = (f_n)^2 + (f_{n+1})^2$ (Problem 9)
    (e) $(f_1)^2 + \cdots + (f_n)^2 = f_n f_{n+1}$ (Problem 10)
    (f) $f_{n+1} f_{n-1} - (f_n)^2 = (-1)^n$ (Problem 11)
    (g) $(f_n)^3 + (f_{n+1})^3 - (f_{n-1})^3 = f_{3n}$
    (h) $f_n = \dfrac{1}{\sqrt{5}} \left[\left(\dfrac{1 + \sqrt{5}}{2} \right)^n - \left(\dfrac{1 - \sqrt{5}}{2} \right)^n \right]$
    (i) $\dfrac{f_{n+1}}{f_n} \to \dfrac{1 + \sqrt{5}}{2}$ as $n \to \infty$
    (j) $(f_{n+2})^2 - (f_{n+1})^2 = f_n f_{n+3}$

20.* Are there infinitely many Fibonacci numbers that are prime? This is the most famous unsolved problem involving the Fibonacci Sequence.

Suggested Reading

For more information on the Fibonacci numbers, we recommend the *Fibonacci Quarterly* published by the Fibonacci Association, especially the February 1963 and April 1963 issues. To learn more about the golden ratio, see Martin Gardner's column, "Mathematical Games," *Scientific American*, August 1959, and David Bergamini and the editors of *Life, Mathematics*, Time-Life Science Library, New York, 1963.

Chapter 5
Programming a Computer

If every instrument could do its own work, if the shuttle should weave of itself and the plectrum play the harp unaided, then managers would not need workers and masters would not need slaves.
ARISTOTLE

Standing Them on Their Heads

Matilda, a learned mathematician, and her husband Mortimer sometimes amused their children with the following trick. They would place eight pennies in a row, all showing an upright Lincoln. After Matilda had left the room, Mortimer would ask the children to choose a number between 0 and 250. Then he would rotate a few pennies so that Lincoln was standing on his head. When Matilda reentered the room, she would scan the pennies and announce the chosen number—right every time. The arrangement above corresponds to 46. What is the principle behind this trick?

5.1 / Binary Arithmetic

The long, arduous search for an efficient way to write numbers has been described in many books. Some of it is conjecture, since it predates recorded history. We can speculate that people first used simple tallies (scratch marks on a stick), that they later invented special symbols to represent smaller numbers, and that they then devised ways of combining these symbols to represent larger numbers. Most of the early systems have long since disappeared from use. One, the Roman numeral system, still makes occasional ceremonial appearances.

This we know. About 1200 A.D. a remarkably useful numeration system began to be used in western Europe. It had its roots in India and was brought to the West by the Arabs. We call it **Hindu-Arabic notation.** It is the numeration system 1, 2, 3, . . . learned in every grade school.

HINDU-ARABIC NOTATION

Two facts about this system are worth commenting on. First, it uses base-ten, probably because humans have ten fingers. Thus ten and its powers (hundred, thousand, etc.) are the basic numbers of the system. In contrast, the Roman numeral system (I, II, III, IV, V, . . .) is a mixed base-five and base-ten system. The ancient Babylonians used a system in which 60 played the prominent role, and their system has influenced the way we measure time and angles.

Much more significant than the base that is used is the concept of place value. Thus, in the Hindu-Arabic system, the numeral 3 has a different value in 36, in 345, and in 3465, depending on the place it occupies; in the first it stands for 3 tens, in the second 3 hundreds, and in the third 3 thousands. This allows us to write any number in a very compact way using only the ten symbols.

$$0, 1, 2, 3, 4, 5, 6, 7, 8, 9$$

The reader can compare our 3268 which stands for

$$3 \text{ thousands} + 2 \text{ hundreds} + 6 \text{ tens} + 8$$

with the Roman numeral equivalent

MMMCCLXVIII

The Hindu-Arabic system is remarkable not only because of the conciseness with which numbers can be written, but because it lends itself to simple and systematic methods of computation. These systematic methods are called **algorithms.** You will begin to appreciate their significance if you try to perform multiplication using Roman numerals.

Multiplication is easy in our system partly because of the exceptional role played by 0. Thus, as we all know, multiplication by 10 simply adjoins a 0, by 100 two 0's, etc.

$$345 \times 10 = 3450$$
$$345 \times 100 - 34,500$$

Recall another notational device we use constantly in mathematics. To indicate how many times a number is to be multiplied by itself, we attach a superscript called an exponent. Thus

$$10^2 = 10 \times 10 = 100$$
$$10^3 = 10 \times 10 \times 10 = 1000$$

Accordingly, we can write

$$3268 = 3(10^3) + 2(10^2) + 6(10) + 8$$

This suggests how we should interpret a numeral when a base other than ten is used, a subject we discuss next.

EIGHTLAND

To help us better understand our place value system, we take a brief glimpse at an imaginary world where all residents have just four fingers on each hand. Once over their hands correspond to our eight. They use only the symbols

$$0, 1, 2, 3, 4, 5, 6, 7$$

The multiplication algorithm

```
  345
   26
 2070
 6900
 8970
```

and their numbers are written in what is called base-eight. Thus 452 in Eightland means

$$4(8^2) + 5(8) + 2$$

which to us is

$$4(64) + 5(8) + 2 = 256 + 40 + 2 = 298$$

Counting the same number of objects as we count, they record the answer differently:

Tally System	*Our System*		*Eightland*	
	Tens	Units	Eights	Units
‖‖‖‖‖‖‖‖‖‖‖‖	2	3	2	7
‖‖‖‖‖‖‖‖‖‖‖‖‖	2	8	3	4

Children in their school system memorize sums and products from flash cards, just as ours do; but they get gold stars for different-looking answers:

4	6	4	6	7
+7	+3	×7	×5	×6
13	11	34	36	52

Fortunately, the usual algorithms for addition, subtraction, multiplication, and division still work. It is, in fact, a good exercise to do the same problem in both base-eight and base-ten, and then check that the answers are really the same.

Base-Eight			*Base-Ten*
134	=	$1(8^2) + 3(8) + 4 =$	92
256	=	$2(8^2) + 5(8) + 6 =$	174
123	=	$1(8^2) + 2(8) + 3 =$	83
535	=	$5(8^2) + 3(8) + 5 =$	349
254	=	$2(8^2) + 5(8) + 4 =$	172
27	=	$2(8) + 7 =$	23
2264			516
530			344
7564	=	$7(8^3) + 5(8^2) + 6(8) + 4 =$	3956

Actually, youngsters in Eightland have a decided advantage; they need to learn their tables only through sevens, not through nines as we do. If this is so, why not move to Sixland, Fourland, or best of all Twoland?

TWOLAND

In this best of all worlds, we need only two symbols, 1 and 0. There are only six flash cards to learn:

| $\begin{array}{r} 0 \\ +0 \\ \hline 0 \end{array}$ | $\begin{array}{r} 0 \\ +1 \\ \hline 1 \end{array}$ | $\begin{array}{r} 1 \\ +1 \\ \hline 10 \end{array}$ | $\begin{array}{r} 0 \\ \times 0 \\ \hline 0 \end{array}$ | $\begin{array}{r} 0 \\ \times 1 \\ \hline 0 \end{array}$ | $\begin{array}{r} 1 \\ \times 1 \\ \hline 1 \end{array}$ |

Of course, there is a price to pay; even relatively small numbers take a lot of room to write down.

Tally System	*Our System*		*Twoland*				
	Tens	Units	2^4	2^3	2^2	2	Units
|||||||[IIIIIIIIIIIIIIIIIIIII]	2	3	1	0	1	1	1
|||||||||||||[IIIIIIIIIIIIIIIIIIIIIIIIIII]	2	8	1	1	1	0	0

Now we can understand what is behind the penny game at the beginning of this section. An upright Lincoln stands for 0, and a Lincoln on its head stands for 1. The display

is just another way of writing

$$00101110$$

When translated to base-ten, it is

$$0(2^7) + 0(2^6) + 1(2^5) + 0(2^4) + 1(2^3) + 1(2^2) + 1(2) + 0$$
$$= \qquad\qquad 32 \qquad + 8 + 4 + 2 = 46$$

You may think that, in writing about bases other than ten, the authors have gone off on just the kind of esoteric flight that is to be expected of mathematicians. We do have our reasons, however. Besides its usefulness in certain games and puzzles (see Problem Set 5.1), base-two notation, often called **binary notation,** is just what one needs to understand how computers do arithmetic. We see why this is so in the next section.

SUMMARY

The human need to record and manipulate numbers has led us to invent many notational systems. Most have disappeared in favor of Hindu-Arabic notation which uses base-ten and features the notion of place value. Thus, in this system,

$$3624 = 3(10^3) + 6(10^2) + 2(10) + 4$$

In fact, any number can be written in this compact way, using only the 10 symbols, 0, 1, 2, 3, 4, 5, 6, 7, 8, and 9. Much more importantly, it allows us to do the four operations of arithmetic (addition, subtraction, multiplication, and division) by means of simple systematic procedures called algorithms.

We use base-ten, but there is nothing in mathematics or nature (except perhaps our ten fingers) that makes this choice of base compelling. There are in fact times when it is better to use a different base, the most notable instance being the case of a modern digital computer in which base-two is employed. Fortunately, the algorithms for arithmetic are valid for any base. The following problems are designed to give you practice in verifying this and in changing from one base to another.

PROBLEM SET 5.1

1. Write the number 36 in
   (a) Base-eight   (b) Base-two   (c) Roman numerals
2. Write the number 51 in
   (a) Base-eight   (b) Base-two   (c) Base-five
3. Let 347_{eight} mean that we are using base-eight notation. Change each of the following to base-ten.
   (a) 23_{eight} (b) 63_{eight} (c) 1234_{eight}
   (d) 10101_{two} (e) 111101_{two} (f) 234_{five}
4. Write the following in base-ten.
   (a) 31_{eight} (b) 43_{eight} (c) 567_{eight}
   (d) 11101_{two} (e) 100011_{two} (f) 567_{nine}
5. Write each of the following base-ten numbers in base-eight.
   (a) 41   (b) 55   (c) 362   (d) 409
6. Write each of the following base-ten numbers in base-eight.
   (a) 36   (b) 98   (c) 513   (d) 777
7. Write each of the numbers in Problem 5 in base-two.
8. Write each of the numbers in Problem 6 in base-two.
9. Look at your answers to Problems 5 and 7. You should be able to see an easy way of going from base-eight to base-two. Use this easy method to write each of the following in base-two.
   (a) 57_{eight} (b) 63_{eight} (c) 324_{eight}
10. Write each of the following in base-eight.
    (a) 110110_{two} (b) 111001_{two} (c) 1111001_{two}
11. Perform the following operations, assuming that all the numbers are written in base-eight. Then check by changing everything to base-ten and doing the operations again.
    (a)   41   (b)   21   (c)   76
       +67   +66   −34
        35   13

(d)   544
    −355

(e)   62
    ×23

(f)   123
    × 42

12. Follow the same directions as in Problem 11, but assume all numbers are written in base-nine.

13. In base-twelve we use A for ten and B for eleven. Perform the indicated operations.

(a)   A3
  + 89
    1B

(b)   AB4
 − 126

(c)   4B
  ×23

14. In base-sixteen, we use A, B, C, D, E, and F for ten, eleven, twelve, thirteen, fourteen, and fifteen respectively. Perform the indicated operations.

(a)   BC
 + 92

(b)   D99
 − 37A

(c)   3B
  ×26

15.* A certain kind of puzzle, known as an alphametric, poses a problem in which letters represent digits in the number system (base-ten unless otherwise specified). The game is to assign values to each letter so as to make the indicated arithmetic correct. Understand that the leftmost letter is not zero, and that distinct letters always represent distinct digits.

(a)   SEND
 +MORE
 MONEY

(b)   TEN
   TEN
 +FORTY
  SIXTY

(c)   WRONG
 +WRONG
  RIGHT

16.* See Problem 15 for directions.

(a)   FLY
 + FOR
 YOUR
 LIFE

(b)   WATER
 − HEAT
   ICE

17.* Ask someone to pick a number between 1 and 31 and to tell you on which cards (below) it appears. By adding the numbers in the upper left-hand corner of each card, you can tell what number was chosen. For example, if you are told that the numbers appear on cards B, D, and E, you can "guess" that the number is $8 + 2 + 1 = 11$.

16	24
17	25
18	26
19	27
20	28
21	29
22	30
33	31

A

8	24
9	25
10	26
11	27
12	28
13	29
14	30
15	31

B

4	20
5	21
6	22
7	23
12	28
13	29
14	30
15	31

C

2	18
3	19
6	22
7	23
10	26
11	27
14	30
15	31

D

1	17
3	19
5	21
7	23
9	25
11	27
13	29
15	31

E

(a) See if you can discover how the cards were constructed. Hint: Express the numbers 1 to 31 in binary notation, using 0's as necessary to express each with five digits; thus $11_{ten} = 01011_{two}$.

(b) Explain how to make a set of seven cards which include all numbers from 1 to 100. They can be used to "guess" a person's age by asking him to tell you on which cards he finds his age.

18* The game of Nim is played with matchsticks as follows. Arrange any number of matchsticks in any number of piles. The first player may take any number of matchsticks from any pile but may choose from only one pile. His opponent then chooses some matchsticks according to the same rules. The winner is the person who picks up the last matchstick. We illustrate with a game between the Poor Nut and Mr. Swift in which the courteous Mr. Swift has offered to let the Poor Nut go first.

	1	2	3	4
Beginning state	111	11	1111	11111
Poor Nut chooses three from pile 4	111	11	1111	11
Mr. Swift chooses one from pile 3	111	11	111	11
Poor Nut chooses two from pile 2	111		111	11
Mr. Swift chooses two from pile 1	1		111	11
Poor Nut chooses two from pile 3	1		1	11
Mr. Swift chooses two from pile 4	1		1	

The Poor Nut now sees that he has lost. Pity on the chap, for had he known the secret of the game, it would have been clear to him that, in accepting Mr. Swift's invitation to go first, he had already lost.

To win at Nim, proceed as follows. Before the game begins, and after each move, list in binary notation the number of matchsticks in each pile. Then note (here you can think in terms of the number system most familiar to you, base-ten) the number of 1's in each column. If the number of 1's in every column is even, call the arrangement unlucky. If any column has an odd number of 1's, call the arrangement lucky. Check on the following statements:

1. The winner's last move can be made only from a lucky position.
2. If you play from an unlucky position, you can only leave your opponent in a lucky position.
3. If you play from a lucky position, you can always manage to leave your opponent in an unlucky position; and you should.

If you see that the original position is unlucky (as Mr. Swift did in our example), you should try to get your opponent to move first. Otherwise you will have to watch for an opportunity to get him in an unlucky position (unless of course he also knows the key to playing, in which case he will beat you if you agree to move first from an unlucky position). Once you get your opponent in an unlucky position, keep him there, which the outline above ensures, and he will lose.

Note that Mr. Swift always moved from a lucky position and left the Poor Nut to move from an unlucky position.

> The calculation at the beginning of our example above
>
> $$\begin{array}{r} 011 \\ 010 \\ 100 \\ \underline{101} \\ 222 \end{array}$$

> It is always better in a game of chance if you don't have to leave things to chance.

5.2 / Computing Machines

Unlike faucets in a water line that frequently leak, switches in an electric circuit are wonderfully decisive devices; they are either on or off. And while we have seen frustrated students of computer programming who were ready to push the computer into a lake, we have not heard anyone suggest hooking it up to a water line. Computers work best when connected to a source of electricity, and internal to a computer are myriads of electric circuits, each with a switch that is on or off. Any attempt to explain how these electric circuits can manipulate numbers should begin with a bit of history.

THE HISTORICAL BACKGROUND

The ancient Chinese abacus was perhaps the first mathematical machine. It consisted of a rigid frame holding rods with moveable rings on them. Numbers were represented by configurations of the rings on the rods; and the operator performed calculations by moving the rings one way or the other.

Blaise Pascal, a man we will meet several times in this book, built the first mechanical adding machine in 1642. It featured a system of interlocking cogwheels which automatically "carried" numbers from one wheel to the next much as in an automobile odometer. The wheels were turned by hand with a stylus.

Pascal's device was gradually improved, culminating in the familiar mechanical adding machine widely used until the 1960s. Numbers were represented by the positions of small wheels on which the digits 0 through 9 were painted. The operator entered numbers by punching buttons, calculations were performed by turning wheels, and the results were read off a register that displayed one digit from each of the wheels. Power was provided by a handcrank or perhaps an electric motor.

The engineering achievement that made the modern digital computer possible was the development of sophisticated electronic circuitry. This, together with the recognition that the state of a circuit could represent a number, meant that the transmission of electric impulses could substitute for the movement of wheels and gears. Suddenly the speed of calculation was limited only by the rate of travel of electric signals. Since these signals travel at close to the speed of light, that seemed like little limitation at all.

The hand-held electronic calculator which became widely available beginning in the 1970s, illustrates the use of electronic circuits to represent numbers and gives some idea of the speed with which calculations can now be done. But even these amazing devices are not true computers, at least in present-day terminology. Why not?

All the devices mentioned so far have a feature in common. They depend on a human operator to tell them what to do at each of the major stages of a calculation. The operator controls by external manipulation what happens inside the machine. The great idea that distinguishes a computer from a mere calculating machine is the notion of an internally stored program, a program that controls the operation of the computer from the beginning of a problem to its end. This idea, suggested by John Von Neumann in 1946, has been developed to the place where a programmer can compose a few instructions, enter them into the computer's memory, and then sit idly by while it performs millions of calculations completely on its own and reports the answers—all in a matter of a

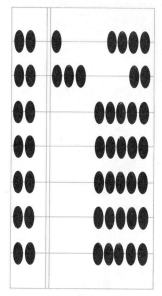

The abacus

Adding machine

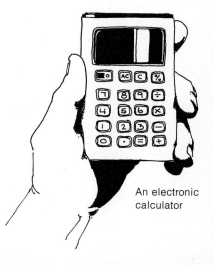

An electronic calculator

John Von Neumann (1903–1957)
Von Neumann and Einstein were perhaps the most brilliant of the host of scientists who emigrated from Europe to the United States during the 1930s. Both joined the Institute for Advanced Study at Princeton. Johnny, as he was affectionately known, made notable contributions in both pure and applied mathematics.

few minutes. And this is possible because, contrary to a player piano which gets its cues from a slowly moving paper tape, the computer taps the instructions stored in its electronic memory by means of high-speed electric impulses.

COMPONENTS OF A COMPUTER

An electronic computer always has five main components: an input device, storage or memory, an accumulator or arithmetic unit, a control unit, and an output device. A relatively simple system involving typewriter input and output is shown below.

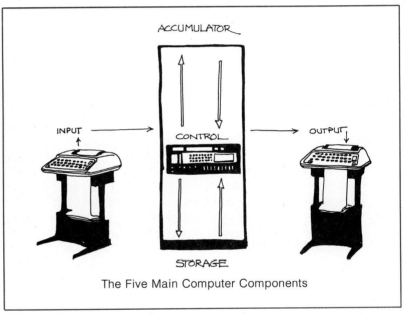

The Five Main Computer Components

A large system accepts input in several forms (punched cards, paper tape, magnetic tape, etc.), has a huge memory, and employs a complicated control system, but the principles are the same. A user writes a program (set of instructions) and enters it with the necessary data into the memory through an input device. The control unit follows these instructions, calling on the accumulator to do the required calculations, and eventually spews out the results through an output device (see page 145).

The fact that there is no need for input-output devices to be located near the rest of the computer has led to another important innovation. A large computer can accept information and instructions from 100 or more input devices at widely scattered locations via telephone lines. Thus many users can share the same basic computer, an arrangement called **time-sharing.** Many high schools, col-

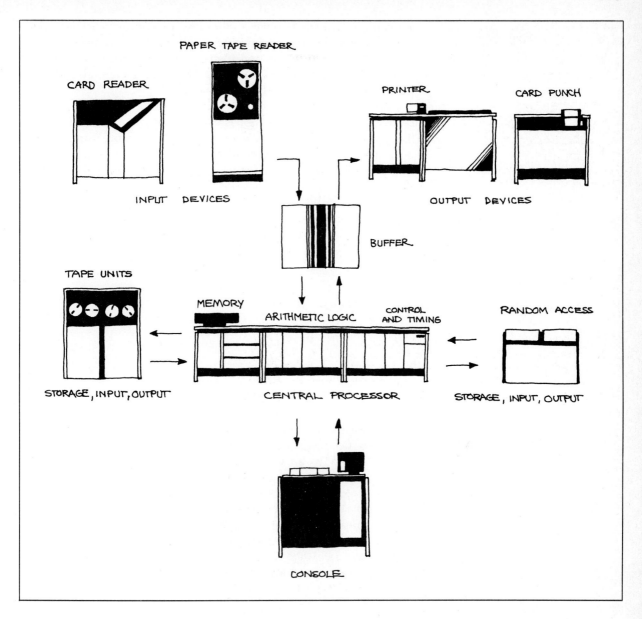

PAPER TAPE READER

CARD READER

PRINTER

CARD PUNCH

INPUT DEVICES

OUTPUT DEVICES

BUFFER

TAPE UNITS

MEMORY

ARITHMETIC LOGIC

CONTROL AND TIMING

RANDOM ACCESS

STORAGE, INPUT, OUTPUT

CENTRAL PROCESSOR

STORAGE, INPUT, OUTPUT

CONSOLE

leges, and businesses thereby gain the benefits of a large, high-speed computer without the necessity of having it on their own premises.

COMPUTER CALCULATIONS

It is easy to build electric or magnetic devices with two stable states. A switch can be open or closed, a light can be on or off, and

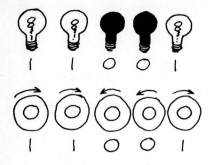

a magnetic ring can have a clockwise or counterclockwise field. Thus all modern digital computers use binary notation (recall Section 5.1). Both numbers and instructions are written as strings of 1's and 0's. Machine language is a binary language, and all early programmers had to learn it.

Modern computers actually handle numbers with 32 or more binary digits. (A binary digit is called a bit.) But for purposes of illustration, let us consider a computer that accepts numbers with only eight binary digits. One scheme for entering numbers into the memory of a computer uses tiny magnetic rings, each having an electrically induced clockwise or counterclockwise magnetic field. Thus, the number 46, expressed in binary notation as 101110, is stored by eight rings with fields as indicated by the arrows.

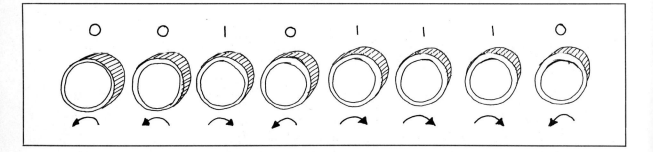

This is not unlike the arrangement of pennies in Section 5.1.

When two numbers are to be added, the first is transferred from memory to an eight-bit register in the accumulator. Then the second is transferred and automatically added to the first by means of electronic wizardry that need not concern us. Finally, the result is transferred back to an appropriate location in the memory of the computer to await further use.

But how does the computer know which numbers to add? There may be 1000 numbers stored at different locations in its memory. This is accomplished by giving each location in the memory an address (also a binary number). Thus an instruction to the computer could read: Take the number stored at address 1101 and add it to the number stored at address 1010. Store the result in the location 1101.

```
  00101110
+ 00100011
  01010001
```

NEW COMPUTER LANGUAGES

To write programs in binary language (machine language) is time-consuming and requires the programmer to become proficient in binary arithmetic. Beginning in the late 1950s, several new langauges (FORTRAN, COBOL, ALGOL, BASIC, and PASCAL) intermediate between ordinary English and machine language were developed. A computer doesn't understand them directly but is programmed to automatically translate them into its own machine language. BASIC is a particularly good example. It uses ordinary base-ten notation, standard mathematical symbols, and a few English words (which, however, must be strung together according to strict grammatical rules). We study this language in Section 5.5.

SUMMARY

Two main ideas made possible the modern high-speed computer: (1) the recognition that (binary) numbers could be represented by the state of an electronic circuit, (2) the notion of storing within the internal memory of the computer, as a set of electronically coded messages, the instructions the computer uses in solving a problem. For the earliest computers, it was necessary to write all numbers and instructions in binary, that is, as strings of 1's and 0's. But since the late 50s, special languages more like ordinary English have been developed, which computers can decipher and translate into their own language.

Many people who do not intend to become programmers will nevertheless have to communicate with those who are. For this it is helpful to be familiar with a certain vocabulary. One should certainly know what is meant by: input, accumulator, control, memory, output, machine language, BASIC language, a program, and time-sharing.

PROBLEM SET 5.2

1. What engineering achievement made possible the modern high-speed computer?
2. What was John Von Neumann's major contribution to the development of computers?
3. What distinguishes a calculator from a computer?
4. Why is binary notation used within a computer?
5. What is a program?
6. What is meant by time-sharing?
7. What are the five main components of a computer and what are their functions?
8. What is machine language? What is the purpose of languages like FORTRAN, BASIC, and COBOL?

In Problems 9 through 16, assume that each number is displayed in an eight-bit register. When two such numbers are added, we may get an overflow, resulting in a wrong answer:

$$
\begin{array}{r}
1\,1\,0\,0\,0\,1\,0\,1 \\
+\,1\,0\,0\,0\,1\,1\,1\,1 \\
\hline
\text{Overflow} \rightarrow \boxed{1}\,0\,1\,0\,1\,0\,1\,0\,0
\end{array}
$$

The 1 in the box is lost. This is a nuisance in addition, though most computers alert the operator to an overflow. But what is a nuisance in addition is actually the basis for subtraction in a computer. The following problems are designed to show why.

9. Perform the indicated subtraction by the usual process, borrowing when necessary.

(a)   1 0 1 1 0 0 1 1
     − 0 0 1 0 0 0 0 1

(b)   1 1 0 1 1 1 1 0
     − 0 1 1 0 1 0 1 1

(c)   1 0 0 0 0 0 1 1
     − 0 1 0 1 1 1 0 0

10. Follow the instructions in Problem 9.

(a)   1 1 0 1 1 1 1 0
     − 0 1 0 1 0 1 1 0

(b)   1 0 1 1 0 0 1 1
     − 1 0 0 0 1 1 0 1

(c)   1 0 0 0 0 0 1 1
     − 0 1 1 1 0 1 1 0

11. The **two's complement** of a binary number is obtained by changing all 1's to 0's and 0's to 1's. Find the two's complement of the subtrahends (bottom numbers) in Problem 9.

12. Find the two's complement of the subtrahends in Problem 10.

13. Perform the indicated addition (discarding the overflow).

(a)   1 0 1 1 0 0 1 1
     + 1 1 0 1 1 1 1 0

(b)   1 1 0 1 1 1 1 0
     + 1 0 0 1 0 1 0 0

(c)   1 0 0 0 0 0 1 1
     + 1 0 1 0 0 0 1 1

14. Add, discarding the overflow.

(a)   1 1 0 1 1 1 1 0
     + 1 0 1 0 1 0 0 1

(b)   1 0 1 1 0 0 1 1
     + 0 1 1 1 0 0 1 0

(c)   1 0 0 0 0 0 1 1
     + 1 0 0 0 1 0 0 1

15. Look at Problems 9, 11, and 13 and state a new rule for binary subtraction.

16. Compare Problems 10, 12, and 14.

5.3 / Step-by-Step Directions

Parents are often heard to complain about sons or daughters who refuse to follow instructions. But a programmer is more likely to complain that a computer follows instructions too closely. One who was particularly frustrated wrote the verse that appears in the

Mrs. Brown's Waffles
(with optional blueberries)

Ingredients
2 eggs
1½ cups buttermilk
1 tsp. soda
1¾ cups flour
2 tsp. baking powder
6 tbsp. melted shortening

Directions
Beat eggs. Beat in other ingredients. Pour batter on iron. Sprinkle on blueberries if desired. Bake. Make eight.

margin. Anyone who has tried to write a computer program knows this frustration. It turns out that giving complete, unambiguous, step-by-step instructions is far more difficult than we might have imagined.

MRS. BROWN'S WAFFLE RECIPE

We don't have to use a technical example to understand some of the problems. Consider Mrs. Brown's recipe. An experienced cook may find it adequate, but an amateur may need more help. So would a computer when faced with an analogous situation involving calculations.

First, the recipe uses some special terminology (e.g., tsp., tbsp.) that the cook may need to have translated. A computer simply stops when it encounters a term it doesn't understand. Second, the recipe leaves out steps (preheat the iron, crack the eggs, and throw away the shells, etc.). Most cooks, using common sense, fill in these steps; but a computer doesn't have common sense. Third, the recipe leaves out important information (temperature setting, baking time, etc.). A person can make an intelligent guess, but a computer must be told everything.

Most of us can comprehend a picture more easily than a set of words. Let's rewrite Mrs. Brown's recipe using a schematic diagram called a flowchart. Those who want to communicate with computers have learned that this is the best way to start writing a set of instructions.

FLOWCHARTS

A **flowchart** is a set of boxes connected by arrows. Each box contains a command (or perhaps a question). The arrows tell the reader what to do next. In the margin, we have indicated our first attempt at a flowchart for Mrs. Brown's recipe. Note that we use circles to indicate the start and stop commands, and rectangles for other commands.

A careful reader will raise an immediate objection. Our first flowchart offers no opportunity to add blueberries. This, however, calls for a decision. Do we want blueberries or not? For this question we use a diamond-shaped box, and in it we write a question that can be answered yes or no. An arrow emanates from the box corresponding to each of these answers. Our second flowchart incorporates this improvement (page 150).

Another reader is sure to object to this flowchart too. He points out that a cook following it will make only 1 waffle, while the recipe calls for 8. We could repeat the instructions eight times, but that would be very inefficient (and result in enough batter for

A Programmer's Lament

I really hate this
    damned machine;
I wish that they
    would sell it.
It never does quite
    what I want,
But only what
    I tell it.
Dennie L. Van Tassel,
The Compleat Computer
SPA, 1976, p. 44

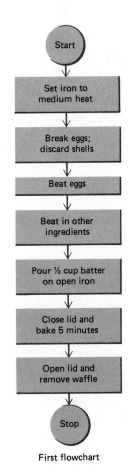

First flowchart

64 waffles). Instead we introduce another decision box with the question, Have 8 waffles been baked? If the answer is yes, we stop. If it is no, we send the cook back to near the beginning of the flowchart with an upward arrow, forming what is called a loop. The

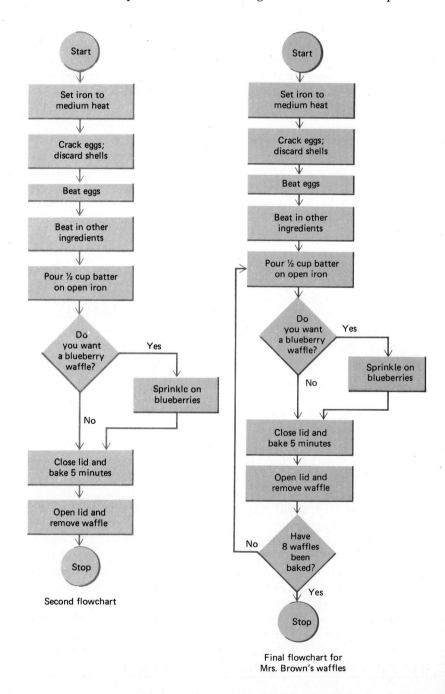

Second flowchart

Final flowchart for
Mrs. Brown's waffles

complete flowchart for Mrs. Brown's recipe is shown on page 150. The reader should study it carefully, paying special attention to what is accomplished by this loop.

SUMMARY

A flowchart is a diagram that provides clear, complete instructions for an activity that must be carried out in a sequence of steps. Each command is enclosed in a rectangle. When a decision is called for, a question is asked demanding a yes or no answer. Diamond-shaped boxes are used for questions. Arrows indicate the next step. If the activity involves a repetitive task, one can ask, Is the job finished? If the answer is no, an arrow directs the reader back to an earlier point in the instructions, forming a loop.

PROBLEM SET 5.3

1. Construct a flowchart for using a pay phone. Allow for the possibility that the line is busy.
2. Construct a flowchart for using a soft-drink machine. Assume that it offers seven flavors but that it may be out of some of them.
3. It may be that 5 minutes is not long enough to bake a waffle. Make an addition to the flowchart for Mrs. Brown's waffles, allowing for a check of whether or not the waffle is brown every 30 seconds beginning at 3 minutes.
4. Construct a complete flowchart for taking a standard 20-question multiple-choice test in a 60-minute time period.
5. Below is shown an island with four lakes and several walking paths, together with a flowchart which tells you how to proceed. The ocean shoreline is not a path.

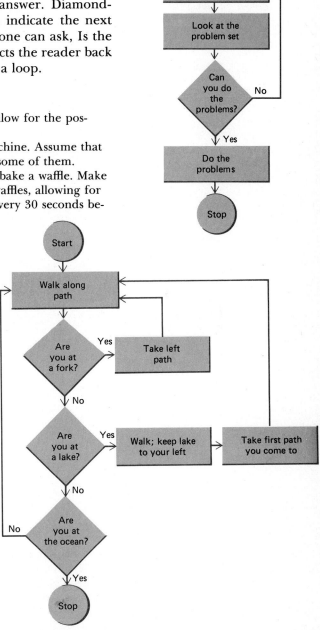

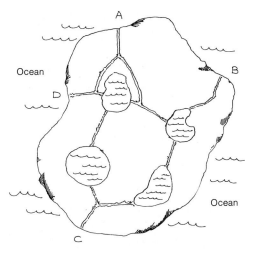

(a) If you start at A, where do you end up?
(b) If you start at B, where do you end up?
(c) If you start at C, where do you end up?

6. Redo Problem 5, replacing the word "left" with "right" in the flow-chart.
7. Follow the instructions in the flowcharts below, starting with the number 7. Then try it with 17 and 47.

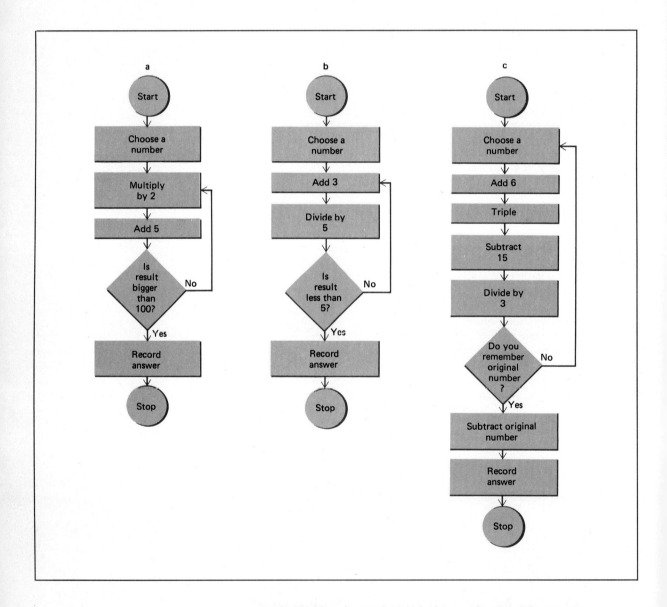

8. Redo Problem 7, starting with the number 10. Then try 20.
9. What happens in Problem 7a if you start with −5?
10. What choices of starting number (other than −5) will cause you to keep going in circles in Problem 7a?
11. Follow the instructions at the top of the next page.

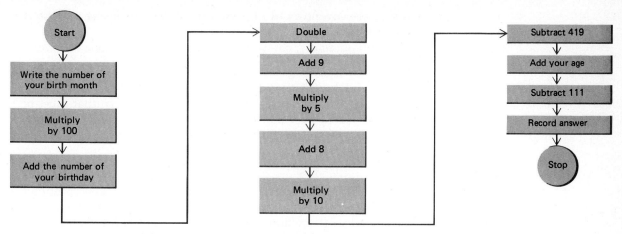

Your answer should be (reading from left to right) your birth
month, birthday, and age.

12.* Why does the flowchart in Problem 11 work?

13.* Consider Problem 7c again. Why does it always give 1 as the
answer?

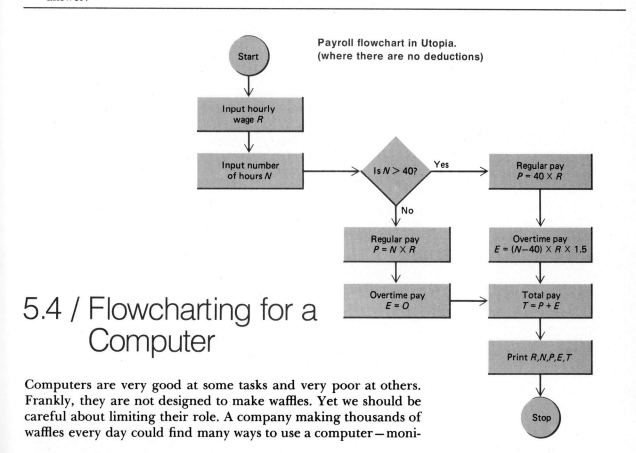

Payroll flowchart in Utopia.
(where there are no deductions)

5.4 / Flowcharting for a Computer

Computers are very good at some tasks and very poor at others.
Frankly, they are not designed to make waffles. Yet we should be
careful about limiting their role. A company making thousands of
waffles every day could find many ways to use a computer—moni-

toring the inventory of ingredients, billing customers, writing the payroll, etc.

Fundamentally, computers are good at two things:

1. Remembering and manipulating huge amounts of numerical data, for example, the inventory of a large company.
2. Performing calculations that follow a pattern that must be repeated over and over. Examples are figuring the monthly payroll for a large company and calculating the orbit of a satellite.

At both tasks, they are far superior to the human brain. They can remember better, and they can calculate faster. The computer's advantage with respect to speed is overwhelming. Some computers can perform 10 million additions in 1 second, more than a person could do in a lifetime. Even more amazing, a computer can continue making calculations at this speed for days or weeks without making a single error.

In this section, we consider problems that require a large number of calculations of the same or very similar nature. We intend to flowchart these problems in a form that can be easily translated into the language of the computer. That translation is the subject of Section 5.5.

In order to make such flowcharts, we need to know precisely what operations a computer can perform. Fundamentally, it can add, subtract, multiply, divide, and raise to a power. In addition, it can make simple decisions. It can decide whether one number is larger than, equal to, or smaller than another. This is enough to allow us to do some pretty sophisticated problems.

EXAMPLE 1 (COMPOUND INTEREST)

Recall from Section 4.5 that, if a principal of $1000 is invested at 8% interest compounded quarterly, it will be worth

$$A = 1000 \ (1 + 0.02)^{40}$$

at the end of 40 periods (i.e., 10 years). We want to instruct the computer to make this calculation for us and then print the answer. To make the problem a bit more interesting, let's allow for an arbitrary principal P, a yearly interest rate R (expressed as a decimal), a number of periods M in a year, and a total number of periods N. Then the required formula is

$$A = P\left(1 + \frac{R}{M}\right)^{N}$$

Note that we use capital letters for all the variables; that's because only capitals are available in most computer languages.

An appropriate flowchart is shown in the margin. It allows for input, it calculates the answer, and it prints the result. And as we will see in Section 5.5, it is easily translated into computer language. Note that we use only one box for the calculation step; that's because one instruction is enough to tell the computer to make such a calculation.

That was all very simple — no decision boxes or loops. We illustrate these ideas in the next example.

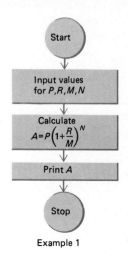

Example 1

EXAMPLE 2 (ADD $1 + 2 + 3 + \cdots + N$)

For the moment, we'll ignore the fact that Carl Gauss found a simple way to do this sum by the time he was 10 years old (Section 4.2). We choose rather to calculate it by straightforward repeated addition. That is, we follow the step-by-step procedure of adding 1 to 0, then 2 to the result, and so on until we finally add N.

Step	K-Amount Added	S-New Sum
1	1	$0 + 1 = 1$
2	2	$1 + 2 = 3$
3	3	$3 + 3 = 6$
4	4	$6 + 4 = 10$
5	5	$10 + 5 = 15$
⋮	⋮	⋮

To construct a flowchart for this problem, we'll make use of three variables, N, K, and S. Here N denotes the largest number to be added; it has to be specified, so that the computer knows when to stop. We use K to denote the amount to be added at each stage, and S to keep track of the accumulated sum.

The flowchart in the margin accomplishes the desired task. The reader should check it carefully, going around the loop several times to see that it really does what it is supposed to do. Also check that there is a stopping mechanism and that the final answer is printed.

The flowchart we have constructed is rather typical of computer flowcharts. It has four main features:

1. Input of the data.
2. Setting of the initial values for the variables.
3. Instructions involving a loop through which repeated calculations can be made.
4. Output of the answer.

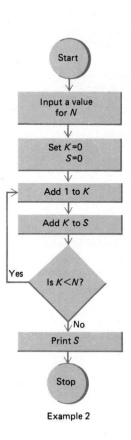

Example 2

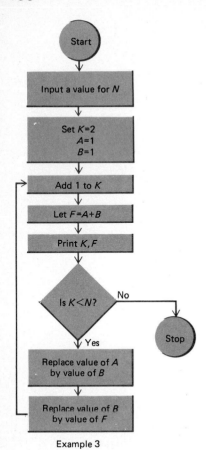

Example 3

EXAMPLE 3 (FIRST *N* FIBONACCI NUMBERS)

Section 4.6 offered a rather complete treatment of the Fibonacci Sequence

$$1, 1, 2, 3, 5, 8, 13, \ldots$$

All we need to know right now is that it starts with two 1's and then that any term is obtained by adding the previous two terms. In symbolic language, we have the recursion formula

$$F_k = F_{k-2} + F_{k-1}$$

(We are using F_k rather than f_k as in Section 4.6). Thus

$$F_3 = F_1 + F_2 = 1 + 1 = 2$$
$$F_4 = F_2 + F_3 = 1 + 2 = 3$$
$$F_5 = F_3 + F_4 = 2 + 3 = 5$$

and so on.

Our aim is to generate these numbers on a computer—in fact, to make a table of the first N of these numbers, N to be specified by the user. Subscripts are hard to manage on a computer, so we avoid them. We use K as any counting number and F as the corresponding Fibonacci number, and we ask the computer to print both K and F at each stage so we can keep track of which F goes with which K.

The flowchart in the margin offers one way to meet these requirements. The reader should study it, going around the loop several times to see what happens to each of the variables. Then compare it with the table below. Note that the shaded portion is the actual computer printout.

TABLE 5-1				
Loop Number	**K**	**F**	**A**	**B**
Initial value	2		1	1
1	3	2	1	2
2	4	3	2	3
3	5	5	3	5
4	6	8	5	8
5	7	13	18	13
6	8	21	13	21
7	9	34	21	34
8	10	55	34	55
⋮	⋮	⋮	⋮	⋮

SUMMARY

Computers are particularly good at tasks involving numerical calculations of the same type over and over. To tell a computer how to do such a task requires first of all a flowchart with step-by-step instructions. Usually these instructions fall in four categories:

1. Input the data.
2. Set the initial values of the variables.
3. Build a loop through which repeated calculations can be made. This includes provisions for giving the variables their new values and a way for the computer to get out of the loop.
4. Print out the answers.

PROBLEM SET 5.4

1. What two things do computers do particularly well?
2. Give an indication of how fast computers really are.
3. Make a flowchart for calculating and printing the value of W if

$$W = \left[\frac{X+Y}{Z}\right]^2$$

Your flowchart should allow the input of any values for X, Y, and Z.

4. Do the same for

$$W = \frac{X - Y + Z}{Y^3}$$

5. Study Flowchart A in the margin. What numbers will be printed?
6. Study Flowchart B. If we input $N = 5$, what numbers will be printed?
7. Make a flowchart for calculating and printing each value of A for $N = 1, 2, \ldots, 60$ if

$$A = 1000(1 + 0.02)^N$$

8. Redo your flowchart for Problem 7 so that it will calculate and print each value of

$$A = P(1 + 0.02)^N$$

for $N = 1, 2, \ldots, 60$ with provision for inputting any value for P.

9. Study Example 1 (compound interest) again. Modify the flowchart in the text so that it will calculate and print each value of A for $N = 1, 2, \ldots, 60$. Recall that

$$A = P\left(1 + \frac{R}{M}\right)^N$$

10. Redo the flowchart in Problem 9 so that it will calculate and print A for $N = 1, 2, \ldots, J$, where J is a counting number whose value can be inputted.

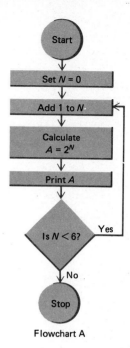

Flowchart A

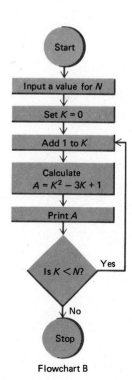

Flowchart B

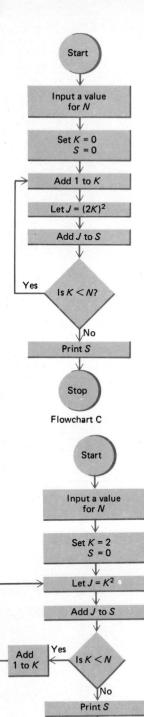

Flowchart C

Flowchart D

11. If in Flowchart C in the margin the value $N = 5$ is used as input, what value for S will be printed?

12. If the value $N = 6$ is used as input in Flowchart D, what value for S will be printed?

13. The flowchart in Example 2 of the text was designed to calculate $S = 1 + 2 + 3 + \cdots + N$. What expression is Flowchart C designed to calculate?

14. What expression is Flowchart D designed to calculate?

15. Make a flowchart for calculating and printing the value of

$$S = 2 + 4 + 6 + \cdots + 2N$$

for any specified N.

16. Make a flowchart for calculating and printing the value of

$$S = 1 + 2 + 4 + 8 + \cdots + 2^{N-1}$$

for any specified N.

17. Make a flowchart for calculating and printing the value of

$$S = 1 + \frac{1}{2} + \frac{1}{3} + \cdots + \frac{1}{N}$$

for any specified value of N.

18. Make a flowchart for calculating and printing each value of

$$S = 1 + \frac{1}{2} + \frac{1}{3} + \cdots + \frac{1}{N}$$

for $N = 1, 2, \ldots, 100$.

19. Make a flowchart for calculating and printing the first N so-called Tribonacci numbers T_k. They are determined by the formula

$$T_k = T_{k-3} + T_{k-2} + T_{k-1}$$

together with $T_1 = T_2 = T_3 = 1$.

**A sample BASIC program: A payroll in Utopia
(see flowchart at beginning of Section 5.4)**

```
1Ø   INPUT R
2Ø   INPUT N
3Ø   IF N > 4Ø, THEN 7Ø
4Ø   LET P=N*R
5Ø   LET E=Ø
6Ø   GO TO 9Ø
7Ø   LET P=4Ø*R
8Ø   LET E=(N−4Ø)*R*1.5
9Ø   LET T=P+E.
1ØØ  PRINT "HOURLY WAGE IS"; R
11Ø  PRINT "NUMBER OF HOURS IS"; N
12Ø  PRINT "REGULAR PAY IS"; P
13Ø  PRINT "OVERTIME PAY IS"; E
14Ø  PRINT "TOTAL PAY IS"; T
15Ø  END
```

20.* Here is a way to calculate $\sqrt{A}$ for any positive number A: Make any reasonable first guess and call it X_1. Then use the recursion formula

$$X_N = \frac{1}{2}\left(X_{N-1} + \frac{A}{X_{N-1}}\right)$$

(a) Try this for $A = 3$, using $X_1 = 2$. Calculate X_2, X_3, X_4, and X_5.
(b) Make a flowchart for calculating X_N for $N = 1, 2, \ldots, 20$.

5.5 / BASIC Programming

We are finally ready to do some actual programming for a computer. The language we've selected is called BASIC, an acronym for *Beginner's All-purpose Symbolic Instruction Code*; it is easy to learn, and computers that understand it are widely available. While we cannot cover everything in one lesson, you may be pleasantly surprised at how much you can do with just the introduction provided here.

THE SYMBOLS USED IN BASIC

The most common type of input device is a modified typewriter with:

Capital letters: A, B, C, D, . . ., X, Y, Z
Digits: Ø, 1, 2, . . ., 9
Punctuation marks: , . : ; " ? ( )
Mathematical symbols: $+ - * / \uparrow = < >$

Note that we use Ø for zero in order to distinguish it from the letter O.

Numbers are written as usual, for example,

$$1763, \ 143.Ø1, \ -1.2357$$

Variables are denoted by letters of the alphabet (always capitals). Arithmetic operations are symbolized in the familiar way with a slight variation for multiplication and raising to a power (exponentiation).

TABLE 5-2				
Operation	Usual Symbol	Example	BASIC Symbol	Example
Addition	$+$	$4 + 5$	$+$	$4 + 5$
Subtraction	$-$	$4 - 5$	$-$	$4 - 5$
Multiplication	$\cdot$ or $\times$	$4 \cdot 5$	$*$	$4 * 5$
Division	$/$	$4/5$	$/$	$4/5$
Exponentiation	Superscript	4^5	$\uparrow$	$4 \uparrow 5$

Parentheses are used in the standard way to indicate the order in which operations are to be done

Usual	*BASIC*
$X(Y + Z)$	$X*(Y + Z)$
$(X + Y)^2$	$(X + Y) \uparrow 2$
$(X \cdot Y)/Z$	$(X*Y)/Z$
$[X + 2(Y - Z)]/3$	$(X + 2*(Y - Z))/3$

It is important to explain how the symbol "=" is used in BASIC. When you use a variable, like X, in a program, the computer automatically picks out a storage location in its memory and gives it the name X. The instruction, LET X = 172, stores the number 172 in the location named X. Sometimes you may want to add a number, say 3, to the value that X presently has. This is accomplished by the instruction, LET X = X + 3, which makes no sense in ordinary algebra but is interpreted by the computer to mean that it should replace the number stored in X by a number 3 larger. This can be very confusing to beginning programmers. For this reason, we suggest that you think of "=" as being equivalent to the phrase "be replaced by."

SIMPLE COMMANDS IN BASIC

Suppose that we want the computer to calculate

$$A = \frac{(1.56)(65)}{78}$$

and print the result. A program that will do this job uses the LET and PRINT commands as follows:

```
10   LET A = (1.56*65)/78
20   PRINT A
30   END
```

Several things should be noted. Each command is given a line number, and the computer executes commands in order of increasing line number unless told otherwise. It is traditional, but not necessary, to use multiples of 10 as line numbers. (This allows you at a later date to modify the program by inserting new commands without having to renumber the old ones.) Every program must end with the command END, thus notifying the computer that it has reached the last line of the program.

In order to get the computer to execute this program, you should type it exactly as shown above, then type the word RUN on a new line, and then press the carriage return button. Almost immediately the computer will type out the answer which happens to be 1.3$\emptyset$.

Next suppose we want a program that will calculate

$$A = \frac{X \cdot Y}{Z}$$

for any given values of X, Y, and Z. For this, we make use of the INPUT command. Here is the program:

    1$\emptyset$ INPUT X
    2$\emptyset$ INPUT Y
    3$\emptyset$ INPUT Z
    4$\emptyset$ LET A=(X*Y)/Z
    5$\emptyset$ PRINT A
    6$\emptyset$ END

The operator types the program exactly as shown, types RUN, and presses the carriage return. Here is what happens; the explanation is given in parentheses:

RUN   (Operator types instruction for computer to execute, and then presses the return key.)

? 1.56   (Computer reads line 10 and types "?" to indicate it wants a value for X. Operator types a value, say 1.56, and presses the return key.)

? 65   (Computer reads line 20 and types "?." Operator types "65" and presses the return key.)

? 78   (Computer reads line 30 and types "?." Operator types "78" and presses the return key.)

1.3$\emptyset$ (Computer executes the remaining commands and prints the answer, 1.30.)

The great strength of a computer lies not in its ability to do arithmetic calculations (which can be done on a calculator), but to understand instructions to do a calculation over and over again for different inputs. Such instructions are given by setting up the loops of which we spoke in Section 5.4, and there are three BASIC commands that are especially useful for this purpose.

The simplest is the GO TO command. If the computer comes to line 60 and finds

$$6\emptyset \quad \text{GO TO} \quad 4\emptyset$$

it next executes line 40. Consider, for instance, a BASIC program for Example 2 of Section 5.4 in which we were to add $1 + 2 \cdots + N$. The flowchart is reproduced in the margin. Part of the desired program may be written as follows.

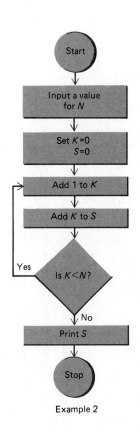

Example 2

```
2∅   LET K=∅
3∅   LET S=∅
4∅   LET K=K+1
5∅   LET S=S+K
6∅   GO TO 4∅
```

The commands from 40 through 60 form the desired loop. After the variables K and S have been given initial values in lines 20 and 30, the execution proceeds as follows.

Line	K	S	Next Command
40	1	0	50
50	1	1	60
60	1	1	40
40	2	1	50
50	2	3	60
60	2	3	40
40	3	3	50
50	3	6	60

The problem with this program so far is, of course, that it will never stop. No provision has been made to break out of the loop. For this, we must allow for the insertion of a value of N at which we want to stop, say

$$1∅ \text{ INPUT N}$$

and the command

$$55 \text{ IF K=N, THEN 7∅}$$

This command has the following effect. If the value of K is not N, the computer will go on to line 60 and thence back to 40 as before. But if $K = N$, it will go to a new command on line 70 which might be

$$7∅ \text{ PRINT S}$$

The completed program is shown in the margin.

Another way to achieve the same end would be to use the IF-THEN command with $<$ rather than $=$, as follows

```
1∅   INPUT N
2∅   LET K=∅
3∅   LET S=∅
4∅   LET K=K+1
5∅   LET S=S+K
6∅   IF K<N, THEN 4∅
7∅   PRINT S
8∅   END
```

**BASIC PROGRAM
FOR EXAMPLE 2**

```
1∅   INPUT N
2∅   LET K=∅
3∅   LET S=∅
4∅   LET K=K+1
5∅   LET S=S+K
55   IF K=N, THEN 7∅
6∅   GO TO 4∅
7∅   PRINT S
8∅   END
```

Finally, we mention the FOR-NEXT command which can be used to shorten the program just written. Consider the program

```
1∅   INPUT N
2∅   LET S=∅
3∅   FOR K=1 TO N
4∅   LET S=S+K
5∅   NEXT K
6∅   PRINT S
7∅   END
```

By using the FOR-NEXT command, the programmer is relieved of setting an initial value for K, increasing its value by 1, and testing its new value. All this is done automatically by the computer.

TABLE 5-3
COMMONLY USED BASIC COMMANDS

Command	Example	Explanation
SCR	SCR	Scratch (erase) whatever is in memory
LIST	LIST	List program in memory
	LIST 3∅	List line 30
RUN	RUN	Execute the program
PRINT	PRINT A,B	Print the values of A and B
	PRINT A+B	Print the value of $A + B$
	PRINT " "	Print the words between quotes
LET	LET A=5	Assign the value 5 to A
	LET A=A+5	Replace the present value of A by $A + 5$
INPUT	INPUT A	Computer types "?" and waits for operator to type a value
GO TO	GO TO 9∅	Transfer control to line 90
IF-THEN	IF A<12, THEN 5∅	If A has value less than 12, control is transferred to line 50; if not, it continues in the regular sequence
	IF A=12, THEN 5∅	As above, with "less than" replaced by "equal to"
FOR-NEXT	For N=4 TO 9 · · NEXT N	Performs commands between the two lines, first for $N = 4$, then for $N = 5, \ldots,$ finally for $N = 9$. Control then passes to next line number.
STOP	STOP	Stop execution
END	END	Must be last command in program

EARLIER EXAMPLES

In Section 5.4, we made flowcharts for three examples, one of which (Example 2) we have just used in the illustration above. The

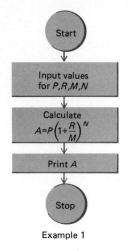

Example 1

reproduced flowcharts together with the corresponding BASIC programs for the other two examples follow.

Example 1. (Compound Interest)

```
10   INPUT P
20   INPUT R
30   INPUT M
40   INPUT N
50   LET  A=P*((1+R/M)↑N)
60   PRINT A
70   END
```

Example 3. (Fibonacci Numbers)

```
10    INPUT N
20    LET  K=2
30    LET  A=1
40    LET  B=1
50    LET  K=K+1
60    LET  F=A+B
70    PRINT K,F
80    IF  K<N, THEN 100
90    STOP
100   LET  A=B
110   LET  B=F
120   GO TO 50
130   END
```

SUMMARY

A program is a set of instructions for a computer. A BASIC program is a set of instructions written in a special language which, however, uses ordinary typewriter symbols. In order to write a BASIC program, one must learn the BASIC vocabulary and some simple gramatical rules, which takes 2 or 3 hours. We hasten to add that real proficiency in writing programs comes only after months or years of practice. For readers who want to study BASIC programming in more depth we recommend John Kemeny and Thomas Kurtz, *BASIC Programming*, 2d ed., Wiley, New York, 1971.

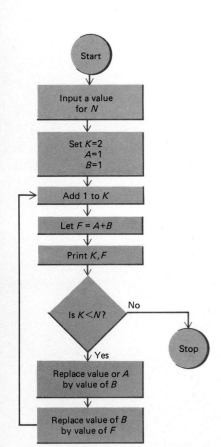

Example 3

PROBLEM SET 5.5

1. Write each of the following in BASIC notation.
   (a) $3X^2 - 2X + 7$ (b) $6(X + Y)^2$
   (c) $(X^2 + 3)/5$ (d) $(X + Y)^3/(Z + 3)$

2. Do the same for
   - (a) $3X^3 - X + 11$
   - (b) $(X^2 - 1)^3$
   - (c) $(3X - 9)/(X + 2)$
   - (d) $((X + Y)/Z)^2$

3–12. Write BASIC programs for the flowcharts of Problems 3 through 12 of Section 5.4. If a loop is involved, use the IF-THEN command.

13. Redo Problem 11 using the FOR-NEXT command.

14. Redo Problem 12 using the FOR-NEXT command.

15–20. Write BASIC programs for the flowcharts of Problems 15 through 20 of Section 5.4.

Chapter 6
Systematic Counting

The deriving of shortcuts from basic principles covers some of the finest achievements of the greatest mathematicians.
M. H. A. NEWMAN

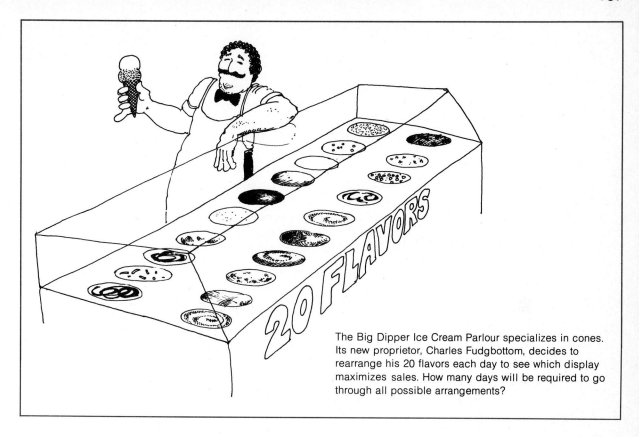

The Big Dipper Ice Cream Parlour specializes in cones. Its new proprietor, Charles Fudgbottom, decides to rearrange his 20 flavors each day to see which display maximizes sales. How many days will be required to go through all possible arrangements?

6.1 / Fundamental Counting Principles

The fact is that the proposed experiment will take

$$2,432,902,008,176,640,000$$

days. Somebody ought to tell poor Charlie before he starts; he will surely lose his taste for the job before it is completed.

To see how this huge number arises, we heed the advice of Section 1.3 and begin with a simple case. Suppose there were only three flavors (almond crackle, banana sherbet, and charcoal licorice) to be put in a case with room for three containers. The six possible arrangements are shown at the right. A similar listing with four flavors and four positions would yield 24 arrangements. And with five flavors and five positions it would be 120. Is a pattern apparent? Some readers may already see one. If so, what is the principle behind the pattern?

Let's go back to the situation with three flavors. There are three

A	B	C
A	C	B
B	A	C
B	C	A
C	A	B
C	B	A

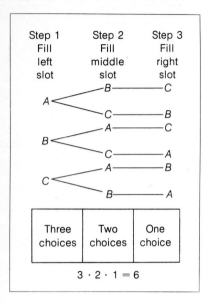

Step 1 Fill left slot	Step 2 Fill middle slot	Step 3 Fill right slot

Three choices	Two choices	One choice

$$3 \cdot 2 \cdot 1 = 6$$

Sherbet

Ice Cream

19   10,000 Lakes   81
AAT 019
MINNESOTA

slots to fill; let's call them the left, middle, and right slots. For the left slot, we have *three* choices; we can fill it with A, B, or C. Once it is filled, there are *two* choices of flavors to fill the middle slot. And this leaves no choice at all for the right slot; we must fill it with the *one* remaining flavor. What shall we do with the three numbers 3, 2, and 1? It is evident from the diagram in the margin that there are $3 \cdot 2 \cdot 1 = 6$ ways to fill the slots. If there were four flavors, the same reasoning would lead us to $4 \cdot 3 \cdot 2 \cdot 1 = 24$; for five it would be $5 \cdot 4 \cdot 3 \cdot 2 \cdot 1 = 120$. Moreover, $20 \cdot 19 \cdot 18 \cdots 3 \cdot 2 \cdot 1$ is the huge number we claimed was the answer to our opening question.

A double-dip ice cream cone serves to round out our discussion. If there are 14 flavors of ice cream and 6 of sherbet, how many different double-dip cones can be made with ice cream on the bottom and sherbet on top? The answer is simply $14 \cdot 6 = 84$. We summarize these insights in an important principle.

MULTIPLICATION PRINCIPLE

If event M can occur in m ways and after it has occurred event N can occur in n ways, the joint event M *and* N (that is, M followed by N) can occur in $m \cdot n$ ways.

Naturally, this principle of multiplication extends to the case of three events or in fact to any finite number of events.

We offer two more illustrations. Homer has seven shirts, nine ties, two pairs of trousers, and nine pairs of socks. How many different outfits can he wear? That is, in how many ways can the four events (choosing a shirt, a tie, a pair of trousers, and a pair of socks) occur? Answer: $7 \cdot 9 \cdot 2 \cdot 9 = 1134$.

Minnesota license plates consist of three letters followed by three digits. How many different plates can be made? Answer: $26 \cdot 26 \cdot 26 \cdot 10 \cdot 10 \cdot 10$

There is another principle, more obvious, that we must know.

ADDITION PRINCIPLE

If event M can occur in m ways and event N in n ways, and if M and N cannot occur simultaneously (i.e., M and N are disjoint), the event M *or* N can occur in $m + n$ ways.

Consider the following hypothetical example. Whisk Airlines offers four flights from Seattle to St. Louis; Zoom Airways has three. In how many ways can we fly from Seattle to St. Louis? We can fly Whisk or Zoom, but not both. There are $4 + 3 = 7$ ways to fly.

But be careful. If there are five dinner flights and six evening flights from St. Louis to New York and these are the only flights, can we conclude that there are $5 + 6 = 11$ flights from St. Louis to New York? No. A dinner flight can also be an evening flight. Without more information, we can't answer the question. The addition principle, as stated, doesn't apply.

Suppose, however, that we know that exactly two of the flights are both evening and dinner flights. The number of St. Louis–New York flights is (see diagram at right) $5 + 6 - 2 = 9$. You may recognize that the situation is completely analogous to that of counting the number of points in the union of two sets. Recall that the **union** of two sets, denoted $A \cup B$, is the set of points in A or B. If $n(A)$ is the number of points in A,

$$n(A \cup B) = n(A) + n(B)$$

provided A and B are disjoint. But if A and B overlap, we have

$$n(A \cup B) = n(A) + n(B) - n(A \cap B)$$

Here $A \cap B$, the **intersection** of A and B, is the set of points in both A and B.

Consider a final question about flights. In how many ways can we fly from Seattle to New York via St. Louis? By the multiplication principle, it must be $7 \cdot 9 = 63$ ways.

TO ADD OR MULTIPLY?

Let us recapitulate. Suppose M and N designate events that can occur in m and n ways, respectively. If the occurrence of M does not change the number of ways in which N can occur, then the combined event M *and* N can occur in mn ways. If M and N are disjoint events (events that cannot occur simultaneously), then the event M *or* N can occur in $m + n$ ways. The rule of thumb, then, is that we multiply when interested in M *and* N and add when interested in M *or* N.

Suppose we are to select a president, vice-president, and secretary from a class consisting of 5 boys and 4 girls. If there are no restrictions as to sex, this can be done in $9 \cdot 8 \cdot 7 = 504$ ways. But suppose the president is to be of one sex and the other two officers of the opposite sex. That is, suppose we are to select a female president and male vice-president and male secretary or a male president and female vice-president and female secretary. This can be done in

$$4 \cdot 5 \cdot 4 + 5 \cdot 4 \cdot 3 = 140$$

ways (see diagram in the margin).

THE FACTORIAL SYMBOL

Let's dip into the ice cream problem again. Our brilliant proprietor can arrange his 20 ice cream containers in $20 \cdot 19 \cdot 18 \cdots 3 \cdot 2 \cdot 1$ ways. If he had n containers to put in n positions, he could do it as

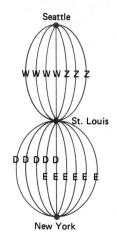

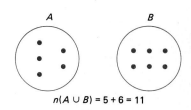

$n(A \cup B) = 5 + 6 = 11$

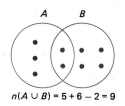

$n(A \cup B) = 5 + 6 - 2 = 9$

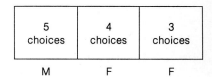

Pres.	VP	Sec.
4 choices	5 choices	4 choices
F	M	M

or

5 choices	4 choices	3 choices
M	F	F

TABLE 6-1

n	$n!$
0	1
1	1
2	2
3	6
4	24
5	120
6	720
7	5,040
8	40,320
9	362,880
10	3,628,800

the product of the numbers from 1 to n. This expression occurs so often that we give it a special name (n factorial) and denote it by a special symbol ($n!$) Thus

$$7! = 7 \cdot 6 \cdot 5 \cdot 4 \cdot 3 \cdot 2 \cdot 1$$
$$n! = n(n-1)(n-2) \cdots 3 \cdot 2 \cdot 1$$

Factorials are tiresome to calculate; so we have provided the table at left. Note that we have defined 0! to be 1, a convenience for later work. Also observe how rapidly the sequence $a_n = n!$ grows. Just as geometric growth ultimately surpasses arithmetic growth, so factorial growth will ultimately conquer geometric growth. For example, $n!$ will ultimately get bigger than 100^n. Can you see why?

It is important for us to be able to manipulate factorials with ease. Be sure you understand the following calculations.

$$\frac{6!}{4!} = \frac{6 \cdot 5 \cdot 4 \cdot 3 \cdot 2 \cdot 1}{4 \cdot 3 \cdot 2 \cdot 1} = 6 \cdot 5 = 30$$

$$\frac{8!}{5!} = \frac{8 \cdot 7 \cdot 6 \cdot 5 \cdot 4 \cdot 3 \cdot 2 \cdot 1}{5 \cdot 4 \cdot 3 \cdot 2 \cdot 1} = 8 \cdot 7 \cdot 6 = 336$$

$$\frac{n!}{(n-2)!} = \frac{n(n-1)(n-2)(n-3) \cdots 3 \cdot 2 \cdot 1}{(n-2)(n-3) \cdots 3 \cdot 2 \cdot 1} = n(n-1)$$

$$\frac{9!}{4! \, 5!} = \frac{9 \cdot 8 \cdot 7 \cdot 6 \cdot 5 \cdot 4 \cdot 3 \cdot 2 \cdot 1}{4 \cdot 3 \cdot 2 \cdot 1 \cdot 5 \cdot 4 \cdot 3 \cdot 2 \cdot 1} = \frac{9 \cdot \overset{2}{8} \cdot 7 \cdot 6}{4 \cdot 3 \cdot 2 \cdot 1} = 9 \cdot 2 \cdot 7 = 126$$

On the other hand, the factorial symbol can be used to condense certain expressions:

$$7 \cdot 6 \cdot 5 \cdot 4 = \frac{7 \cdot 6 \cdot 5 \cdot 4 \cdot 3 \cdot 2 \cdot 1}{3 \cdot 2 \cdot 1} = \frac{7!}{3!}$$

$$2 \cdot 4 \cdot 6 \cdot 8 \cdot 10 = (2 \cdot 1)(2 \cdot 2)(2 \cdot 3)(2 \cdot 4)(2 \cdot 5) = 2^5 \cdot 5!$$

Some calculators have a factorial button. Note that this should not be used indiscriminately. Try, for example, to evaluate 25!/21! using this button. What happens? Why is it preferable to do part of this calculation by hand?

SUMMARY

Correct counting depends on two principles which we now state in a slightly new way. *Multiplication principle:* If task M can be performed in m ways and after it is done task N can be performed in n ways, the joint task M followed by N can be performed in $m \cdot n$ ways. *Addition principle:* If task M can be performed in m ways and task N in n ways, and if we wish to do either M or N (but not both), we can proceed in $m + n$ ways.

PROBLEM SET 6.1

In the following problems, you need not do all the arithmetic. For example, 11 · 10 · 9 · 8 would be an acceptable answer rather than 7920.

1. Podunk Pizzeria offers 3 choices of salad, 10 kinds of pizza, and 4 different desserts. How many different three-course meals can Homer order?
2. The Fair Game Dating Service has cards for 60 men and 80 women. How many different dates can it arrange?
3. Horace plans to drive from Posthole to Klondike but will make an intermediate stop at Pitstop to pick up his friend Hugo. There are four roads from Posthole to Pitstop, and five from Pitstop to Klondike. How many different routes can Horace choose?
4. A club with 20 members wants to choose a president and a vice-president. In how many ways can this be done?
5. Homer has six bow ties, eight regular ties, and four shirts. How many different tie-shirt combinations can he wear?
6. Hugo can go from Springfield to Chicago by one of three busses or one of two trains. From Chicago to New York he will pick one of four air flights. How many different choices does he have for his trip from Springfield to New York?
7. There are five different roads from Podunk to Tubville. In how many ways can Homer make a roundtrip between the two cities? A roundtrip returning by a different road?
8. A combination lock with 10 digits on its face is opened by turning clockwise to a digit, counterclockwise to a second digit, and then clockwise to a third digit (but digits may be repeated, for example, 3-3-1). How many different combination locks of this type are possible?
9. The letters RATS are written one on each of four cards. How many code words can be made with these cards? Hint: Code words can have one letter, two letters, three letters, or four letters.
10. The letters STUPID are written one on each of six cards. How many code words can be made with these cards?
11. The Tooterville Toads have 22 players (3 catchers, 9 pitchers, 5 infielders, and 5 outfielders). In how many ways can the manager fill the nine positions (assuming, for example, that an infielder can play any of the four infield positions)?
12. The manager of the Podunk Possums has chosen the nine players to start today's game but must turn in the batting order. How many different batting lineups are possible? How many if the pitcher bats last?
13. Telephone numbers consist of seven digits (e.g., 641-2256). Assuming any choice of digits can be used, how many different telephone numbers are possible? How many if the first digit cannot be 0?
14. A man wants to have a picture taken of himself with his four boys and two girls all lined up in a row. In how many ways can the photographer line them up if the man insists on being in the middle with a girl adjacent to him on either side?

Order Form

Salad _____

Pizza _____

Dessert _____

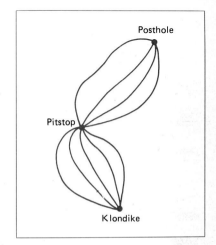

Posthole

Pitstop

Klondike

Tooterville Lineup

Catcher _____

Pitcher _____

1st base _____

2nd base _____

SS _____

3rd base _____

LF _____

CF _____

RF _____

15. How many three-digit numbers are there? (Be careful; the first digit cannot be 0). How many are there without repeated digits?

16. How many even three-digit numbers are there? How many of these have no repeated digits?

17. How many three-digit numbers are there ending in 3, 4, or 5 that are greater than 400? Greater than 432?

18. How many numbers greater than 5000, with no repeated digits, can be formed using only, 0, 2, 3, 4, 5, and 6?

19. Calculate the easy way (i.e., first cancel between the numerator and denominator).

    (a) $\dfrac{7!}{2!}$ (b) $\dfrac{8!}{4!}$

    (c) $\dfrac{7!}{4!3!}$ (d) $\dfrac{9!}{5!2!2!}$

20. Calculate:

    (a) $\dfrac{8!}{5!}$ (b) $\dfrac{9!}{5!4!}$

    (c) $\dfrac{100!}{98!}$ (d) $\dfrac{10!}{5!3!2!}$

21. Simplify:

    (a) $\dfrac{(n-1)!}{n!}$ (b) $\dfrac{(n!)^2}{(n-1)!(n-2)!}$

22.* How many seating arrangements are possible for a class of 15 students if there are 30 chairs in a room, arranged in 6 rows of 5 each and bolted to the floor?

23.* In how many ways can nine different books be divided among three people
    (a) If each person is to get three books?
    (b) If each person is to get at least one book?
    (c) If there are no restrictions on the number of books each person gets?

[c] 24. Minnesota license plates used to consist of two letters followed by four digits before being changed to the system described in this section of three letters followed by three digits. How many more plates can be produced using the new system.

Problems 25–28 give you a chance to experiment with your calculator but require that it have a y^x key and a factorial key. In each case, try $n = 5, 10, 15, \ldots$ and continue until you are convinced of the answer.

[c] 25. Which grows faster as n increases, $n!$ or 10^n?

[c] 26. Which grows faster as n increases, $n!$ or n^{10}?

[c] 27. Which grows faster, $n!$ or $(n/2)^n$?

[c] 28. Which grows faster, $n!$ or $(n/3)^n$?

6.2 / Permutations

To permute a set of objects is to rearrange them. And a **permutation** of a set of objects is just an (ordered) arrangement of the objects. Thus RHMEO is a permutation of HOMER, as are MOREH and ORMEH. How many permutations are there of HOMER? We can consider this a problem of filling five positions.

$$\square \quad \square \quad \square \quad \square \quad \square$$

We can fill the first position in five ways. Having done this we can fill the second in four ways, etc. From the multiplication principle, the answer is $5! = 5 \cdot 4 \cdot 3 \cdot 2 \cdot 1 = 120$.

But now suppose we ask how many three-letter words we can make from the letters HOMER. We include nonsense words like EOM and HMR here, as well as more familiar words like HOE and

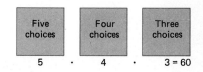

$$5 \quad \cdot \quad 4 \quad \cdot \quad 3 = 60$$

HEM. Again we are faced with a problem of filling positions, three of them. We can fill the first position in five ways, the second in four ways, and the third in three ways. The answer (by the multiplication principle) is $5 \cdot 4 \cdot 3 = 60$.

This is such a common problem that we introduce special language for it. When we have n distinguishable objects and we select r of them to arrange in a row, we call the resulting arrangement a **permutation of n things taken r at a time.** The question of interest is: How many of these permutations are there? The answer is designated by the symbol $_nP_r$. Thus $_5P_3$ denotes the number of permutations of five things taken three at a time; in particular, it represents the number of three-letter words that can be made from the five-letter word HOMER. It has the value

$$_5P_3 = 5 \cdot 4 \cdot 3 = 60$$

Similarly,

$$_6P_4 = 6 \cdot 5 \cdot 4 \cdot 3 = 360$$
$$_{10}P_2 = 10 \cdot 9 = 90$$

In general, $_nP_r$ is the product of the numbers starting with n and working down until there are r factors. We can even give a formula:

$$_nP_r = n(n-1)(n-2) \cdots (n-r+1)$$

You should check that this formula gives the right answer in each of the examples immediately above. Note that, when $r = n$,

$$_nP_n = n(n-1)(n-2) \cdots 3 \cdot 2 \cdot 1 = n!$$

There is another formula for $_nP_r$ which we will need later. Let's see if we can work up to it by means of examples. Note that

$$_5P_3 = 5 \cdot 4 \cdot 3 = \frac{5 \cdot 4 \cdot 3 \cdot 2 \cdot 1}{2 \cdot 1} = \frac{5!}{2!}$$
$$_7P_3 = 7 \cdot 6 \cdot 5 = \frac{7 \cdot 6 \cdot 5 \cdot 4 \cdot 3 \cdot 2 \cdot 1}{4 \cdot 3 \cdot 2 \cdot 1} = \frac{7!}{4!}$$

The general formula that these special cases suggest is

$$_nP_r = \frac{n!}{(n-r)!}$$

It is correct, as you can verify by writing out both numerator and denominator and canceling.

Here is a good problem dealing with permutations. How many words of any length can be made from HOMER? We translate this into: How many five-letter words or four-letter words or three-letter words or two-letter words or one-letter words can be made? The answer (using the addition principle) is

$$_5P_5 + {_5P_4} + {_5P_3} + {_5P_2} + {_5P_1}$$
$$= 5 \cdot 4 \cdot 3 \cdot 2 \cdot 1 + 5 \cdot 4 \cdot 3 \cdot 2 + 5 \cdot 4 \cdot 3 + 5 \cdot 4 + 5$$
$$= 120 + 120 + 60 + 20 + 5 = 325$$

In how many ways can we scramble SCRAMBLE? The answer is $_8P_8 = 8 \cdot 7 \cdot 6 \cdot 5 \cdot 4 \cdot 3 \cdot 2 \cdot 1$. How many four-letter words can be made from SCRAMBLE? We can make $_8P_4 = 8 \cdot 7 \cdot 6 \cdot 5$ or 1680 of them.

It is important to notice that the letters of HOMER and SCRAMBLE are all different. In fact, in our discussion thus far, we have assumed that we were talking about permutation of distinguishable objects. But what happens if some of the objects are alike? This is our next subject.

PERMUTATIONS WITH SOME OBJECTS ALIKE

Homer's girlfriend was christened HESTER AMANDA GOOD-ROAD. Understandably, she prefers being known as Amy. But there are things to do with even the worst of names. Take HESTER, for example. With just the slightest bit of juggling, it becomes ESTHER which isn't really bad at all. And that leads us to the question we wanted to ask. How many different names (words) can be made from HESTER using all six letters exactly once? If it weren't for the fact that there are two E's, it would be easy. The answer would be $6! = 720$. But there are two E's, and unfortunately HESTER and HESTER (the second has the two E's interchanged) spell exactly the same word.

Let's pretend we can distinguish between the two E's. We could write one in black ink and the other in red but, better yet, let's tag the two E's with subscripts, so that we have E_1 and E_2. Now HE_1STE_2R and HE_2STE_1R look different. In fact, if we maintained this distinction, we could make $6! = 720$ different arrangements. But of course we have really counted every word twice corresponding to the interchange of the two E's. There are only 720/2 or 360 genuinely different words that can be made.

Take AMANDA next. Suppose we tag the three A's with subscripts so that we can distinguish between them. Then $A_1MA_2NDA_3$ can be arranged in $6! = 720$ different ways. But every word is counted $3! = 6$ times, because the letters $A_1A_2A_3$ can be arranged in $3 \cdot 2 \cdot 1$ ways. Thus the number of different six-letter words that can be made from AMANDA is

$$\frac{6!}{3!} = 6 \cdot 5 \cdot 4 = 120$$

GOODROAD is even more interesting, for it has two D's and three O's. If we pretend that we can distinguish among the two

HESTER

$\frac{6!}{2!}$ words

AMANDA

$\frac{6!}{3!}$ words

GOODROAD

$\frac{8!}{2!3!}$ words

D's and three O's, we can make 8! words. But of course we really can't, so we must divide 8! by 2! and by 3! to compensate for the repeated letters.

In each of the above cases, we supposed that each letter was available for use only the number of times it appeared in the given word. The words "permute" and "scramble" imply this. But what happens if each letter is available in unlimited supply? This is our final subject.

ARRANGEMENTS WITH UNLIMITED REPETITIONS ALLOWED

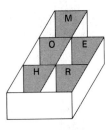

We suppose now that we have five bins filled with the letters H, O, M, E, and R, respectively. Thus we can make words like ROMER, or even RRRRR. How many five-letter words can we make altogether? Again we are faced with filling five positions. But after filling the first position with one of the five letters, we still have five choices for the second position, etc. By the multiplication principle, we can create $5 \cdot 5 \cdot 5 \cdot 5 \cdot 5 = 3125$ words. How many four-letter words can we make? Answer: $5 \cdot 5 \cdot 5 \cdot 5 = 625$.

The general situation is this. We have n items, each available in unlimited supply. How many different ordered sets of length m can we make? The answer is $n \cdot n \cdot n \cdots n$ (m times); that is, it is n^m.

Homer is learning how to make signals from his new sailboat. He has plain flags of six colors (red, blue, green, yellow, white, and black) which he can run up a pole as shown at the left. How many different signals can he make using four flags? We are faced with an immediate interpretation problem. Does he have just one flag of each color? If so, it is a permutation problem and the answer is $_6P_4 = 6 \cdot 5 \cdot 4 \cdot 3 = 360$. Does he have several (at least four) flags of each color? In this case the answer is $6 \cdot 6 \cdot 6 \cdot 6 = 1296$.

This example nicely illustrates one of the chief difficulties in counting problems, namely, the difficulty in interpreting the problem correctly. And it reinforces the maxim of Section 1.1. Before you try to solve a problem, *be sure you know precisely what the problem is.* Until this is settled, you are just spinning your wheels.

Yet you should not only think of your problem in some vague way, you should face it, you should see it clearly, you should ask yourself: *What do you want?*
George Polya

SUMMARY

We have considered three quite different arrangement problems in this section: (1) permutations of distinct objects, (2) permutations with some objects alike, and (3) arrangements with unlimited repetitions allowed. Let's put them together using the idea of bins. Suppose there is just one letter in each bin on page 177. How many three-letter words can we make? Equivalently, how many permutations of BAN are there? Answer: $3! = 6$. Suppose there are three

A's, two N's, and one B in the bins. How many six-letter words can we make; equivalently, how many permutations are there of BANANA? Answer: $6!/3!2! = 60$. Finally, suppose the bins are full of letters. How many four-letter words can we make? Answer: $3 \cdot 3 \cdot 3 \cdot 3 = 81$.

PROBLEM SET 6.2

1. How many different words can be made using all the letters of
   (a) MEAT   (b) CREAM   (c) ORANGES

2. In how many ways can we permute the letters of
   (a) FORD   (b) HORNET   (c) MUSTANG

3. Suppose the letters MAVERICK are written on eight cards. How many four-letter words can be made? Five-letter words?

4. The letters PONTIAC are written on seven cards. How many six-letter words can be made? Five-letter words?

5. How many different words can be made using all the letters of
   (a) NIXON   (b) HOOVER
   (c) KENNEDY   (d) EISENHOWER

6. How many different words can be made using all the letters of
   (a) MAINE   (b) KANSAS
   (c) WISCONSIN   (d) MISSISSIPPI

7. How many words of all lengths can be made from PONIES if letters cannot be repeated?

8. How many three- or four- or five-letter words can be made from MONKEYS if letters cannot be repeated?

9. How many six-letter words can be made from CHEESE if:
   (a) Letters can be used only as often as they appear in CHEESE?
   (b) Letters can be used an unlimited number of times?

10. Homer has 12 poker chips; 5 are white, 4 are red, and 3 are blue. In how many different-looking stacks of 12 chips can he pile them?

11. If 10 horses are entered in a race, in how many ways can the first three places (win, place, and show) be taken?

12. Thirty people enter a contest offering scholarships to the first- and second-place winners. In how many ways can the two winners be chosen?

13. A coin is to be tossed 10 times. How many possible results are there? Hint: Think of this as making 10-letter words using H's and T's.

14. A die is to be tossed five times. How many possible results are there?

15. Homer has nine flags: four red, three green, and two white. How many nine-flag signals can he run up a pole?

16.* With the flags in Problem 15, how many seven-flag signals can Homer run up a pole?

17.* The World Series is won by the team that first gets four wins in a maximum of seven games. In a Dodgers-Twins series, count all possible outcomes (e.g., DDDD and DDTTDD are two possible ways for the Dodgers to win).

[c] 18. How many code words of length 10 or less can be made from the 26 letters of the alphabet if letters cannot be repeated in a word?

[c] 19. Answer the question of Problem 18 if letters can be repeated.

Charlie is especially proud of his triple-treat sundae. It consists of three scoops of ice cream of different flavors selected from the 20 available, topped with whipped cream and a cherry. If the order in which the scoops are put in the dish doesn't matter, how many different triple-treat sundaes can Charlie produce?

6.3 / Combinations

In Sections 6.1 and 6.2, order played a major role, and we considered arrangements of things (ice cream containers, letters, and flags). Now we face a counting problem in which order is irrelevant. The whipped cream and cherry are also irrelevant, so we set them aside and regard each of the six sundaes at the top of page 179 as the same; they have exactly the same three flavors (almond crackle, banana sherbet and charcoal licorice).

If we count the number of arrangements (permutations) of three different scoops of ice cream, since there are 20 flavors available, we get $20 \cdot 19 \cdot 18$. But this counts each sundae six times. Thus there are only $(20 \cdot 19 \cdot 18)/6 = 1140$ different possible triple-treat sundaes. If a sundae contained four different scoops of ice cream, the answer would be the number of permutations $_{20}P_4$ divided by the number of ways of arranging four things, namely, 4!; that is,

$$\frac{_{20}P_4}{4!} = \frac{20 \cdot 19 \cdot 18 \cdot 17}{4 \cdot 3 \cdot 2 \cdot 1} = 4845$$

An unordered subset of a given set of objects is called a **combination** of the objects. If we select r objects from a set of n distinguishable objects, the resulting subset is called a **combination of n things taken r at a time.** We use the symbol $_nC_r$ to denote the number of such combinations. Thus

$$_{20}C_3 = \frac{_{20}P_3}{3!} \qquad _{20}C_4 = \frac{_{20}P_4}{4!} \qquad _{10}C_6 = \frac{_{10}P_6}{6!}$$

and, in general,

$$_nC_r = \frac{_nP_r}{r!} = \frac{n(n-1)(n-2) \cdots (n-r+2)(n-r+1)}{r(r-1)(r-2) \cdots 2 \cdot 1}$$

The formula above is both correct and useful. We remember how to use it by the following device. First write out the denominator. Then fill in the same number of factors in the numerator, starting with n and working down. Finally, cancel common factors. Thus

$$_{15}C_4 = \frac{15 \cdot \overset{7}{\cancel{14}} \cdot 13 \cdot \cancel{12}}{\cancel{4} \cdot \cancel{3} \cdot \cancel{2} \cdot 1} = 15 \cdot 7 \cdot 13 = 1365$$

and

$$_{10}C_6 = \frac{10 \cdot \cancel{9} \cdot \cancel{8} \cdot 7 \cdot \overset{3}{\cancel{6}} \cdot \cancel{5}}{\cancel{6} \cdot \cancel{5} \cdot \cancel{4} \cdot \cancel{3} \cdot \cancel{2} \cdot \cancel{1}} = 210$$

For theoretical purposes, another formula is very useful. Recall from Section 6.2 that

$$_nP_r = \frac{n!}{(n-r)!}$$

Thus

$$_nC_r = \frac{_nP_r}{r!} = \frac{n!}{r!(n-r)!}$$

We turn to several applications. Keep in mind that combinations are just subsets of a given collection of objects, subsets in which the order of the objects does not matter. A class of 15 students wants to choose a social committee consisting of 4 of its members. In how many ways can this be done? Is order important in this problem? No. Therefore the answer is $_{15}C_4 = 1365$ (see calculation above). The same class wishes to choose a president, vice-president, secretary, and treasurer. In how many ways can this be done? Is order important? Yes. Now there are four specific slots to fill. Therefore the answer is

$$_{15}P_4 = 32{,}760$$

A 4-card hand is to be dealt from a suit of 13 cards. In how many ways can this be done? Is order important in a hand of cards? No. It is the cards that are held, not the order in which they are held, that is significant. Thus the answer is

$$_{13}C_4 = \frac{13 \cdot \cancel{12}^{\,5} \cdot 11 \cdot \cancel{10}}{\cancel{4} \cdot \cancel{3} \cdot \cancel{2} \cdot 1} = 715$$

UNORDERED SETS WITH UNLIMITED REPETITIONS ALLOWED

Consider again Charlie's triple-treat sundaes which we discussed earlier. Every sundae contains three scoops of ice cream of *different* flavors; the customer makes his or her own choice.

On Monday of the second week in his new position, our enterprising proprietor, Charles Fudgbottom, was frozen in his tracks by two completely unexpected requests. One particularly obstreperous customer insisted on a triple-treat sundae with one scoop of almond crackle and two scoops of charcoal licorice. And the very next customer demanded three scoops of banana sherbet. Fortunately, Charlie had the good sense to realize that three scoops of ice cream was a triple treat even if the scoops were not of different flavors. And suddenly it occurred to him. Not only could he make the 1140 sundaes he had been advertising, but many many more. How many more? How many triple-treat sundaes can Charlie make if repetitions of flavors are allowed?

Charlie has 20 flavors to work with. That's too many for clear thinking. Let's suppose there are only 7 flavors called A, B, C, D, E, F, and G. A customer can order three scoops of anything on the list. Imagine that Sally, one of Charlie's best customers, has a terrible case of laryngitis and can't speak. Naturally, Sally nods her head for yes and shakes it for no. Here is the way she orders a sundae. When Charlie points to A, Sally may shake her head, which translates no, go on to the next flavor. Or she may nod twice, and then shake once, which means two scoops of A, and go to the next flavor. With Y for yes and N for no her signals become words. YNYNNNYNN translates into one A, one B, and one E, whereas NNNNYNNYY means one E and two G. Sally's order is always a word, albeit a strange one, consisting of six N's and three Y's. Each triple-treat sundae can be translated into a nine-letter NY word, and every such word corresponds to a sundae.

How many three-scoop sundaes can Charlie make using seven flavors and allowing repetitions? Translated into our new language, the question reads, how many nine-letter words can be made using six N's and three Y's? But this is a familiar problem (Section 6.2). The answer is 9!/6!3! With 20 flavors, it would be

22!/19!3! or 1540. And in general if Charlie has n flavors, he can make

$$\frac{(n+2)!}{(n-1)!3!} = {}_{n+2}C_3$$

sundaes. Suppose there are n categories of objects, each available in unlimited supply, from which we wish to select r objects, not caring about order of selection but with repetitions allowed. We can do this in ${}_{n+r-1}C_r$ ways.

SUMMARY

We can summarize much of what we have learned in Sections 6.2 and 6.3. In a typical selection problem, begin by asking two questions:

1. Is the order of selection important?
2. Are unlimited repetitions allowed?

If you can answer both questions unambiguously yes or no, refer to the following chart. Keep in mind that limited repetition problems must be handled separately.

Problem
Select r objects from n categories (bins). How many ways can this be done?

Solution
It can be done in the number of ways indicated in the table below.

Method consists entirely in properly ordering and arranging things to which we should pay attention.
René Descartes

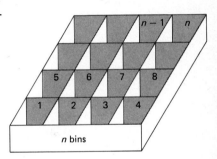

n bins

FOUR COUNTING PROBLEMS		
	Unlimited Repetitions	**No Repetitions**
Order Relevant	n^r	${}_nP_r$
Order Irrelevant	${}_{n+r-1}C_r$	${}_nC_r$

Charlie's newest concoction is a super five-deck cone. How many different five-deck cones can he make? Is order important? Probably not (though any soda jerk who has served a 5-year-old may disagree). Are repetitions allowed? Yes. The answer is

$$_{24}C_5 = \frac{24 \cdot 23 \cdot 22 \cdot 21 \cdot 20}{5 \cdot 4 \cdot 3 \cdot 2 \cdot 1} = 34{,}504$$

If order is considered relevant, the answer is a whopping $20^5 = 2^5 \cdot 10^5 = 3{,}200{,}000$.

PROBLEM SET 6.3

1. Recall the easy way to calculate $_nC_r$. First write out $r!$ in the denominator. Then put the same number of factors in the numerator, starting with n and working down. Cancel and calculate. Thus

$$_8C_3 = \frac{8 \cdot 7 \cdot \cancel{6}}{\cancel{3} \cdot \cancel{2} \cdot 1} = 56$$

Evaluate each of the following.
   - (a) $_{10}C_3$ (b) $_9C_4$ (c) $_{15}C_3$
   - (d) $_{100}C_2$ (e) $_8C_5$ (f) $_8P_5$

2. Evaluate:
   - (a) $_7C_3$ (b) $_5C_4$ (c) $_{50}C_3$
   - (d) $_{11}C_2$ (e) $_9C_3$ (f) $_9P_3$

3. The local union wants to select a committee of 3 to represent it in contract negotiations. If there are 50 people in the union, in how many ways can this be done?

4. The faculty of 30 at Podunk University plans to select a 4-member educational policies committee. In how many ways can it make its selection?

5. From a penny, a nickel, a dime, a quarter, and a half-dollar, how many different sums can be formed of:
   - (a) Three coins each?
   - (b) Four coins each?
   - (c) At least three coins each? Hint: Use the addition principle.

6. Given eight points on a sheet of paper, no three on the same line, how many triangles can be drawn using three of these points as vertices?

7. From a suit of 13 playing cards, how many different 5-card hands can be dealt? How many of these hands include the queen?

8. In how many ways can Clement select 4 of 12 friends to invite to lunch? In how many ways can she do it if she doesn't want to invite Euodias and Syntyche together?

9. From a group of six girls and nine boys, how many five-member committees can be formed involving
   - (a) Three boys and two girls? Hint: First select boys, then girls, and use the multiplication principle.
   - (b) Three girls and two boys?
   - (c) Five boys or five girls? Hint: Use the addition principle.
   - (d) At least three boys?

10. From a bag containing six white balls and ten black balls, in how many ways can we draw a group of six balls consisting of
    - (a) Three white balls and three black balls?
    - (b) Six white balls or six black balls?
    - (c) At least three white balls?

    Note: Assume the balls are distinguishable; for example, they may have numbers on them.

11. In how many ways can a group of 12 people be divided up so that 4 play tennis, 4 play golf, and 4 go hiking?

12. In how many ways can the 52 cards in a standard deck be dealt to

four players? Write the answer in combination symbols. It's too big to evaluate.

13. Hester won the big prize on a television quiz show — three brand-new cars. She may choose from Ford, Chevrolet, Plymouth, Volvo, and Saab in any combination, or even select three of the same kind. How many different choices can she make?

14. Professor Witquick went to the department storeroom to pick up five new pencils. There he found black, red, and green pencils available in abundant supply. He quickly calculated the number of choices available. What was his answer?

15. Jan has 10 different dresses. In how many different ways can she dress for school this week (Monday through Friday) if
    (a) She can wear a dress over and over again?
    (b) She wants to wear a different dress every day?
    (c) She wants to wear a different dress every day but decides that the order in which she wears them is really unimportant?

16.* Professor Witquick plans three vacations this year and will choose from Europe, Mexico, the Caribbean, Japan, and Hawaii. How many choices does he have? Analyze this problem under various assumptions. (Is order significant? Would he go to the same place more than once?)

17.* Using only pennies, nickels, and dimes, in how many ways can we make 16 cents? Twenty cents? Thirty-five cents? Sixty cents? Begin by trying to fill in the table below (p = pennies; n = nickels; d = dimes). Look for a pattern.

TABLE 6-2

Amount	1	2	3	4	5	6	7	8	9	10	11	12	13	14	15	16	20	35
Number of ways with p and n	1	1	1	1	2	2	2	2	2	3							4	5
Number of ways with p, n, and d										4								

If a_n is the number of ways to make n cents using pennies and nickels, b_n the number of ways to make n cents using pennies, nickels, and dimes, find a formula connecting a_n and b_n.

18.* In how many ways can you get a sum of 16 in a throw of five dice? Assume the dice are of different colors.

[c] 19. Consider a standard deck of 52 cards consisting of 4 suits (spades, clubs, hearts, diamonds) each with 13 cards (2, 3, 4, . . . , 10, jack, queen, king, ace). A bridge hand consists of 13 cards.
    (a) How many different bridge hands are there?
    (b) How many of them have exactly 3 aces?
    (c) How many of them have no aces?
    (d) How many of them have cards from just 3 suits?
    (e) How many of them have only honor cards (aces, kings, queens, jacks)?
    (f) How many of them have one card of each kind (one ace, one king, one queen, etc.)?
    (g) How many of them have all cards from just one suit?

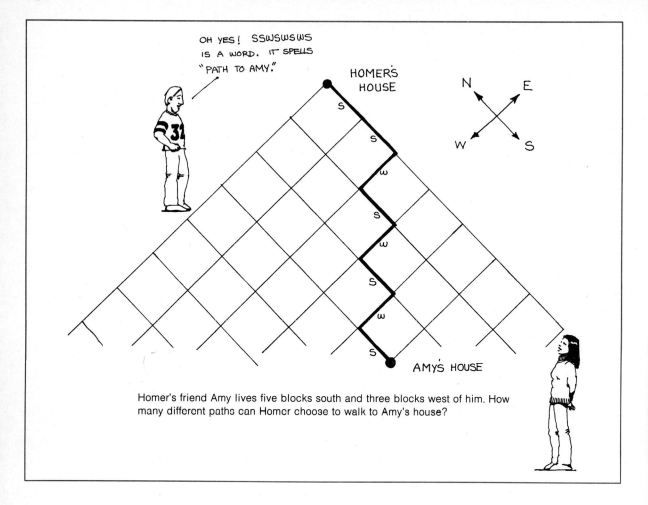

Homer's friend Amy lives five blocks south and three blocks west of him. How many different paths can Homer choose to walk to Amy's house?

6.4 / The Binomial Theorem

First we need to sharpen the question. What is a path? Can Homer cut across backyards? No; the neighbors have had their fill of Homer's shortcuts. Does he wander around the neighborhood on his way to Amy's? No. Homer has had his fill of visiting with the neighbors. He is looking for the shortest route to Amy's house, but there are many such paths. In fact, Homer must walk exactly five blocks south and three blocks west, but he can do this in any order. If S denotes south and W west, SSWSWSWS does spell "path to Amy"; so does WWWSSSSS.

How many paths are there from Homer's house to Amy's house? Exactly the number of eight-letter words consisting of five S's and three W's. We have just dressed a new subject in a set of old clothes. And in the old clothes, we recognize it. It is the problem of permutations with some objects alike (Section 5.2). The answer is

$$\frac{8!}{5!\,3!} = \frac{8 \cdot 7 \cdot 6}{3 \cdot 2 \cdot 1} = 56$$

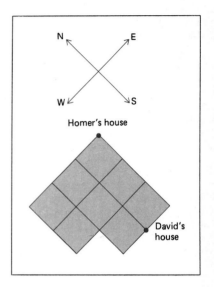

Homer's house

Homer can travel a different path on every Saturday night date for over a year. Another observation is more relevant. The answer can be given as $_8C_5$, since its value is $8!/(5!3!)$.

We can obtain the answer $_8C_5$ by another kind of reasoning. In his eight-block walk, Homer must select five blocks on which he will walk south (which can be done in $_8C_5$ ways); on the other three, he will have to walk west.

How many paths are there from Homer's house to David's house (three blocks south, one block west)? As many as there are four-letter words consisting of three S's and one W. And this is $4!/(3!1!)$ which is of course also equal to $_4C_3$; both have the value 4. This time the answer is easy to check. The four paths correspond to the words SSSW, SSWS, SWSS, and WSSS.

Our general claim is this. The number of paths from Homer's house to any intersection is $_{m+n}C_m$, where m is the number of blocks south and n the number of blocks west. Here it must be understood that $_nC_0 = 1$. We can nicely summarize our discussion in a triangular arrangement. The number at an intersection is the number of paths from the vertex to that point.

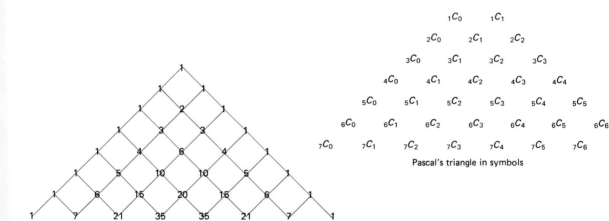

Pascal's triangle

Pascal's triangle in symbols

PASCAL'S TRIANGLE

The triangular array which appears above has been called **Pascal's Triangle** after the great mathematician-philosopher Blaise Pascal.

Blaise Pascal (1623–1662)
In 39 short years, this French mathematician discovered an important theorem in geometry, helped found the theory of probability, planted the seeds of the calculus, wrote a treatise on the "arithmetical triangle," invented a computing machine and authored *Penses,* a profound religious work.

No other array of numbers has so intrigued mathematicians. It is loaded with nuggets awaiting discovery. Let's uncover a few of them.

First, note the symmetry of the array. If we fold the triangle across its altitude, the numbers match. For example,

$$_6C_2 = {_6C_4}$$
$$_7C_3 = {_7C_4}$$
$$_7C_1 = {_7C_6}$$

and, in general,

$$_nC_r = {_nC_{n-r}}$$

There is a simple logical explanation for this formula. When we select r objects out of n objects to form a group, we simultaneously select another group of $n - r$ objects, namely, those left out.

Second, note that any number in the interior of the triangle is the sum of the two neighbors immediately above it. For example,

or, equivalently, reading from the right diagram,

The general fact, written symbolically, is

$$_nC_r = {_{n-1}C_{r-1}} + {_{n-1}C_r}$$

It is a recursion formula. We explain how to establish it algebraically in Problem 16.

Look at the second diagonal now (see figure titled Features of Pascal's Triangle). It is the sequence of counting numbers n. The third diagonal gives the triangular numbers t_n (see Section 4.1), and the fourth the summed triangular numbers $T_n = t_1 + t_2 + \cdots + t_n$ (often called tetrahedral numbers. Why?)

Next, add the horizontal rows. We get powers of 2. Algebraically, this means that (cf. Problems 9 and 10)

$$_nC_0 + {_nC_1} + {_nC_1} + \cdots + {_nC_n} = 2^n$$

This has an important interpretation. Recall that $_nC_r$ was originally introduced as the number of combinations of n things taken r at a time. Thus the formula says that the total number of subsets

Note diagonals

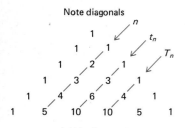

Add horizontals

			1				1
		1 —		1			2
	1 —		2 —		1		4
1 —		3 —		3 —		1	8
1 —	4 —		6 —		4 —	1	16

Features of Pascal's triangle

> Look around when you have got your first mushroom or made your first discovery: they grow in clusters.
> *George Polya*

of all sizes of a set with n elements is 2^n. For example, the set {A, B, C, D} ought to have $2^4 = 16$ subsets. Let's check by listing them (see the table below).

TABLE 6-3
SUBSETS OF {A, B, C, D}: $1 + 4 + 6 + 4 + 1 = 2^4$

Zero-Member Set	One-Member Subsets	Two-Member Subsets	Three-Member Subsets	Four-Member Subsets
The empty set { }	{A} {B} {C} {D}	{A, B} {A, C} {A, D} {B, C} {B, D} {C, D}	{A, B, C} {A, B, D} {A, C, D} {B, C, D}	{A, B, C, D}
$_4C_0 = 1$	$_4C_1 = 4$	$_4C_2 = 6$	$_4C_3 = 4$	$_4C_4 = 1$

Can you find the Fibonacci numbers hidden in Pascal's Triangle? They are there, though we must admit that they are somewhat disguised (Problem 11).

THE BINOMIAL FORMULA

By far the most important fact related to Pascal's triangle is an algebraic theorem dealing with raising a binomial, say $x + y$, to an integral power.

$$(x + y)^0 = 1$$
$$(x + y)^1 = x + y$$
$$(x + y)^2 = x^2 + 2xy + y^2$$
$$(x + y)^3 = x^3 + 3x^2y + 3xy^2 + y^3$$
$$(x + y)^4 = x^4 + 4x^3y + 6x^2y^2 + 4xy^3 + y^4$$
$$\vdots \qquad\qquad \vdots$$
$$(x + y)^n = {_nC_0}x^ny^0 + {_nC_1}x^{n-1}y^1 + \cdots + {_nC_{n-1}}x^1y^{n-1} + {_nC_n}x^0y^n$$

The last displayed formula is the **Binomial Formula.** We have obtained it by observing the pattern of coefficients we get as we raise $x + y$ to higher and higher powers. An observation is hardly a proof. Nevertheless it is correct (we suggest a way to prove it in Problem 15); and we shall feel free to use it from now on. Because of this formula the numbers $_nC_r$ are often called the binomial coefficients.

In the Binomial Formula, we may substitute any numbers for x and y. For example, the substitution $x = 1$, $y = 1$ leads to the formula displayed near the bottom of page 186.

The Binomial Formula provides a good way of calculating numbers near 1, such as the numbers that occur in compound interest tables (see Section 4.5). Since

$$(x + y)^{12} = x^{12} + 12x^{11}y + 66x^{10}y^2 + 220x^9y^3 + \cdots$$
$$(1.01)^{12} = 1^{12} + 12(1)^{11}(0.01) + 66(1)^{10}(0.01)^2 + 220(1)^9(0.01)^3$$
$$+ \cdots$$
$$= 1 + 0.12 + 0.0066 + 0.000220 + \cdots$$
$$\approx 1.12682$$

Note that, if we are interested in only five-decimal-place accuracy, there is no need to calculate all the terms in the expansion. Only the first few contribute to the first five decimal places.

SUMMARY

To specify a path from A to B (B being southwest of A) on a rectangular grid, one need only say how many blocks south and west, in some order, a person should walk. The number of such paths is then the number of words that can be formed with S's and W's used the appropriate number of times. If B is m blocks south and n blocks west of A, there are $_{m+n}C_m$ paths from A to B.

The numbers $_nC_r$ can be displayed in an array called Pascal's triangle which exhibits many beautiful and surprising patterns. These numbers also appear in the Binomial Formula which tells us how to expand $(x + y)^n$. Once again, we offer an example to remind the reader of a simple way to calculate them.

$$_{15}C_3 = \frac{\overset{5}{\cancel{15}} \cdot \overset{7}{\cancel{14}} \cdot 13}{\cancel{3} \cdot \cancel{2} \cdot \cancel{1}} = 35 \cdot 13 = 455$$

PROBLEM SET 6.4

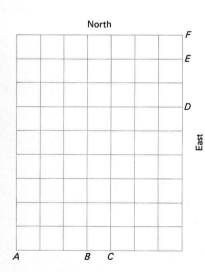

North

East

F

E

D

A   *B* *C*

1. Evaluate:
   (a) $_5C_3$ (b) $_9C_3$
   (c) $_{10}C_4$ (d) $_{12}C_4$
2. Evaluate:
   (a) $_6C_3$ (b) $_{12}C_3$
   (c) $_9C_4$ (d) $_{40}C_3$
3. List all the five-letter words that can be made with three E's and two N's. Are there $_5C_3$ of them?
4. List all the six-letter words that can be made with three E's and three N's. Are there $_6C_3$ of them?
5. How many paths (shortest routes) are there, in the diagram at the left, from
   (a) C to D? Hint: Three E's and six N's.
   (b) C to F?
   (c) A to D?
6. How many paths are there in the diagram from
   (a) B to E?
   (b) B to D?
   (c) B to F?

7. Calculate:
   (a) $_{11}C_2$ and $_{11}C_9$.
   (b) $_8C_3$ and $_8C_5$.
   What formula do these examples illustrate?
8. Recall the formula

$$_nC_r = \frac{n!}{r!(n-r)!}$$

Use it to show that

$$_nC_r = {_nC_{n-r}}.$$

9. Consider the set $\{a, b, c, d, e\}$. Finish filling in the table below.

TABLE 6-4		
Zero-member subset	{ }	$_5C_0 = 1$
One-member subsets	$\{a\}$ $\{b\}$ $\{c\}$ $\{d\}$ $\{e\}$	$_5C_1 = 5$
Two-member subsets		$_5C_2 = 10$
Three-member subsets		$_5C_3 = 10$
Four-member subsets		$_5C_4 = 5$
Five-member subsets		$_5C_5 = 1$

Notice that the total is
$$_5C_0 + {_5C_1} + {_5C_2} + {_5C_3} + {_5C_4} + {_5C_5} = 32 = 2^5$$

10. Again consider the set $\{a, b, c, d, e\}$. Suppose we wish to pick a subset out of it. We have five decisions to make, each with two choices. Will a go in the subset or not? Will b go in the subset or not? And so on. By the multiplication principle, we can make these decisions in $2 \cdot 2 \cdot 2 \cdot 2 \cdot 2$ ways. Looked at from this perspective, how many subsets does $\{a, b, c, d, e, f, g\}$ have?
11. Fill in the boxes.

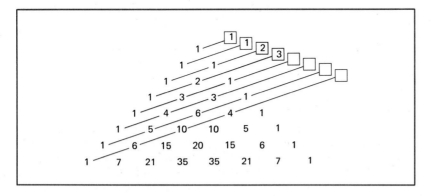

What do you observe?
12. Find 11^2, 11^3, 11^4 and 11^5. What do you observe?
13. Expand each of the following:
   (a) $(a + b)^4$
   (b) $(c - d)^5$ Hint: $c - d = c + (-d)$.
   (c) $(u + 2v)^6$

14. Expand:
    (a) $(1+x)^7$
    (b) $(2+y)^5$
    (c) $(2x+y)^5$

15.* Here is an outline of a proof of the Binomial Theorem. Try to understand it.
    a. $(x+y)^n = (x+y)(x+y)(x+y) \cdots (x+y)$, ($n$ factors).
    b. When expanded, each term arises from picking x's from r of the factors and y's from the remaining factors and then multiplying these x's and y's together.
    c. Many of the terms are alike and can be collected together. When this is done, each resulting term has the form $Ax^r y^{n-r}$.
    d. The coefficient A is the number of ways of picking r x's out of the n factors, i.e., $A = {}_nC_r$.

16.* Remember that

$$ {}_nC_r = \frac{n!}{r!(n-r)!} \qquad {}_{n-1}C_r = \frac{(n-1)!}{r!(n-1-r)!} \qquad \cdots $$

Use these facts and some algebra to show that
$$ {}_nC_r = {}_{n-1}C_{r-1} + {}_{n-1}C_r $$

17.* Convince yourself that
$$ {}_nC_r = {}_{n-1}C_{r-1} + {}_{n-1}C_r $$
is correct by filling in the details in the following argument. Consider a set of n objects, one of which we label A. To form a subset consisting of r objects we may choose to include A or not include it. In how many ways can we form the subset in each of these two cases?

18.* (Project.) Try to discover other patterns in Pascal's Triangle. Here is one you can try. Take any number a in the interior of the triangle and consider its six neighbors.

$$
\begin{array}{ccc}
 & s & t \\
x & a & u \\
 & w & v \\
\end{array}
$$

Calculate suw and tvx. Do this for several choices of a and then guess a result. Can you prove it?

[c] 19. Use your calculator to check the accuracy of the estimate of $(1.01)^{12}$ obtained at the end of this section.

[c] 20. Use the first four terms of the Binomial Formula to estimate $(1.02)^{15}$. Then use your calculator to evaluate this number directly.

[c] 21. Some banks compound interest *continuously*. To understand what this means, imagine putting 1 dollar in the bank today at 12% interest compounded n times per year where n is very very large. Estimate the value at the end of the year, namely $(1 + 0.12/n)^n$, by using the first four terms of the Binomial Formula. Note, for example, that $n(n-1)/n^2$ is approximately 1 for large n.

Chapter 7
The Laws of Chance

There are laws of chance. We must avoid the philosophically intriguing question as to why chance, which seems to be the antithesis of all order and regularity, can be described at all in terms of laws.
WARREN WEAVER

Probability is the very guide of life.
Cicero.

7.1 / Equally Likely Outcomes

If we can be sure of anything, we can be sure that our two college students are not going to study. There may be three outcomes of this experiment, but a coin that balances on one edge, when weighed on the scales of justice, is likely to be found wanting. A fair coin landing on a flat surface will show either heads or tails.

More is true. If the coin has not been doctored, it is just as likely to fall heads as tails. By this we mean that, if it were tossed over and over again, say millions of times, it would fall heads approximately one-half of the time and tails about one-half of the time. How do we know that this is true for a particular coin? We don't; but it is an assumption that most of us are willing to make, at least if the coin appears to be perfectly round and unworn.

There are many situations in which the assumption of equal likelihood seems plausible. Toss a standard die. Is there any reason to think that one side has an edge in the battle to show its face? We think not. Draw a card from a well-shuffled deck. Nature is perfectly democratic. It gives each card the same chance of being drawn.

But one should be careful. If a stockbroker tries to convince you that the market can do only one of three things—go up, stay the same, or go down—he is of course right. But, if he further insists that these three possibilities are equally likely and therefore that the odds are two to one against losing, hang on to your wallet.

Now we are ready to introduce the main notion of this chapter, the *probability* of an event. Consider the experiment of tossing a die, which can result in six equally likely outcomes. Of these six outcomes, three show an even number of spots. We therefore say that the probability that an even number will show is $\frac{3}{6}$.

Or consider the experiment of picking a card from a well-shuffled deck. Thirteen of the 52 cards are diamonds. Thus we say that the probability of getting a diamond is $\frac{13}{52}$.

These examples illustrate the general situation. If an experiment can result in any one of n equally likely outcomes and if exactly m of them are included in the event E, we say that the **probability of E** is m/n. We write this as

$$P(E) = \frac{m}{n}$$

> Logically cautious readers may have noticed a disturbing aspect of this definition of probability. Since it speaks of "equally likely," i.e., equally probable, events, the definition sits on its own tail, so to speak, defining probability in terms of probability.
> *Warren Weaver*

A PAIR OF DICE

We use two dice, one shaded and the other white, to illustrate the major ideas of probability theory. Since one die can fall in 6 ways, two dice can fall in $6 \times 6 = 36$ ways. The 36 equally likely outcomes are shown below.

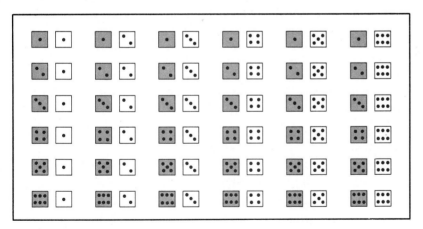

Most questions about a pair of dice have to do with the total number of spots showing after a toss. For example, what is the probability of getting a total of 7? Six of the 36 outcomes result in this event (see margin). Therefore the required probability is $\frac{6}{36} = \frac{1}{6}$. Similarly,

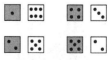

Six ways to get 7.
P(getting 7) = 6/36.

Two ways to get 11
P(getting 11) = 2/36

P(getting a total of 11) $= \frac{2}{36} = \frac{1}{18}$
P(getting a total of 12) $= \frac{1}{36}$
P(getting a total greater than 7) $= \frac{15}{36} = \frac{5}{12}$
P(getting a total less than 13) $= \frac{36}{36} = 1$
P(getting a total of 13) $= \frac{0}{36} = 0$

The last two are worthy of comment. The event "getting a total less than 13" is certain to occur; we call it a **sure event.** The probability of a sure event is always 1. However, the event "getting a total of 13" cannot occur; it's called an **impossible event.** The probability of an impossible event is 0.

Consider the event "getting 7 or 11" which is important in the dice game called craps. From the picture, we note that 8 of the 36 outcomes give a total of 7 or 11, so the probability of this event is $\frac{8}{36}$. But note that we could have calculated this probability by adding together the probability of getting 7 and the probability of getting 11:

$$P(7 \ or \ 11) = P(7) + P(11)$$
$$\frac{8}{36} = \frac{6}{36} + \frac{2}{36}$$

It appears that we have found a very useful property of probability. However, a different example dampens our enthusiasm. Note that

$$P(\text{odd } or \text{ over } 7) \neq P(\text{odd}) + P(\text{over } 7)$$

since

$$\frac{27}{36} \neq \frac{18}{36} + \frac{15}{36}$$

Why is it that we can add probabilities in one case and not in the other? The reason is a simple one. The events "getting 7" and "getting 11" are disjoint (they can't both happen). But the events "getting an odd total" and "getting over 7" overlap (a number such as 9 satisfies both conditions).

Considerations like these lead us to the main properties of probability:

1. P(impossible event) $= 0$; P(sure event) $= 1$.
2. $0 \leq P(A) \leq 1$ for any event A.
3. $P(A \ or \ B) = P(A) + P(B)$, provided A and B are disjoint, i.e., can't both happen at the same time.

From these properties, another follows. The events "A" and "not A" are certainly disjoint. Thus, from property 3,

$$P(A \ or \text{ not } A) = P(A) + P(\text{not } A)$$

But the event "A or not A" must occur; it is a sure event. Hence, from property 1,

$$1 = P(A) + P(\text{not } A)$$

or

4. $P(A) = 1 - P(\text{not } A)$.

Let us illustrate this fourth property. If A is the event "getting a total less than 11," not A is the event "getting a total of at least 11." Thus

$$P(\text{total} < 11) = 1 - P(\text{total} \geq 11) = 1 - \tfrac{3}{36} = \tfrac{33}{36} = \tfrac{11}{12}$$

RELATION TO SET LANGUAGE

Most American students are introduced to the language of sets in the early grades. Even so, a brief review may be helpful. In everyday language, we talk of a bunch of grapes, a class of students, a herd of cattle, a flock of birds, or perhaps even a team of toads, a passel of possums, or a gaggle of geese. Why are there so many words to express the same idea? Mathematicians use one word — **set.** The objects that make up a set are called **elements** or **members.**

Sets can be put together in various ways. We have $A \cup B$ (read "A **union** B"), which consists of the elements in A or B. There is $A \cap B$ (read "A **intersection** B"), which is made up of the elements in both A *and* B. We have the notion of an **empty set** $\emptyset$ and of a **universe** S (the set of all elements under discussion). Finally, we have A' (read "A **complement**"), which is composed of all elements in the universe that are *not* in A.

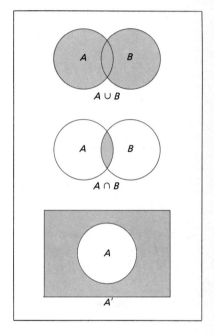

The language of sets and the language used in probability are very closely related, as the list below suggests.

Probability Language	Set Language
Outcome	Element
Event	Set
Or	Union
And	Intersection
Not	Complement
Impossible event	Empty set
Sure event	Universe

If we borrow the notation of set theory, we can state the laws of probability in a very succinct form:

1. $P(\emptyset) = 0; P(S) = 1$.
2. $0 \leq P(A) \leq 1$.
3. $P(A \cup B) = P(A) + P(B)$, provided $A \cap B = \emptyset$.
4. $P(A) = 1 - P(A')$.

POKER HANDS

We assume the reader is familiar with a standard deck of 52 cards. There are 4 suits (spades, clubs, hearts, and diamonds), each with

Difficulties with probability are usually caused, not by the simple rules that are used, but by the problem of trying to count the total number of outcomes and also those outcomes that lead to the event in question. This is one reason for the inclusion of a chapter on counting in this book (chapter 6).

13 cards (2, 3, ..., 10, jack, queen, king, and ace). A poker hand consists of 5 cards. Now recalling that $_{52}C_5$ is the number of combinations of 52 things taken 5 at a time, we see that there are

$$_{52}C_5 = \frac{52 \cdot 51 \cdot 50 \cdot 49 \cdot 48}{5 \cdot 4 \cdot 3 \cdot 2 \cdot 1} = 2{,}598{,}960$$

different possible poker hands. Suppose we plan to deal a 5-card poker hand from a well-shuffled deck.

What is the probability of a diamond flush (i.e., 5 diamonds)? Of the $_{52}C_5$ possible outcomes, $_{13}C_5$ give a diamond flush. Thus

$$P(\text{diamond flush}) = \frac{_{13}C_5}{_{52}C_5} \approx .0005$$

What is the probability of a flush of any kind? By property 3,

$$
\begin{aligned}
P(\text{flush}) = &\ P(\text{diamond flush}) + P(\text{heart flush}) + P(\text{club flush}) \\
&+ P(\text{spade flush}) \\
\approx &\ .0005 + .0005 + .0005 + .0005 \\
= &\ .002
\end{aligned}
$$

What is the probability of getting at least 1 ace? Here it is much easier to look at the complementary event "get no aces" and apply property 4. Note that there are 48 nonaces in a deck.

$$P(\text{at least 1 ace}) = 1 - P(\text{no aces})$$

$$= 1 - \frac{_{48}C_5}{_{52}C_5} \approx 1 - .66 = .34$$

SUMMARY

If an experiment can result in n equally likely outcomes of which m are in the event A, the probability of A is given by

$$P(A) = \frac{m}{n}$$

From this definition follow the four main properties or laws of probability which were stated twice in this section. Some problems can be worked by appealing directly to the definition. To calculate a probability is then a matter of counting all the outcomes and counting those that lead to the event in question. Often this is difficult, and it is better to break the problem into disjoint parts. Then, one of the properties of probability allows us to add together the probabilities of these pieces. Occasionally, it is easier to calculate the probability of the complementary event and then subtract it from 1. This is justified by another of the properties. In any case, it is important to have a clear notion of the experiment to be performed. This means, in particular, that one must be able to enumerate the (equally likely) outcomes in a systematic way.

PROBLEM SET 7.1

1. An ordinary die is tossed. What is the probability that the number of spots on the upper face will be
   (a) Three?   (b) Greater than 3?   (c) Less than 3?
   (d) An even number?   (e) An odd number?

2. Nine balls, numbered 1, 2, . . . , 9, are in a bag. If one is drawn at random, what is the probability that its number is
   (a) Nine?   (b) Greater than 5?   (c) Less than 6?
   (d) Even?   (e) Odd?

3. A penny, a nickel, and a dime are tossed. List the eight possible outcomes of this experiment. What is the probability of
   (a) Three heads?   (b) Exactly two heads?
   (c) More than one head?

4. A coin and a die are tossed. Suppose that one side of the coin has a 1 on it, and the other a 2. List the 12 possible outcomes of this experiment, e.g., (1, 1), (1, 2), (1, 3). What is the probability of
   (a) A total of 4?   (b) An even total?   (c) An odd total?

5. Two ordinary dice are tossed. What is the probability of
   (a) A double (both showing the same number)?
   (b) The number on one of the dice being twice that on the other?
   (c) The numbers on the two dice differing by at least 2?

6. A letter is chosen at random from the word PROBABILITY. What is the probability that it will be (treating Y as a vowel)
   (a) P?   (b) B?   (c) M?   (d) A vowel?

7. Two regular tetrahedra (tetrahedra have four identical equilateral triangles for faces) have their faces numbered 1, 2, 3, and 4. Suppose they are tossed and we keep track of the outcome by listing the numerals on the bottom faces, e.g., (1, 1), (1, 2).
   (a) How many outcomes are there?
   (b) What is the probability of a sum of 7?
   (c) What is the probability of a sum less than 7?

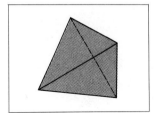

8. Two regular octahedra (polyhedra having eight identical faces) have their faces numbered 1, 2, . . . , 8. Suppose they are tossed and we record the outcomes by listing the numerals on the bottom faces.
   (a) How many outcomes are there?
   (b) What is the probability of a sum of 7?
   (c) What is the probability of a sum less than 7?

9. What is wrong with each of the following statements?
   (a) Since there are 50 states, the probability of being born in Wyoming is $\frac{1}{50}$.
   (b) The probability that a person smokes is .45, and that he drinks, .54; therefore the probability that he smokes or drinks is .54 + .45 = .99.
   (c) The probability that a certain candidate for president of the United States will win is $\frac{3}{5}$, and that she will lose, $\frac{1}{4}$.
   (d) Two football teams A and B are evenly matched; therefore the probability that A will win is $\frac{1}{2}$.

10. During the past 30 years, Professor Witquick has given only 100 A's and 200 B's in Math 13 to the 1200 students who registered for the

class. Based on these data, what is the probability that a student who registers next year

(a) Will get an A or a B?

(b) Will not get either an A or a B?

11. A poll was taken at Podunk University on the question of coeducational dormitories, producing the following results.

TABLE 7-1				
	Administrators	Faculty	Students	Total
For	4	16	100	120
Against	3	32	100	135
No opinion	3	2	40	45
TOTAL	10	50	240	300

On the basis of this poll, what is the probability that

(a) A randomly chosen faculty member will favor coed dorms?

(b) A randomly chosen student will be against coed dorms?

(c) A person selected at random at Podunk University will favor coed dorms?

(d) A person selected at random at Podunk University will be a faculty member who is against coed dorms?

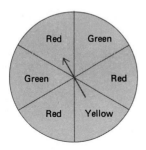

12. The well-balanced spinner shown at the left is spun. What is the probability that the pointer will stop at

(a) Red?　　　　　　　　(b) Green?

(c) Red or green?　　　　(d) Not green?

13. Four balls numbered 1, 2, 3, and 4 are placed in a bag, mixed, and drawn out, one at a time. What is the probability that they will be drawn in the order 1, 2, 3, 4?

14. A five-volume set of books is placed on a shelf at random. What is the probability they will be in the right order?

15. From an ordinary deck of 52 cards, 1 card is drawn. What is the probability that it will be

(a) Red?　　(b) A spade?　　(c) An ace?

Note: Two of the suits are red, and two are black.

16. From an ordinary deck of cards, two cards are drawn. This can be done in $_{52}C_2$ ways. What is the probability that both will be

(a) Red?　　(b) Of the same color?　　(c) Aces?

17. From an ordinary deck of 52 cards, 3 cards are drawn. What is the probability that

(a) All will be red?　　(b) All will be diamonds?

(c) Exactly 1 will be a queen?　　(d) All will be queens?

18. If three men and two women are seated at random in a row, what is the probability that

(a) Men and women will alternate?

(b) The men will be together?

(c) The women will be together?

19. A die has been doctored so that the probabilities of getting 1, 2, 3, 4, 5, and 6 are, respectively, $\frac{1}{3}$, $\frac{1}{4}$, $\frac{1}{6}$, $\frac{1}{12}$, $\frac{1}{12}$, and $\frac{1}{12}$. Assuming the rules for probabilities are still valid, find the probability of throwing
    (a) An even number.
    (b) A number less than 5.
    (c) An even number or a number less than 5.

20. On a history exam, eight events are to be matched with eight dates, with each item used just once. Homer is sure of four dates and matches the others at random. What is the probability that he will be right on
    (a) All eight dates? (b) At least six dates?
    (c) Just four dates?

21. The third law of probability is $P(A \cup B) = P(A) + P(B)$, provided $A \cap B = \varnothing$. Show that the following extension always holds.
    $P(A \cup B) = P(A) + P(B) - P(A \cap B)$

22* Three dice are tossed. What is the probability that 1, 2, and 3 all will appear?

23* In poker, what is the probability that a player will be dealt a straight flush (five consecutive cards in the same suit; an ace counts both high and low)?

24* A careless secretary typed four letters and four envelopes, and then inserted the letters randomly into the envelopes. Find the probability that
    (a) None went into the right envelope.
    (b) At least one went into the right envelope.
    (c) All went into the right envelope.
    (d) Exactly two went into the right envelope.

[c] 25. If a poker hand (5 cards) is dealt from a standard deck (52 cards), the probability of a diamond flush (that is, all diamonds) is

$$\frac{_{13}C_5}{_{52}C_5} = \frac{\dfrac{13 \cdot 12 \cdot 11 \cdot 10 \cdot 9}{5 \cdot 4 \cdot 3 \cdot 2 \cdot 1}}{\dfrac{52 \cdot 51 \cdot 50 \cdot 49 \cdot 48}{5 \cdot 4 \cdot 3 \cdot 2 \cdot 1}}$$

Calculate this value but be sure to do all the cancelling you can before using your calculator.

[c] 26. Calculate the probability of getting a poker hand consisting of all honor cards (aces, kings, queens, jacks).

[c] 27. Calculate the probability of getting a poker hand consisting of all kings and queens.

[c] 28. Calculate the probability of getting a bridge hand (13 cards) consisting completely of honor cards.

[c] 29. Calculate the probability of getting a poker hand with no honor cards.

There is no other branch of mathematics in which intelligent people make such foolish mistakes.

7.2 / Independent Events

A perfectly balanced coin has shown nine tails in a row. What is the probability that it will show a tail on the tenth flip? Some people argue that a mystical law of averages makes the appearance of a head practically certain. It is as if the coin had a memory and a conscience; it must atone for falling on its face nine times in a row. Such thinking is pure, unadulterated nonsense. The probability of showing a tail on the tenth flip is $\frac{1}{2}$, just as it was on each of the previous nine flips. The outcome of any flip is independent of what happened on previous flips.

Here is a different question, not to be confused with the one just answered. If one plans to flip a coin 10 times, what is the probability of getting all tails? To answer, we reason that there are two possibilities on the first flip, two on the second, etc. By the multiplication principle for counting (Chapter 6), there are $2 \cdot 2 \cdot 2 \cdot 2 \cdot 2 \cdot 2 \cdot 2 \cdot 2 \cdot 2 \cdot 2$ or 1024 possible outcomes of this experiment, only one of which consists of all tails. The probability of 10 tails is 1/1024, a very unlikely event indeed.

Here are the same questions stated for a game with higher stakes. A couple already has 9 girls. What is the probability that its tenth child will also be a girl? Assuming that boys and girls are equally likely (which is not quite true), the answer is $\frac{1}{2}$. But suppose a newly married couple decides, with great passion and little thought, to have 10 children. Assuming that nature assents to their decision, what is the probability they will have all girls? It is 1/1024.

Consider now the worried patient in the opening cartoon. Most of us would not (and should not) find the doctor's logic very comforting. In fact, if we make the assumption that the outcome of the tenth operation is completely independent of the first nine, the probability of a tenth failure is still $\frac{9}{10}$. The assumption of independence may be questioned. Perhaps doctors improve with experience; but this particular patient had better hope for a miracle.

DEPENDENCE VERSUS INDEPENDENCE

Perhaps we can make the distinction between dependence and independence clear by describing an experiment in which we again toss two dice—one brown and the other white. Let A, B, and C designate the following events:

    A: brown die shows 6
    B: white die shows 5
    C: total on the two dice is greater than 7

Consider first the relationship between B and A. It seems quite clear that the chance of B occurring is not affected by our knowledge that A has occurred. In fact, if we let $P(B|A)$ denote the **probability of B, given that A has occurred,** then (see illustration in margin)

$$P(B|A) = \tfrac{1}{6}$$

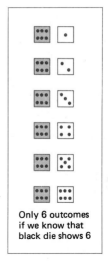

Only 6 outcomes if we know that black die shows 6

But this is equivalent to the answer we get if we calculate $P(B)$ without any knowledge of A. For then we look at all 36 outcomes for two dice and note that 6 of them have the white die showing 5, that is,

$$P(B) = \tfrac{6}{36} = \tfrac{1}{6}$$

We conclude that $P(B|A) = P(B)$, just as we expected.

The relation between C and A is very different; C's chances are greatly improved if we know that A has occurred. From the marginal illustration, we see that

$$P(C|A) = \tfrac{5}{6}$$

However, if we have no knowledge of A, and calculate $P(C)$ by looking at the 36 outcomes (page 193) for two dice, we find

$$P(C) = \tfrac{15}{36} = \tfrac{5}{12}$$

Clearly $P(C|A) \neq P(C)$.

This discussion has prepared the way for a formal definition. If $P(B|A) = P(B)$, we say that A and B are **independent events.** If $P(B|A) \neq P(B)$, A and B are **dependent events.**

And now, recalling that $A \cap B$ means A and B, we can state the multiplication rule for probabilities:

$$P(A \cap B) = P(A)P(B|A) = P(B)P(A|B)$$

In words, the probability of both A and B occurring is equal to the probability that A will occur multiplied by the probability that B will occur, given that A has already occurred. In the case of independence, this takes a particularly elegant form:

$$P(A \cap B) = P(A)P(B)$$

To illustrate these rules, consider first the problem of drawing two cards one after another from a well-shuffled deck. What is the probability that both will be spades? Based on previous knowledge (i.e., Section 7.1), we respond:

$$P(\text{two spades}) = \frac{{}_{13}C_2}{{}_{52}C_2} = \frac{\dfrac{13 \cdot 12}{2 \cdot 1}}{\dfrac{52 \cdot 51}{2 \cdot 1}} = \frac{13 \cdot 12}{52 \cdot 51}$$

But here is another approach. Consider the events:

A: getting a spade on the first draw
B: getting a spade on the second draw

Our interest is in $P(A \cap B)$. According to the rule above, it is given by

$$P(A \cap B) = P(A)P(B|A) = \tfrac{13}{52} \cdot \tfrac{12}{51}$$

which naturally agrees with our earlier answer.

Here is a different but related problem: Suppose we note the character of the first card, replace it, shuffle the deck, and draw a second card. Now A and B are independent.

$$P(A \cap B) = P(A)P(B) = \tfrac{13}{52} \cdot \tfrac{13}{52}$$

URNS AND BALLS

For reasons not entirely clear, teachers have always illustrated the central ideas of probability by talking about urns (vases) containing colored balls. Most of us have never seen an urn containing

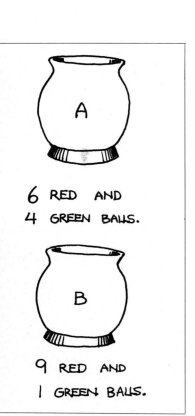

A

6 RED AND
4 GREEN BALLS.

B

9 RED AND
1 GREEN BALLS.

colored balls, but it won't hurt to use a little imagination.

Consider two urns labeled A and B, A containing six red balls and four green balls, and B containing nine red balls and one green ball. If a ball is drawn from each urn, what is the probability that both will be red?

$$P(\text{red from A } and \text{ red from B}) = P(\text{red from A}) \cdot P(\text{red from B})$$
$$= \tfrac{6}{10} \cdot \tfrac{9}{10} = \tfrac{54}{100} = \tfrac{27}{50}$$

As a second problem, suppose we choose an urn at random and then draw one ball. What is the probability that it will be red? In other words, what is the probability of the event

"choose A *and* draw red" *or* "choose B *and* draw red"?

The events in quotation marks are disjoint, so their probabilities can be added. We obtain the answer

$$\tfrac{1}{2} \cdot \tfrac{6}{10} + \tfrac{1}{2} \cdot \tfrac{9}{10} = \tfrac{15}{20} = \tfrac{3}{4}$$

The reader should note the procedure we use. We describe the event using the words "and" and "or." When we determine probabilities, "and" translates into "times," and "or" into "plus."

A HISTORICAL EXAMPLE

Two French mathematicians, Pierre de Fermat (1601–1665) and Blaise Pascal (1623–1662), are usually given credit for originating the theory of probability. This is how it happened: The famous gambler, Chevalier de Méré, was fond of a dice game in which he would bet that a 6 would appear at least once in four throws of a die. He won more often than he lost for, though he probably didn't know it,

$$P(\text{at least one 6}) = 1 - P(\text{no 6's})$$
$$= 1 - \tfrac{5}{6} \cdot \tfrac{5}{6} \cdot \tfrac{5}{6} \cdot \tfrac{5}{6} = 1 - .48 = .52$$

Growing tired of this game, he introduced a new one played with two dice. Méré then bet that at least one double 6 would appear in 24 throws of two dice. Somehow (perhaps he noted that $\tfrac{4}{6} = \tfrac{24}{36}$) he thought he should do just as well as before. But he lost more often than he won. Mystified, he proposed it as a problem to Pascal who in turn wrote to Fermat. Together they produced an explanation. Here it is:

$$P(\text{at least one double 6}) = 1 - P(\text{no double 6's})$$
$$= 1 - (\tfrac{35}{36})^{24} = 1 - .51 = .49$$

Out of this humble and slightly disreputable origin grew the science of probability.

> He [Méré] is very intelligent but he is not a mathematician: this as you know is a great defect.
> *Pascal, in a letter to P. de Fermat*

SUMMARY

We must carefully distinguish between independent and disjoint events. Two events are independent if the occurrence of one does not influence the occurrence of the other. They are disjoint if they cannot occur simultaneously. The corresponding laws of probability are

$$P(A \cap B) = P(A) \cdot P(B) \qquad \text{(if } A \text{ and } B \text{ are independent)}$$
$$P(A \cup B) = P(A) + P(B) \qquad \text{(if } A \text{ and } B \text{ are disjoint)}$$

Both laws have extensions which are always valid:

$$P(A \cap B) = P(A)P(B|A)$$
$$P(A \cup B) = P(A) + P(B) - P(A \cap B)$$

The last law was introduced in Problem 21 in Section 7.1.

PROBLEM SET 7.2

1. Toss a balanced die three times in succession. What is the probability of all 1's?
2. Toss a fair coin four times in succession. What is the probability of all heads?
3. Spin the two spinners pictured at the left. What is the probability that

    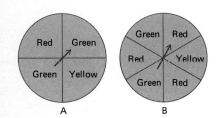

    (a) Both will show red?
    (b) Neither will show red?
    (c) Spinner A will show red and spinner B not red?
    (d) Spinner A will show red and spinner B red or green?
    (e) Just one of the spinners will show green?
4. The two boxes shown below are thoroughly shaken and a ball drawn from each. What is the probability that
    (a) Both will be 1's?
    (b) Exactly one of them will be a 2?
    (c) Both will be even?
    (d) Exactly one of them will be even?
    (e) At least one of them will be even?

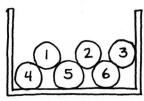
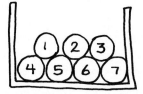

5. An urn contains five red balls and seven black balls. Two balls are drawn in succession. What is the probability of drawing two red balls if
    (a) The first ball is replaced before the second is drawn?
    (b) The first ball is not replaced before the second is drawn?

6. Answer the questions in Problem 5 if the urn contains five red balls, six black balls, and seven green balls.

7. Given that $P(A) = .8$, $P(B) = .5$, and $P(A \cap B) = .4$, find
   (a) $P(A \cup B)$ (b) $P(B|A)$ (c) $P(A|B)$

8. Given that $P(A) = .8$, $P(B) = .4$, and $P(B|A) = .3$, find
   (a) $P(A \cap B)$ (b) $P(A \cup B)$ (c) $P(A|B)$

9. In each case indicate whether or not the two events seem independent to you. Explain.
   (a) Getting an A in physics and getting an A in math.
   (b) Getting an A in physics and winning the tennis match.
   (c) Getting a new shirt for your birthday and stubbing your toe the next day.
   (d) Going to Harvard and being an American Indian.
   (e) Being a woman and being a doctor.
   (f) Walking under a ladder and having an accident the next day.

10. In each case indicate whether the two events are disjoint.
    (a) Getting an A in Physics 101 and getting a B in Physics 101.
    (b) The sun shining on Tuesday and raining on Tuesday.
    (c) In tossing two dice, getting an odd sum and getting the same number on each die.
    (d) In tossing two dice, getting an odd sum and getting a six on one of the dice.
    (e) Getting an A in physics and getting an A in math.
    (f) Not losing the football game and not winning the football game.

11. A machine produces bolts which are put in boxes. It is known that 1 box in 10 will have at least one defective bolt in it. Assuming that the boxes are independent of each other, what is the probability that a customer who ordered 3 boxes will get all good bolts?

12. Suppose the probability of being hospitalized during a year is .20. Assuming independence of family members, what is the probability that no one in a family of five will be hospitalized this year? Do you think the assumption of independence is reasonable?

13. Suppose that 4% of males are color blind, that 1% of females are color blind, and that males and females each make up 50% of the population. If a person is chosen at random, what is the probability that this person will be color blind?

14. Consider two urns, one with three red balls and seven white balls, and the other with six red balls and four white balls. If an urn is chosen at random and then a ball drawn, what is the probability that it will be red?

15. A committee of three is chosen at random from among a group of six boys and four girls. What is the probability that it will consist of all boys?

16. A coin is tossed eight times. What is the probability of getting at least one head?

17.* Let us suppose that, in a World Series, the probability that team A will win any given game is $\frac{2}{3}$. Then $\frac{1}{3}$ is the probability that team B will win a given game. What is the probability that
   (a) A will win the series in four games?
   (b) B will win in four games?

(c) The series will end in four games?

(d) A will win in five games?

(e) The series will end in five games?

18.* Fifteen girls went to the beach. Five got sunburned, eight got bitten by mosquitoes, and seven returned without a mishap. Find the probability that

(a) A girl was both burned and bitten.

(b) A burned girl was bitten.

(c) A bitten girl was burned.

19.* Among families with four children known to have at least one boy, what percentage actually have two boys? What assumptions do you have to make to do this problem?

In the game Yahtzee, which involves rolling five dice, it is very desirable to get several 6's. What is the probability of getting exactly three 6's on the first roll? Four 6's?

7.3 / The Binomial Distribution

The manufacturer of Yahtzee chooses to make all five dice of the same color, usually white. This is a matter of convenience; surely the reader will agree that a little paint on each of the dice does not affect the way they roll. Let's imagine that the five dice have been painted black, green, purple, red, and white, respectively. This will allow us to keep track of each die separately. Now call getting a 6 a success (S), and getting anything else a failure (F). Thus, for each die,

$$P(S) = \tfrac{1}{6} \qquad P(F) = \tfrac{5}{6}$$

If we roll five dice, there are 10 ways of getting exactly three 6's (see chart in the margin). Because of the independence of the five dice, we can calculate the probability of each of these events by multiplication. For example,

$$P(\text{SSSFF}) = \tfrac{1}{6} \cdot \tfrac{1}{6} \cdot \tfrac{1}{6} \cdot \tfrac{5}{6} \cdot \tfrac{5}{6} = (\tfrac{1}{6})^3(\tfrac{5}{6})^2$$

and

$$P(\text{SSFSF}) = \tfrac{1}{6} \cdot \tfrac{1}{6} \cdot \tfrac{5}{6} \cdot \tfrac{1}{6} \cdot \tfrac{5}{6} = (\tfrac{1}{6})^3(\tfrac{5}{6})^2$$

Color of Die				
B	**G**	**P**	**R**	**W**
S	S	S	F	F
S	S	F	S	F
S	S	F	F	S
S	F	S	S	F
S	F	F	S	S
S	F	S	F	S
F	S	S	F	S
F	S	F	S	S
F	S	S	S	F
F	F	S	S	S

In fact, each of the 10 disjoint events has this same probability. Consequently,

$$P(\text{three 6's in rolling five dice}) = 10(\tfrac{1}{6})^3(\tfrac{5}{6})^2 \approx .03$$

We need to know why the number 10 appears in this problem. Look at the chart again. A row is determined as soon as we decide where to put the three S's, that is, as soon as we select from the set {B, G, P, R, W} three dice to classify as S's. Now, we can choose three objects from five in $_5C_3$ ways; and recall that

$$_5C_3 = \frac{5 \cdot 4 \cdot 3}{3 \cdot 2 \cdot 1} = 10$$

There is another way to view this problem. We are really asking for the number of five-letter words that can be made using three S's and two F's. From Section 6.2, we know that there are 5!/3!2! such words. But

$$\frac{5!}{3!2!} = {}_5C_3$$

No matter how we look at it, there are $_5C_3$ ways of getting three 6's in a roll of five dice. Thus

$$P(\text{three 6's in rolling five dice}) = {}_5C_3(\tfrac{1}{6})^3(\tfrac{5}{6})^2$$

and, by similar reasoning,

$$P(\text{four 6's in rolling five dice}) = {}_5C_4(\tfrac{1}{6})^4(\tfrac{5}{6})^1$$

We make one final remark about rolling five dice. Whether we roll five dice at once (as in Yahtzee) or roll one die five times is of no significance in probability questions. From now on we adopt the second point of view.

THE BINOMIAL DISTRIBUTION

The general situation we have in mind is this. Suppose that the outcomes of an experiment fall into two categories. One we call a success (S) and the other a failure (F). The probabilities of S and F are presumed to be known, say

$$P(\text{S}) = p \qquad P(\text{F}) = q$$

It should be clear to the reader that $q = 1 - p$.

For example, if p were .3, then q would be .7. The experiment is repeated n times. What is the probability of getting exactly k successes? The answer, based on the same reasoning as in the five-dice problem, is

$$P(k \text{ successes in } n \text{ trials}) = {}_nC_k p^k q^{n-k}$$

This result is closely related to the Binomial Formula we met in Section 6.4. Recall, for example, that

$$(p + q)^5 = (q + p)^5$$
$$= {}_5C_0 p^0 q^5 + {}_5C_1 p^1 q^4 + {}_5C_2 p^2 q^3 + {}_5C_3 p^3 q^2 + {}_5C_4 p^4 q^1 + {}_5C_5 p^5 q^0$$
$$= q^5 + 5pq^4 + 10p^2q^3 + 10p^3q^2 + 5p^4q + p^5$$

The terms in the last row are, respectively, the probabilities of zero, one, two, three, four, and five successes in an experiment repeated five times (e.g., the Yahtzee problem). For this reason, the number of successes in n trials of an experiment is said to exhibit a **Binomial Distribution.**

The case $p = q = \frac{1}{2}$ is particularly interesting. Suppose that a perfectly balanced coin is tossed nine times. What is the probability of getting exactly one head? Two heads? Three heads?

$$P(\text{one head in nine tosses}) = {}_9C_1(\tfrac{1}{2})^1(\tfrac{1}{2})^8 = 9 \cdot \tfrac{1}{512} \approx .02$$
$$P(\text{two heads in nine tosses}) = {}_9C_2(\tfrac{1}{2})^2(\tfrac{1}{2})^7 = 36 \cdot \tfrac{1}{512} \approx .07$$
$$P(\text{three heads in nine tosses}) = {}_9C_3(\tfrac{1}{2})^3(\tfrac{1}{2})^6 = 84 \cdot \tfrac{1}{512} \approx .16$$

The bar graph below shows the probability for any number of heads from zero to nine. We have superimposed on it the famous normal curve (see Section 8.3), which is often used to approximate the Binomial Distribution.

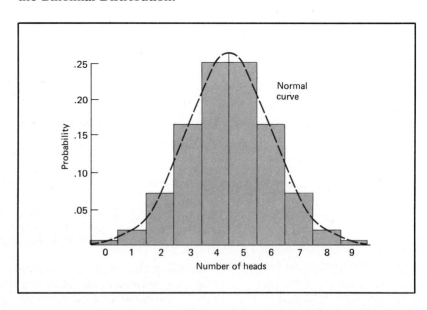

If you toss a coin 9 times, you may of course get anything from zero to nine heads. But if you repeat this experiment 100 or 1000 times and record the proportion of times you get zero, one, two, . . ., nine heads, you should expect a distribution something like that in the chart.

THE WAY THE BALL BOUNCES

A very simple device called a Hexstat allows one to do the equivalent of tossing a coin nine times over and over again with very little effort. It consists of a device in which balls are dropped through a triangular-shaped maze, which should remind you of the path problem and Pascal's Triangle (Section 6.4).

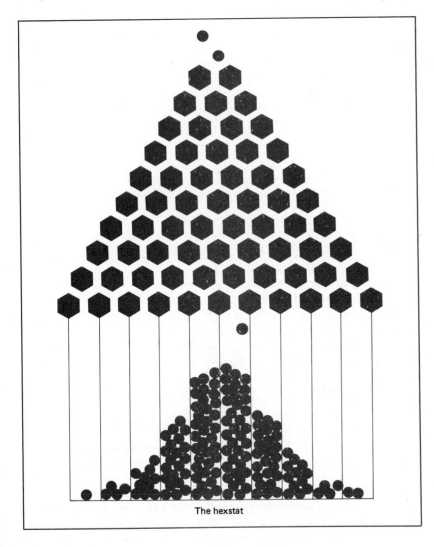

The hexstat

> Notice that a ball dropped through the top slot has two "choices." It can go right or left, each with probability $\frac{1}{2}$. It continues to choose right or left at each of the nine stages of the maze, which is exactly like tossing a coin nine times. Thus the probability is $(\frac{1}{2})^9$ that a ball will wind up in the left column, $_9C_1(\frac{1}{2})^9 = 9(\frac{1}{2})^9$ that it will wind up in the next column, etc. Dropping 100 balls through this maze is equivalent to tossing a coin nine times on 100 occasions.

TWO PRACTICAL APPLICATIONS

A certain rare disease has been studied for over 50 years, and it is known that 30% of the people afflicted with it will eventually recover without treatment; the rest will die. A drug company has discovered what it claims is a miracle cure, citing as evidence the fact that 8 of the 10 people on whom it was tested recovered. Of course, this might have happened by chance even if the drug is absolutely worthless. We would like to know the probability that 8 or more of the 10 patients would have recovered without treatment.

$$P(8 \text{ would have recovered}) = {}_{10}C_8(.3)^8(.7)^2 \approx .0014$$
$$P(9 \text{ would have recovered}) = {}_{10}C_9(.3)^9(.7)^1 \approx .0001$$
$$P(10 \text{ would have recovered}) = {}_{10}C_{10}(.3)^{10}(.7)^0 \approx .0000$$

Therefore the probability of 8 or more recovering naturally is $.0014 + .0001 = .0015$. Either we have observed a very rare event or the drug is useful. The latter conclusion seems likely; thus the drug merits further experimentation.

A company that manufactures $\frac{1}{2}$-inch steel bolts advertises that at most 1% of its bolts will break under a stress of 10,000 pounds. To maintain quality control, it tests a random sample of 100 bolts from each day's output and keeps track of the number of defective ones. A typical record for 10 days is: 0, 2, 0, 1, 0, 3, 0, 0, 0, 1. The question the company faces is to decide when its manufacturing process is out of control. For example, does getting 3 defective bolts on the sixth day establish that something is wrong? Not necessarily. For by chance, one may get 3 or 4 or even 10 defective bolts in a random sample of 100 even if the day's total production has only 1% defective.

Based on many years' experience, the company has established the following rule. When samples from two consecutive days yield 3 or more defective bolts each, the manufacturing process is shut down and the equipment carefully checked. For if the process were under control (i.e., 1% or fewer defective bolts),

> It is remarkable that a science which began with the consideration of games of chance should have become the most important object of human knowledge. . . . The most important questions of life are, for the most part, really only problems of probability.
>
> *Pierre Simon de Laplace*

$$P(0 \text{ defective}) = {}_{100}C_0(.01)^0(.99)^{100} \approx .37$$
$$P(1 \text{ defective}) = {}_{100}C_1(.01)^1(.99)^{99} \approx .37$$
$$P(2 \text{ defective}) = {}_{100}C_2(.01)^2(.99)^{98} \approx .18$$

Thus

$$P(3 \text{ or more defective}) = 1 - P(\text{less than 3 defective})$$
$$\approx 1 - (.37 + .37 + .18) = .08$$

The probability of 3 or more defective bolts being produced on two consecutive days is $(.08)(.08) = .0064$. An event with this small a probability is so rare that it is unlikely to have occurred by chance; it seems likely that something is wrong with the process.

SUMMARY

Consider an experiment in which the outcomes can be put into two categories, one of which we call success and the other failure. Suppose that the probability of success is p, while that of failure is $q = 1 - p$. When this experiment is repeated n times, the probability of exactly k successes occurring is given by

$$P(k \text{ successes in } n \text{ trials}) = {}_nC_k p^k q^{n-k}$$

For example, if we toss a fair die 12 times and consider getting a 6 to be a success, the probability of getting exactly four 6's is

$$_{12}C_4\left(\tfrac{1}{6}\right)^4\left(\tfrac{5}{6}\right)^8$$

PROBLEM SET 7.3

1. Calculate:
   (a) $_6C_1$ (b) $_6C_2$ (c) $_6C_3$
2. Calculate:
   (a) $_8C_1$ (b) $_8C_2$ (c) $_8C_3$
3. A balanced coin is tossed six times. Calculate the probability of getting
   (a) No heads.
   (b) Exactly one head.
   (c) Exactly two heads.
   (d) Exactly three heads.
   (e) More than three heads.
4. A fair coin is tossed eight times. Calculate the probability of getting
   (a) No tails.
   (b) Exactly one tail.
   (c) Exactly two tails.
   (d) Exactly three tails.
   (e) At most three tails.
5. Experiments indicate that, for an ordinary thumbtack, the probability of falling head down is $\frac{1}{3}$ and head up, $\frac{2}{3}$. Write an

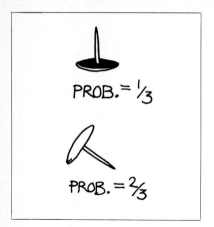

PROB. = $\frac{1}{3}$

PROB. = $\frac{2}{3}$

TABLE 7-2
TABLE OF $_{10}C_k p^k q^{10-k}$

k	p = .25	p = .35	p = .50
0	.0563	.0135	.0010
1	.1877	.0725	.0098
2	.2816	.1757	.0439
3	.2503	.2522	.1172
4	.1460	.2377	.2051
5	.0584	.1536	.2461
6	.0162	.0689	.2051
7	.0031	.0212	.1172
8	.0004	.0043	.0439
9	.0000	.0005	.0098
10	.0000	.0000	.0010

expression (but do not evaluate) for the probability in 12 tosses of its falling
(a) Head up exactly four times.
(b) Head up exactly six times.

6. On a true-false test of 20 items, Homer estimates the probability of his getting any one item right at $\frac{3}{4}$. Write an expression for the probability of his getting
(a) Exactly 19 right.
(b) At least 19 right.

7. Extensive tables have been developed to calculate binomial probabilities. Use the table at the left to calculate
(a) $_{10}C_6(.25)^6(.75)^4$
(b) $_{10}C_3(.35)^3(.65)^7$

8. Veterinarians know that German shepherd pups will die from a certain disease with a probability of $\frac{1}{4}$. In a litter of 10 pups, what is the probability that
(a) Exactly two will die?
(b) At most two will die?

9. Major Electronics sells transistors to the United States government in lots of 1000. The government takes a random sample of 10 from each lot, puts them through a rigorous test, and accepts the lot if no more than 3 of the 10 break down. Major Electronics feels certain that at least three-fourths of its transistors will pass government tests. If they are correct, what is the probability that any given lot will be accepted by the government?

10. Assuming that men and women are equally likely to turn up in Math 412, what is the probability that Professor Witquick will have a class of nine women and one man?

11. Slugger Brown has a batting average of .350. In a three-game series, he expects to have 10 official at-bats. What is the probability that he will get three or more hits? What assumptions did you make in getting your answer?

12. A multiple-choice test has 10 questions, each with 4 alternative answers. If Homer uses the method of pure guessing, what is the probability he will get at least 5 right?

13.* In the game of Yahtzee, five dice are thrown at once. What is the probability on one throw of getting
(a) Five of a kind?
(b) Four of a kind?
(c) Three of a kind?
(d) A full house (two of one kind and three of another)?
(e) A large straight (1, 2, 3, 4, and 5 or 2, 3, 4, 5, and 6)?

14.* A player in Yahtzee has reached his last turn to play and needs at least three 6's to achieve a top score. He has three chances to get it. On the first throw he tosses all five dice, but on the second and third he tosses only those that do not already show 6's. What is the probability that he will succeed?

c 15. A fair die is tossed 20 times. Calculate the probability of getting 3 or less sixes.

c 16. A coin is loaded so that the probability of getting a head is

.469. Calculate the probability of getting exactly 6 heads when this coin is tossed 10 times.

[c] 17. A company's records indicate that the probability it will make a seriously defective TV set is .0124. Calculate the probability that in a production run of 100 sets, more than 2 will be seriously defective.

[c] 18. In a certain country, 104 boys are born for every 100 girls. In that country, what percentage of families with 8 children should have 4 boys and 4 girls?

When no tips were put on the counter, the angry hatcheck boy distributed the hats at random. He hoped that nobody would get the right hat. What is the probability that he succeeded?

7.4 / Some Surprising Examples

The problem posed by the angry hatcheck boy is surprisingly difficult; that's why we are going to follow the advice given in Section 1.3 to begin with a simple case. If there were only three men, their hats could be distributed in 3! = 6 ways (ABC, ACB, BAC, *BCA, CAB,* and CBA). Two of these ways (those in italics) result in wrong hats on every head. The probability of this event is $\frac{2}{6} = .333$. With four men, there are 4! = 24 ways of distributing the four hats and, as the reader can check by making a list, nine of them assign a wrong hat to everyone. The probability has increased to $\frac{9}{24} = .375$. By the time we get to six men, the problem of the ill-fitting hats has become a large headache. But with determined effort, we can list all 6! = 720 ways of distributing the six hats and discover that 265 of them fail to put even one hat on the right head. The probability has dropped slightly to .368.

What if there were 100 men and 100 hats? Surely there is a better method than listing all 100! arrangements. There is. It was discovered by M. de Montmart in 1708, and the answer is easy to understand though the method of deriving it is not. Montmart showed that the probability P_n of no matches in a distribution of n hats to n men is given by

$$P_n = 1 - \frac{1}{1!} + \frac{1}{2!} - \frac{1}{3!} + \frac{1}{4!} - \frac{1}{5!} + \frac{1}{6!} + \cdots \pm \frac{1}{n!}$$

TABLE 7-3	
n	**P_n**
1	.000
2	.500
3	.333
4	.375
5	.367
6	.368
7	.368
8	.368
9	.368
10	.368
100	.368
1000	.368

Most people think that, with 100 people, it would be almost certain that at least one person would get the right hat; that is, P_{100} would be very small. But contrary to intuition $P_{100} = .368$ (correct to three decimal places); it is almost exactly the same as P_6. In fact, P_n gets closer and closer to $1/e$, e being a mysterious number like π which occurs frequently in advanced mathematics. (e is an irrational number with a value of 2.71828....)

Here is a different version of the same problem. A mischievous mailperson has 10 letters, 1 for each of 10 houses. She distributes them at random. What is the probability that at least 1 letter will be delivered to the right house? The event "at least one letter will go to the right house" is complementary to "no letter will go to the right house." The probability of the latter is P_{10} which we find in our table. Thus the probability we desire is

$$1 - P_{10} = 1 - .368 = .632$$

THE BIRTHDAY PROBLEM

Here is an old problem that has confused many a student and an occasional professor. Suppose there are 40 students in Math 13; let's call them $A_1, A_2, \ldots, A_{40}$. What is the probability that at least two of them have the same birthday? With 365 days to choose from, one might think it quite unlikely that there would be a match. Yet we shall show that the probability is .89; it is very likely that at least one match will occur.

To see this, we first make three simplifying assumptions which, if not quite true, are at least approximately correct.

1. There are 365 days in a year.
2. One day is as likely as another for a birthday.
3. Our 40 students were born ("chose" their birthdays) independently of each other.

Now let B be the event that at least two have the same birthday. Consider the complementary event B' that no two have the same birthday. We can think of B' as occurring this way. First A_1 chooses a birthday, then A_2 chooses a birthday different from that of A_1, next A_3 chooses a birthday different from both that of A_1 and A_2, etc. Using our three assumptions, we obtain

$$P(B') = 1 \cdot \frac{364}{365} \cdot \frac{363}{365} \cdot \,\cdots\, \frac{326}{365} \approx .11$$

Consequently,

$$P(B) = 1 - P(B') = 1 - .11 = .89$$

Here is a related problem. How large a class is needed to give a 50% chance (that is, probability $\frac{1}{2}$) of at least one pair of matching birthdays? To answer this and other birthday problems, we let

$Q_n = P$(at least two in a class of n people having the same birthday)

By looking at the complementary event as in our first problem, we can calculate any Q_n we wish. Some values are indicated in the table in the margin. Note that a class of 25 has well over a 50% chance of a match of birthdays (a class of 23 has almost exactly a 50% chance).

TABLE 7-4	
n	Q_n
5	.03
10	.12
15	.25
20	.41
25	.57
30	.71
35	.81
40	.89

PROBABILITY MISUSED

The following story quoted from the April 26, 1968, issue of *Time* magazine illustrates in a vivid way how an uncritical use of probability theory can lead one astray.

After an elderly woman was mugged in an alley in San Pedro, Calif., a witness saw a blonde girl with a ponytail run from the alley and jump into a yellow car driven by a bearded Negro. Eventually tried for the crime, Janet and Malcolm Collins were faced with the circumstantial evidence that she was white, blonde, and wore a ponytail while her Negro husband owned a yellow car and wore a beard. The prosecution, impressed by the unusual nature and number of matching details, sought to persuade the jury by invoking a law rarely used in a courtroom — the mathematical law of statistical probability.

The jury was indeed persuaded, and ultimately convicted the Collinses (*Time,* Jan. 8, 1965). Small wonder. With the help of an expert witness from the mathematics department of a nearby college, the prosecutor explained that the probability of a set of events actually occurring is determined by multiplying together the probabilities of each of the events. Using what he considered "conservative" estimates (for example, that the chances of a car's being yellow were 1 in 10, the chances of a couple in a car being interracial 1 in 1,000), the prosecutor multiplied all the factors together and concluded that the odds were 1 in 12 million that any other couple shared the characteristics of the defendants.

Only one couple. The logic of it all seemed overwhelming, and few disciplines pay as much homage to logic as do the law and math. But neither works right with the wrong premises. Hearing an appeal of Malcolm Collins' conviction, the California Supreme Court recently turned up some serious defects, including the fact that not even the odds were all they seemed.

To begin with, the prosecution failed to supply evidence that "any of the individual probability factors listed were even roughly accurate." Moreover, the factors were not shown to be fully independent of one another as they must be to satisfy the mathematical law; the factor of a Negro with a beard, for instance, overlaps the possibility that the bearded Negro may be part of an interracial couple. The 12 million to 1 figure, therefore, was just "wild conjecture." In addition, there was not complete agreement among the witnesses about the characteristics in question. "No mathematical equation," added the court, "can prove beyond a reasonable doubt (1) that the guilty couple in fact possessed the characteristics described by the witnesses, or even (2) that only one couple possessing those distinctive characteristics could be found in the entire Los Angeles area."

Improbable Probability. To explain why, Judge Raymond Sullivan attached a four-page appendix to his opinion that carried the necessary math far beyond the relatively simple formula of probability. Judge Sullivan was willing to assume it was unlikely that such a couple as the one described existed. But since such a couple did exist—and the Collinses demonstrably did exist—there was a perfectly acceptable mathematical formula for determining the probability that another such couple existed. Using the formula and the prosecution's figure of 12 million, the judge demonstrated to his own satisfaction and that of five concurring justices that there was a 41% chance that at least one other couple in the area might satisfy the requirements.

"Undoubtedly," said Sullivan, "the jurors were unduly impressed by the mystique of the mathematical demonstration but were unable to assess its relevancy or value." Neither could the defense attorney have been expected to know of the sophisticated rebuttal available to them. Janet Collins is already out of jail, has broken parole and lit out for parts unknown. But Judge

Sullivan concluded that Malcolm Collins who is still in prison at the California Conservation Center, had been subjected to "trial by mathematics" and was entitled to a reversal of his conviction. He could be tried again but the odds are against it.

Time, April 26, 1968, p. 41. Reprinted by permission from *Time*, The Weekly News Magazine; © 1968 Time Inc.

SUMMARY

All of us are guided to some extent by intuition; and a well-developed intuition can be invaluable even in mathematics. However, in problems involving probability, most people are easily deceived by intuition. It is therefore especially important to think through such problems in a very careful systematic way. We have tried to do this for two problems that many people find both intriguing and mystifying, the hat check problem and the birthday problem. The first few questions in the problem set are related to these two problems. But we also consider several new problems which many readers may find challenging and surprising.

PROBLEM SET 7.4

1. Consider four bottles labeled A, B, C, and D.
   (a) List all 24 ways of arranging these bottles in a row.
   (b) How many of these arrangements have every letter out of its normal position?
   (c) If the four bottles are arranged in a row at random, what is the probability none of them will be in its normal position?
   (d) Calculate:

$$1 - \frac{1}{1!} + \frac{1}{2!} - \frac{1}{3!} + \frac{1}{4!}$$

   It should agree with your answer to part (c).
2. A playful secretary typed eight different business letters, addressed the eight corresponding envelopes, and then distributed the letters among the envelopes at random. Use one of the tables in this section to calculate the probability that
   (a) No one will get the right letter.
   (b) At least one person will get the right letter.
3. For a room of 20 people, calculate the probability that no 2 have the same birthday. Use a table in this section.
4. For a room of 30 people, calculate the probability that at least 2 of them have the same birthday.
5. A mischievous stock clerk took 24 cans of pineapple and marked them at random using a labeling device which can print any number of cents from 0 to 99. Write an expression for the probability that no 2 cans were marked with the same price?

6. The stock clerk in Problem 5 marked 120 cans of orange juice at random using the same labeling device. What is the probability that at least 2 were marked with the same price? Don't think too long on this one.

7. Who is right in the following problem? Peter and Paul are equally skilled at playing marbles, but Peter has two marbles and Paul only one. They roll to see who comes closer to a stake in the ground.

Peter reasons that there are four possibilities: Both his marbles may be better than Paul's, or his first may be better and his second worse, or his second may be better and his first worse, or both may be worse. Since three of the four cases lead to Peter's winning, Peter wins with a probability of $\frac{3}{4}$.

Paul looks at it this way. Each of the three marbles has an equal chance of winning. Two of the marbles are Peter's. Therefore Peter wins with a probability of $\frac{2}{3}$.

8. Here is an old folklore problem that admittedly has sexist overtones. A sultan wishes to increase the proportion of women (making larger harems possible). He decrees, "As soon as a mother gives birth to a male child, she shall be sterilized. If she bears a female, she may continue to have children." Will this law achieve the desired end?

9. Criticize the following imaginary conversation.

Susan: What is the probability of there being no horses on Mars?
Sally: Not knowing anything about it, I'd have to say $\frac{1}{2}$.
Susan: And what is the probability of no cows on Mars?
Sally: Again, I'd have to say $\frac{1}{2}$.
Susan: And of no dogs?
Sally: $\frac{1}{2}$.

(This conversation continued until Susan had named seven more forms of life.)

Susan: Very well. By the laws of probability, this means that the probability there are no horses, nor cows, nor dogs, nor any of the seven other forms of life is $\frac{1}{2} \cdot \frac{1}{2} \cdot \frac{1}{2} \ldots =$ 1/1024. Thus the probability of life on Mars is at least $1 - (1/1024) = 1023/1024$. It's almost a certainty.

10. Professor Witquick and his wife each play a fair game of chess, but an honest observer would have to admit that the professor usually loses to his wife. One day, their son asked his father for $10 for a Saturday night date. After puffing his pipe for a moment, the professor replied:

"Today is Wednesday. You will play a game of chess tonight, another tomorrow, a third on Friday. Your mother and I will alternate as opponents. If you can win two games in a row you will get the $10."

"Whom do I play first, you or Mom?"

"Take your choice," said the professor.

If you were the son, whom would you choose?

[c] 11. Suppose that, in flight, airplane engines fail independently of each other with a probability of .001 and that an airplane must have at least half of its engines working to fly successfully. Calculate the probability that a plane will make a successful flight if it has
(a) Two engines.   (b) Four engines.

12* Suppose in Problem 11 that the probability of failure for an engine is q. Show that a four-engine plane is safer if $q < \frac{1}{3}$, but a two-engine plane is safer if $q > \frac{1}{3}$.

13* (St. Petersburg paradox.) A coin is tossed until a head appears. When a head appears on the first toss, the bank pays the player $1. When a head appears for the first time on the second toss, the bank pays $2. When a head first appears on the third toss, the bank pays $4, on the fourth toss $8, etc. How much should a player be willing to pay for the privilege of playing this game?

Chapter 8
Organizing Data

*When you can measure what you are speaking about and
express it in numbers you know something about it; but
when you cannot express it in numbers, your knowledge is
of a meagre and unsatisfactory kind.*
LORD KELVIN

*Statistical thinking will one day be as necessary for efficient
citizenship as the ability to read and write.*
H. G. WELLS

"Let us sit on this log at the roadside," says I, "and forget the inhumanity and ribaldry of poets. It is in the columns of ascertained facts and legalized measures that beauty is to be found. In this very log we sit upon, Mrs. Sampson," says I, "is statistics more wonderful than any poem. The rings show that it is sixty years old. At a depth of two thousand feet, it would become coal in three thousand years. The deepest mine in the world is at Killingworth near New Castle. A box four feet long, three feet wide, and two feet eight inches deep will hold one ton of coal. . . ." "Go on Mr. Pratt," says Mrs. Sampson, "Them ideas is so original and soothing. I think statistics are just as lovely as they can be."
O. Henry.

8.1 / Getting The Picture

What are statistics? For many people, including O. Henry, statistics are just numerical facts about this, that, and the other. But unlike Mrs. Sampson, most of us find little of interest or beauty in the mere accumulation of numerical data. It is only when data have been condensed, analyzed, and interpreted that they begin to have significance. Thus by the subject called **statistics,** we mean the science of organizing, interpreting, and drawing inferences from numerical data.

An old proverb says that one picture is worth a thousand words. Statisticians believe this maxim. Faced with a mass of data, a statistician tries to think of a clear way to display it. The picture he draws depends on what he thinks is significant or at least what he wants us to believe is significant about the data. Some commonly used devices are shown below. We call them pie graphs and bar graphs.

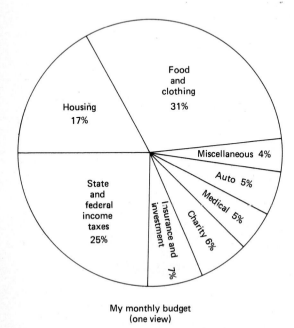

My monthly budget
(one view)

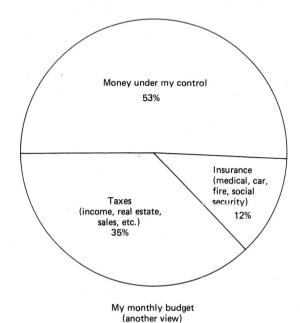

My monthly budget
(another view)

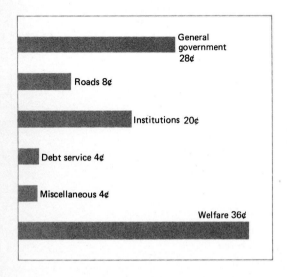

Your state tax dollar (one view)

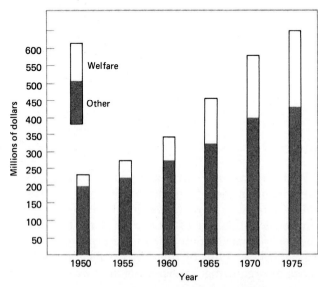

Your state tax dollars (another view)

TABLE 8-1 DATA: WHAT DO THEY MEAN?											
32	34	39	40	42	40	50	55	60	62	60	50
42	44	46	42	48	50	50	58	64	62	62	56
48	51	52	52	60	61	55	57	80	60	60	58
51	53	52	55	60	54	60	60	80	70	70	66

FREQUENCY CHARTS AND BAR GRAPHS

Consider the set of data above. How shall we display it in a pictorial way that will bring out its main features? Clearly one cannot answer this question without some idea of what the data represent.

Let us suppose that Professor Witquick has just received the SMOB (Standard Measure of Brainpower) scores for the 48 students who have registered for his freshmen calculus course. They are recorded in the table above. The professor is delighted to note two 80's, perfect scores on the SMOB, but he also sees several scores in the 30's and 40's. Under his breath, he damns the high schools for failing to teach mathematics as they ought to. But before making further judgments, he decides to analyze the data more carefully.

One good idea is a **frequency chart.** Professor Witquick groups the data into 10 classes (8 to 15 classes are commonly used), keeping track of how many fall into each class by means of tallies. Then he displays the same information on a **bar graph.** Now he can compare this year's class with last year's, for which he has kept a bar graph. When he does this, maybe he will discover that this year's class is actually better than he expected. In any case, the bar graph gives him a much better idea of the ability of his class than the raw data he started with.

FREQUENCY CHART (SMOB SCORES)		
Class Range	**Class Tally**	**Frequency**
31–35	I I	2
36–40	I I I	3
41–45	I I I I	4
46–50	++++ I I	7
51–55	++++ ++++	10
56–60	++++ ++++ II	12
61–65	++++	5
66–70	I I I	3
71–75		0
76–80	I I	2

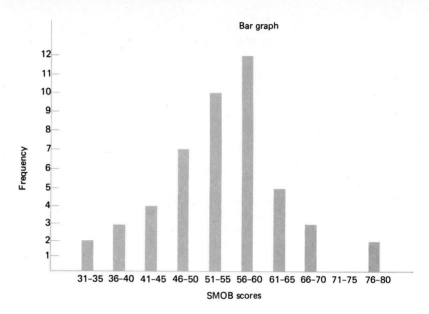

Bar graph

TREND LINES AND COMPARATIVE BAR GRAPHS

Consider the same data once more (page 223). Suppose now that the first row represents the monthly car sales of Chrysler City during 1971, the second row shows sales during 1972, etc. It makes no sense to treat the sales data as we did the SMOB scores. What interests the sales manager is a chart showing trends over time, or perhaps some kind of chart that compares one year with another.

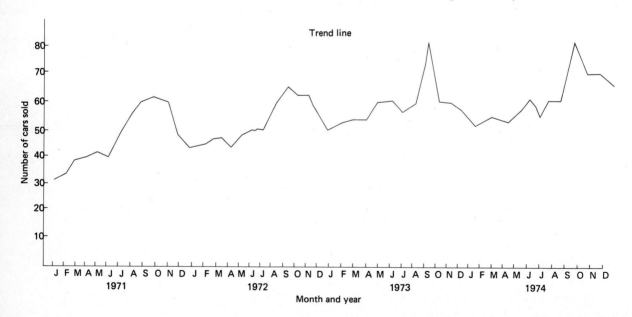

Trend line

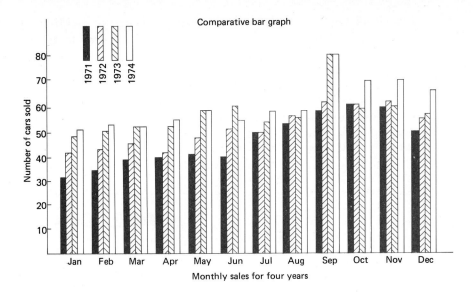

Comparative bar graph

Monthly sales for four years

We can accomplish the first job with a so-called **trend line,** and the second with a **comparative bar graph** (see accompanying diagrams). Note that, while great fluctuations are present, the trend is upward; and this makes the sales manager happy. Can you account for the spurt in sales each year during early fall?

DISTORTING DATA

We have seen that a chart or graph can be used to display data so that their true significance is easily grasped. But it is also possible to use a graph to distort data or to exaggerate the importance of carefully selected bits of information. This is a favorite trick of some advertisers, and most of us have learned to discount their claims even if backed up by fancy charts.

> There are three kinds of lies: lies, damned lies and statistics.
> *Disraeli*

Suppose that you want to win an argument or that you are trying to sell something, say stocks in the ABC Company. Naturally you want to display the stock price history of the company in a favorable way; perhaps you are tempted to use a bit of careful distortion. Some possibilities are shown on page 226.

Or suppose that the ABC Company is engaged in oil refining and has doubled production from 1965 to 1975 and then doubled it again from 1975 to 1980. This performance is good, but there are ways to make it appear even better. Draw a small oil drum representing production in 1965, another twice as high for 1975, and a third twice as high again for 1980. So that the drums will look like real oil drums, you double the diameter as well as the height. You can't quite be accused of cheating; yet you have given

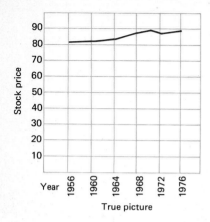

True picture

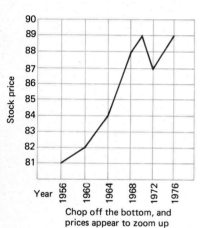

Chop off the bottom, and prices appear to zoom up

Pick only the best years

the visual impression that oil production has grown eightfold at each stage (see Problems 11 and 12). A blazing title encourages the impression.

Of course, you want your clients to extrapolate into the future. If oil production doubled in 10 years, and then again in 5 years, it can be expected to double again in $2\frac{1}{2}$ years, then again in $1\frac{1}{4}$ years, etc. This kind of wishful thinking is plain foolishness, as any mathematician knows. For if production were to maintain that pattern, by 1985 it would be doubling every second.

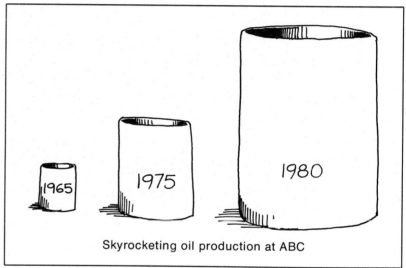

Skyrocketing oil production at ABC

Mark Twain recognized this fallacy and wrote about it in *Life on the Mississippi*, Harper & Row, New York, 1901, p. 136:

In the space of one hundred and seventy-six years the Lower Mississippi has shortened itself two hundred and forty-two miles. That is an average of a trifle over one mile and a third per year. Therefore, any calm person, who is not blind or idiotic, can see that in the Old Oolitic Silurian Period, just a million years ago last November, the Lower Mississippi River was upward of one million three hundred thousand miles long, and stuck out over the Gulf of Mexico like a fishing rod. And by the same token any person can see that seven hundred and forty-two years from now the Lower Mississippi will be only a mile and three-quarters long, and Cairo and New Orleans will have joined their streets together, and be plodding comfortably along under a single mayor and a mutual board of aldermen. There is something fascinating about science. One gets such wholesale returns of conjecture out of such a trifling investment of fact.

SUMMARY

We have sampled a few of the many possibilities for pictorial display of data. A statistician needs both common sense and imagination — common sense to grasp what is important in the data, and imagination to think of a way to display the data clearly and without distortion. Data, clearly displayed and fairly interpreted, can lead to wise decisions. But they can also be manipulated by a deceptive person who wants to justify a course of action already decided on.

PROBLEM SET 8.1

1. At Podunk University the student congress allocated its $40,000 budget as follows: student newspaper $15,000; literary magazine $4000; social committee $8000; student government $3000; student organizations $5000; special projects $2000; reserve fund $1500; miscellaneous $1500. Make a pie graph to display these data.
2. Make a bar graph to display the data in Problem 1.
3. Make a frequency chart for data set A shown in the margin, using the class intervals 24 to 29, 30 to 35, 36 to 41, etc.
4. Follow the directions in Problem 3 for data set B.
5. Make a bar graph based on the frequency chart in Problem 3.
6. Make a bar graph based on the frequency chart in Problem 4.
7. Which do you think is more likely? Data set A represents the scores on a physics exam, or it represents the ages of welfare recipients in Jackson Township. Justify your answer.
8. Answer the question in Problem 7 for data set B.
9. Suppose that the columns in data set A represent the monthly tractor sales of Smith's Farm Implements for 1975, 1976, and 1977, respectively. Display this data in two good ways. What conclusions can you draw?
10. Suppose that the columns in data set B represent the monthly snowmobile sales of the Olson Manufacturing Company for 1975, 1976, and 1977, respectively. Make a comparative bar graph displaying these data. What conclusions can you draw?
11. If you double the dimensions of a square, what happens to its area? If you double the dimensions of a cube, what happens to its volume?
12. If you triple the radius of a circle, what happens to its area? If you triple the radius of a sphere, what happens to its volume?
13. Which of the following statistical arguments seem valid to you? Explain your answer.
    (a) In 1976, 492 men and 301 women drivers had auto accidents in Mudville. Therefore women are safer drivers than men.
    (b) Ohio had more traffic fatalities than Rhode Island in 1975. Therefore it is safer to drive in Rhode Island.
    (c) Ninety percent of all the cars sold in the United States by a certain foreign country are still on the road. No American manufacturer can make this claim. Therefore these foreign cars last longer than American cars.

DATA SET A		
25	24	30
40	50	48
50	52	54
81	81	83
84	88	94
60	62	72
59	64	72
57	59	68
49	54	66
43	50	54
42	45	50
30	35	42

DATA SET B		
80	83	92
62	70	81
63	71	80
50	55	54
36	34	40
32	36	40
24	30	34
25	24	30
30	36	40
55	59	64
81	90	93
92	92	94

DATA SET C	
1966	90
1967	81
1968	90
1969	83
1970	92
1971	82
1972	90
1973	85
1974	89
1975	88

(d) All the students in a class failed a certain physics test. It must be a dumb class.

14. The total sales of a certain company were $10,000, $20,000, $40,000, and $80,000 in 1978, 1979, 1980, and 1981, respectively. A stockbroker claims that it will certainly have total sales of $1 million by 1985. How would you analyze this claim?

15.* Roll a pair of dice 100 times, keeping track of the total number of spots showing after each roll. Make a frequency diagram and a bar graph for these data.

16.* Data set C represents the total car sales of Watson Pontiac during each of the years 1966 to 1975. Make a bar graph that displays this information in an objective manner. Now make a bar graph that distorts this information in such a way as to make the sales appear to be going up dramatically. Suggestions: Combine years, chop off the bottom, etc.

8.2 / On The Average

Few people can look at a mass of data and make sense out of it. This is why statisticians search for ways to condense and summarize data. The goal of a summary is to pick out what is most significant. Summaries do not tell everything; but they ought to give the main idea.

Let's carry the notion of summarizing data to its absolute extreme. For a given set of numerical data, what *one* number best represents these data? That is the theme of this section.

THE MEAN, MEDIAN, AND MODE

In our opening cartoon, President Hornblower of the XYZ Company is explaining the salary situation to his employees. He has computed what he calls the average annual pay. It is $26,000, and he seems immensely pleased with this number. But not his workers. They know that not a one of them makes as much as $26,000. How can this be?

Here are the actual salary data for the XYZ Company, data that Hornblower would much prefer to keep secret. President Hornblower makes $216,000; Vice-President Snodgrass gets $78,000. There are three supervisors at $20,000 each, eight skilled workers at $16,000 each, nine laborers at $12,000 each, and a student assistant at $8,000. One way of displaying these data is simply to list them from highest to lowest with repetitions as they occur:

216,000
 78,000
 20,000; 20,000; 20,000
 16,000; 16,000; 16,000; 16,000; 16,000; 16,000; 16,000;
    16,000
 12,000; 12,000; 12,000; 12,000; 12,000; 12,000; 12,000;
    12,000; 12,000
  8000

We call this a **distribution,** and the actual numbers that occur are called values. Thus the distribution above consists of 23 values.

The **mode** of a distribution is the most frequently occurring value. In the salary example, the mode is 12,000. It is possible for several different values to occur with the maximum frequency. Each of them is then called a mode.

The **median** of a distribution is its middle value. If there is an odd number of values, this makes good sense. For example, in the distribution of 23 salaries, the middle value is the twelfth from the top (or bottom), which happens to be 16,000. When there is an even number of values, two of them will vie for the middle. In this case, we take the median to be the number halfway between these two; that is, we add them and divide by 2. In the distribution of 14 quiz scores shown at the right, the median is $(7 + 6)/2 = 6.5$. Incidentally, this distribution of scores has two modes, namely, 5 and 7. We say it is bimodal.

The **arithmetic mean** (often called simply the mean or average) is the number most commonly used to represent a distribution. If $x_1, x_2, \ldots, x_n$ are the values of the distribution (including repetitions), the mean $\bar{x}$ (read "x bar") is defined by

$$\bar{x} = \frac{x_1 + x_2 + \cdots + x_n}{n}$$

QUIZ SCORES
10
9, 9
8
7, 7, 7
6, 6
5, 5, 5
4, 4

Median: 6.5
Modes: 5 and 7
Mean: 6.57

Thus, for the quiz scores,

$$\bar{x} = \frac{4+4+5+5+5+6+6+7+7+7+8+9+9+10}{14} = \frac{92}{14} \approx 6.57$$

With regard to salaries at the XYZ Company, President Hornblower was right. The mean salary (average salary) is $26,000. But who can blame the workers for thinking that this is a mean way to figure the average pay?

There is another way to calculate the mean, which is especially useful when the data are grouped into categories. Let us suppose that $x_1, x_2, \ldots, x_m$ are *distinct* (i.e., different) values of the distribution and that they occur with frequency $f_1, f_2, \ldots, f_m$, respectively. Then

$$\bar{x} = \frac{f_1 x_1 + f_2 x_2 + \cdots + f_m x_m}{f_1 + f_2 + \cdots + f_m}$$

Looked at this way, the mean of the quiz scores is

$$\bar{x} = \frac{2(4) + 3(5) + 2(6) + 3(7) + 1(8) + 2(9) + 1(10)}{2+3+2+3+1+2+1} = \frac{92}{14} \approx 6.57$$

Of course, the answer is the same as before.

Which of these three numbers—mean, median or mode—should we use to represent a distribution? There is no easy answer to this question, except to say that one should understand what each of them does. The median and mode are easy to comprehend; the mean is quite mysterious. That is why we now take a good deal of space to explain it.

| If you don't understand the average, you're about average. |

THE MEAN AS A BALANCE POINT

Long before we knew anything about statistics, most of us learned something about teeter-totters (or seesaws). In order to balance a big man like your father, you knew you had to sit further from the fulcrum than he did. And by trial and error, you found just the right place to sit or, to put it another way, you found the spot between the two of you where the fulcrum belonged. Finding this place using mathematics is the subject we now investigate.

Consider a teeterboard of negligible weight. Along one edge, mark off a numerical scale using 1-foot intervals (see diagram in margin). Suppose that a 2-pound weight rests at the 1-foot mark and a 5-pound weight at the 8-foot mark. Our question is: Where should the fulcrum be placed in order for the teeterboard to balance? Let's label this point $\bar{x}$ and try to determine its value on the scale we have marked off. Long ago, experimenters learned that a certain condition must be satisfied if the board is to balance; the *first moment* must be 0. What this means is that we must add together the product of each weight times its distance from the fulcrum (this distance being taken as negative to the left of the fulcrum and positive to the right) and get 0 as the result. Thus (see diagram in margin)

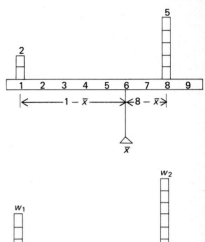

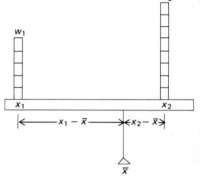

(Weight) times (distance) plus (weight) times (distance) is 0

$$2(1 - \bar{x}) + 5(8 - \bar{x}) = 0$$
$$2 \cdot 1 - 2 \cdot \bar{x} + 5 \cdot 8 - 5 \cdot \bar{x} = 0$$
$$2 \cdot 1 + 5 \cdot 8 = 2 \cdot \bar{x} + 5 \cdot \bar{x}$$
$$2 \cdot 1 + 5 \cdot 8 = (2 + 5)\bar{x}$$
$$\frac{2 \cdot 1 + 5 \cdot 8}{2 + 5} = \bar{x}$$

The answer is $\bar{x} = 6$, but what is more interesting to us is the form of the answer. If the first weight of size w_1 were placed at x_1, and the second weight of size w_2 placed at x_2, the same reasoning would lead us to

$$\frac{w_1 \cdot x_1 + w_2 \cdot x_2}{w_1 + w_2} = \bar{x}$$

The general situation is handled in exactly the same way. If there are m weights of size $w_1, w_2, \ldots, w_m$ spaced along the board at points $x_1, x_2, \ldots, x_m$,

$$\frac{w_1 \cdot x_1 + w_2 \cdot x_2 + \cdots + w_m \cdot x_m}{w_1 + w_2 + \cdots + w_m} = \bar{x}$$

This is the result we were looking for. It tells us how to find the balance point $\bar{x}$ for a whole set of weights placed along a teeterboard.

The formula just displayed should look vaguely familiar. It is exactly the same formula we gave earlier in the section for the mean of a distribution of values, $x_1, x_2, \ldots, x_m$, provided we interpret w_i to be the frequency of x_i. We have arrived at a very important conclusion. *The mean of a distribution is the point at which the distribution balances when we interpret the frequency of a value as a weight at that value.*

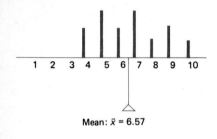

Mean: $\bar{x} = 6.57$

Take the distribution of quiz scores (page 229) as an example. If we draw a bar graph for this distribution and think of the bars as weights, the distribution will balance at the mean, namely, at $\bar{x} = 6.57$.

Or refer back to the example of salaries at the XYZ Company. Now we can understand why there was a mean salary of $26,000 even though all but two employees got less than that amount. President Hornblower and Vice-President Snodgrass with their huge salaries balance off the other 21 employees.

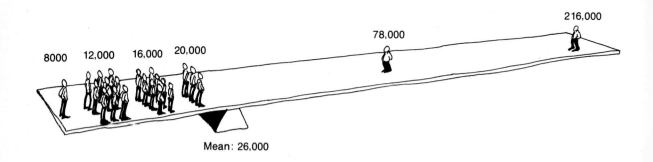

Mean: 26,000

This example serves to emphasize one characteristic of the mean. It is strongly influenced by the presence of an extremely large (or small) value. The median, however, is not influenced at all by such values. Many people prefer to use the median as a measure of the center of a distribution in any situation, such as one involving salaries, where extreme values are likely to occur.

GRADE-POINT AVERAGES

TABLE 8-2		
Course	Credits	Grade
History 11	3	B
Economics 12	3	A
Math 13	5	B
Art 20	3	C
Pol. Sci. 12	3	B
Math 14	5	A
English 15	3	D
Physics 17	5	C
Phys. Ed. 11	1	B
Music 40	2	F

Grading schemes vary from college to college, but the basic principles are the same. At Podunk University, students take semester courses which vary from 1 to 5 credits. In each course a student receives a grade of A, B, C, D, or F. In order to treat grades numerically, A is given the value 4, B the value 3, and so on down to F which has the value 0. The basic unit is the credit; to figure the weighting factor for each grade, one must compute the number of credits at that grade. At the left is Susan Smart's transcript for her first year at Podunk University. She has 2 credits of F, 3 credits of D, 8 credits of C, 12 credits of B, and 8 credits of A. Thus her grade-point average is

$$\bar{x} = \frac{2 \cdot 0 + 3 \cdot 1 + 8 \cdot 2 + 12 \cdot 3 + 8 \cdot 4}{2 + 3 + 8 + 12 + 8} = \frac{87}{33} \approx 2.64$$

SUMMARY

For a distribution of values, we wanted to find a single number that was most representative of these values. Three candidates were offered: the mode (the most frequently occurring value), the median (the middle value), and the mean (the balance point). The last-mentioned is defined by the formula

$$\bar{x} = \frac{x_1 + x_2 + \cdots + x_n}{n}$$

and, even though it is strongly influenced by very large or very small values, it is the most commonly used.

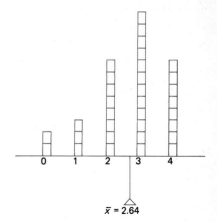

$\bar{x} = 2.64$

PROBLEM SET 8.2

1. Calculate the mean, median, and mode for each of the following sets of data.
   (a) 2, 8, 7, 8, 6, 5, 8, 10, 5
   (b) 10, 20, 18, 16, 20, 16, 13, 16, 17
2. Calculate the mean, median, and mode for each of the following sets of data.
   (a) −4, 8, −2, −1, 5, −1, 6, 2, 1
   (b) 51, 42, 36, 41, 50, 38, 43
3. Find the mean, median, and mode for the lengths (i.e., number of letters) of words in the first paragraph of this section. Begin by making a frequency chart.
4. Find the mean, median, and mode for the lengths of words in the quote from O. Henry at the beginning of Section 8.1. Begin by making a frequency chart.
5. A group of 100 women was weighed, with results as indicated in the margin. What is the mean weight? Use the midpoint of each class in making this computation. For example, assume that each individual in the first class weighs 110 pounds.
6. Faculty salaries at Podunk University are as listed in the table in the margin. Compute the average salary, making the assumption indicated in Problem 5. However, explain why this assumption may cause a large error.
7. The incomes for the five employees of Friend's Furniture are
   $10,000; $12,000; $12,000; $13,000; $63,000
   Calculate the mean and median salaries. In your opinion, which number gives a better indication of salaries at Friend's?
8. Each of 10 students got 50 on a certain exam. What were the mean, median, and mode? Make up an example of 10 scores, not all equal, in which the mean, median, and mode are all 50.
9. A temperature of 25°C is considered ideal for human beings. In Boondocks, the mean temperature is 25°C. Does this mean that the temperature in Boondocks is ideal? Explain.
10. An instructor tells her class that she is going to disregard the lowest of the 10 quiz scores in Math 13 when she determines each

TABLE 8-3

Weight (pounds)	Frequency
105–115	10
115–125	12
125–135	15
135–145	20
145–155	14
155–165	15
165–175	10
175–185	3
185–195	1

TABLE 8-4

Salary (dollars)	Number
Professor	
16,000–20,000	10
Associate professor	
14,000–16,000	20
Assistant professor	
12,000–14,000	40
Instructor	
10,000–12,000	30

student's final grade. Will this necessarily change a student's median score? Mean score? Explain.

11. Last term, Amy enrolled in Math, German, English, History, and Physical Education for 5, 4, 3, 3, and 1 credits, respectively. She got A in Math and German, B in English, and C in History and Physical Education. What was her grade-point average for the term? (A = 4, B = 3, C = 2.)

12. Homer took the same courses as Amy (Problem 11) and got C in Math, B in English, and A in the other three courses. Without calculating, decide whether Homer or Amy has the higher grade-point average. Now calculate Homer's average to make sure.

13. Five people weighing 120, 200, 150, 160, and 130 pounds are standing on a plank at distances 1, 2, 6, 9, and 10 feet, respectively, from one end of the plank. Assuming that the weight of the plank is negligible, where should the fulcrum be placed so that the system will balance?

14. Horace has scores of 60, 75, 40, and 80 on the first four quizzes in Math 13. What must he average on the last two so that his overall average will be 75?

15* Amy drove 90 miles at an average speed of 30 mph. What must she average on the return trip to have an overall average of 50 mph? Of 60 mph?

16* Homer averaged 60 on the quizzes in Math 13 and 70 on the quizzes in Math 14. Yet the average of all his quiz scores in Math 13 and Math 14 was not 65. How can this be?

17* It is possible to have a class in which no student scores below the mean on the final exam. State precisely the conditions under which this can happen. Show that your answer is correct.

X	Y
1.375	1250
6.250	225
4.625	415
3.125	785
4.500	235
8.625	135
9.875	75
3.375	615
4.125	285
6.250	225
5.875	145
5.750	185
4.375	525

[c] 18. Find the mean of the X data shown in the margin. If your calculator has an $\bar{x}$ key, learn how to use it.

[c] 19. Find the mean of the Y data shown in the margin.

[c] 20. Suppose for the data in the margin that X represents the price of a share on the oca exchange and Y the number of shares sold at that price during a certain day. What was the mean price of a share that day?

[c] 21. At Hamline University, grades are assigned numerical values as follows. A(4.00), A$^-$(3.75), B$^+$(3.25), B(3.00), B$^-$(2.75), C$^+$(2.25), C(2.00), C$^-$(1.75), D$^+$(1.25), D(1.00), D$^-$(.75), F(.00). Anne has 4 credits of A$^-$, 6 credits of B$^+$, 8 credits of B$^-$, 9 credits of C, 5 credits of C$^-$. What is her grade point average?

[c] 22. Dale has 3 credits of A$^-$, 6 credits of B$^+$, 9 credits of B, 3 credits of B$^-$, and 11 credits of C$^+$. Using the scale of Problem 21, what is his grade point average?

DON'T WORRY. THAT ROPE
IS ONE INCH THICK ON
THE AVERAGE.

8.3 / The Spread

One can hardly blame our mountain climber if he feels a little insecure. The average doesn't tell all there is to know about a rope, or about a set of numbers. One also wants to know something about the variability of the data. How are they spread out? How are they dispersed?

A picture or a graph gives some idea of the spread of a distribution. But once again, a statistician wants a number that measures this dispersion. One such measure is the **range,** which is defined as the difference between the largest and smallest values in the distribution. It is simple to calculate but has little else to recommend it. It's too coarse a measure; two distributions may have the same range but be very different in character. A much more discriminating measure is the *standard deviation*.

THE STANDARD DEVIATION

Professor Witquick has given two quizzes to his Advanced Calculus class of six students. Here are the results:

Quiz 1	*Quiz 2*
$x_1 = 10$	$x_1 = 7$
$x_2 = 8$	$x_2 = 7$
$x_3 = 7$	$x_3 = 6$
$x_4 = 5$	$x_4 = 6$
$x_5 = 5$	$x_5 = 6$
$x_6 = 1$	$x_6 = 4$

$$\bar{x} = \frac{10 + 8 + 7 + 5 + 5 + 1}{6} = 6 \qquad \bar{x} = \frac{7 + 7 + 6 + 6 + 6 + 4}{6} = 6$$

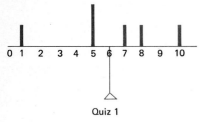

Quiz 1

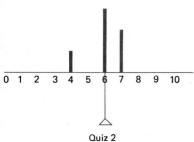

Quiz 2

Note that both distributions have the same mean; their bar graphs balance at the same point. Yet the two distributions are very different. The first one spreads out much more than the second. We want to measure this spread.

One idea that may occur to us is to look at the deviations of the scores from the mean. They are certainly much larger in the first case. Suppose we add up these deviations and use the sum as a measure:

Quiz 1		*Quiz 2*	
$x_1 - 6 =$	4	$x_1 - 6 =$	1
$x_2 - 6 =$	2	$x_2 - 6 =$	1
$x_3 - 6 =$	1	$x_3 - 6 =$	0
$x_4 - 6 =$	-1	$x_4 - 6 =$	0
$x_5 - 6 =$	-1	$x_5 - 6 =$	0
$x_6 - 6 =$	-5	$x_6 - 6 =$	-2
TOTAL	0	TOTAL	0

That wasn't much help. In both cases the sum of the deviations is 0. (We should have known this would happen, since we really just calculated the first moment about the mean, which is 0 by definition.) Of course, we see the difficulty. Some deviations are positive and some are negative; in the sum, the deviations cancel each other out. What can we do to make all the deviations positive? How about squaring them?

Quiz 1		*Quiz 2*	
$(x_1 - 6)^2 =$	16	$(x_1 - 6)^2 =$	1
$(x_2 - 6)^2 =$	4	$(x_2 - 6)^2 =$	1
$(x_3 - 6)^2 =$	1	$(x_3 - 6)^2 =$	0
$(x_4 - 6)^2 =$	1	$(x_4 - 6)^2 =$	0
$(x_5 - 6)^2 =$	1	$(x_5 - 6)^2 =$	0
$(x_6 - 6)^2 =$	25	$(x_6 - 6)^2 =$	4
TOTAL	48	TOTAL	6

Now we have a reasonable measure of the spread. But there are still two things wrong. First, the sum may be large simply because there was a huge number of values. We can take care of that by dividing by the number of values in each case (that is, by averaging the squared deviations). Second, the units of the original data (feet, pounds, amperes, or whatever) have been squared. To get back to the original units, we take the square root. All of this discussion has led us to the following formula for the **standard deviation** s:

Think of the standard deviation as measuring the average distance the values lie from the mean.

$$s = \sqrt{\frac{(x_1 - \bar{x})^2 + (x_2 - \bar{x})^2 + \cdots + (x_n - \bar{x})^2}{n}}$$

Admittedly the formula is complicated; so we'll describe how to calculate it by means of four rules:

1. Subtract the mean from each value (i.e., compute the deviation).
2. Square each deviation.
3. Average the squared deviations (add them up and divide by the number of values).
4. Take the square root of the result.

For quizzes 1 and 2, we get

Quiz 1

$$s = \sqrt{\tfrac{48}{6}} = \sqrt{8} \approx 2.83$$

Quiz 2

$$s = \sqrt{\tfrac{6}{6}} = 1$$

Note that the first value is much larger than the second, as expected. In the table below, we have worked out the standard deviation for a distribution of 15 values, using a convenient format.

TABLE 8-5			
Value, x_i	Deviation, $x_i - \bar{x}$	Squared Deviation, $(x_i - \bar{x})^2$	Calculation of s
10	3	9	$\bar{x} = \tfrac{105}{15} = 7$
9	2	4	$s = \sqrt{\tfrac{36}{15}} \approx 1.55$
8	1	1	
8	1	1	
8	1	1	
8	1	1	
7	0	0	
7	0	0	
7	0	0	
7	0	0	
6	−1	1	
6	−1	1	
5	−2	4	
5	−2	4	
4	−3	9	
105		36	

THE NORMAL DISTRIBUTION

To get a better idea of the significance of the standard deviation, we consider a curve that pervades all theoretical discussions in statistics, the **normal curve.** Its precise definition is complicated, involving ideas from calculus. For us, it is enough to say that its graph is a smooth, bell-shaped curve something like the curve shown on page 238. Distributions that occur in real life are never exactly normal, but there is a remarkable tendency for naturally occurring distributions to be approximately normal. This is especially true when they arise from measurements of something, say weight, height, or scores on an intelligence test.

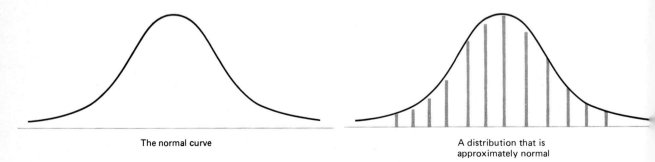

The normal curve

A distribution that is
approximately normal

In the left-hand illustration below we show (approximately) how normally distributed values arrange themselves about the mean. For example, 68% of the values lie within 1 standard deviation of the mean, 96% lic within 2 standard deviations, and almost 100% are within 3 standard deviations. These percents are obtained by means of calculus; they correspond to areas under the normal curve.

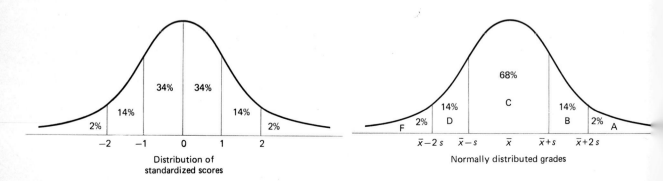

Distribution of
standardized scores

Normally distributed grades

It is sometimes said that course grades ought to be normally distributed. In the right-hand figure above we show what some people mean by this. For example, to get an A a student has to score at least 2 standard deviations above the mean. This happens only about 2% of the time. Students who encourage their teachers to grade on a normal curve may want to rethink their position. And teachers who announce their intention to do so also should consider their position, making sure it can be defended. In fairness, it should be admitted that many teachers take grading on a normal curve to mean something less stringent than what a strict interpretation implies.

STANDARDIZED SCORES

Facts by themselves sometimes convey little information. It is when facts are put in context that they begin to take on real meaning. To say that Susan Smart got an 8 on a quiz is a worthless piece of information. To say that she got an 8 on a quiz that had a mean of 6 and a standard deviation of 2.83 says a great deal more.

When it comes to comparing scores on tests in two different classes, the importance of context becomes especially significant. Let us suppose that Susan scored 8 in her Advanced Calculus class of six students where the mean was 6 with a standard deviation of 2.83, and that she also got 8 in British History where the mean was 7 with a standard deviation of 1.55, there being 15 students in that class. Did she do better in Advanced Calculus or in British History? How can we compare scores in two different classes? There is a way.

Our aim is to put all scores on a common scale. To do this, we compute the **standardized score** or **Z score,** defined by

$$Z = \frac{x - \bar{x}}{s}$$

Here x is a given score, and $\bar{x}$ and s are the mean and standard deviation of all the scores, respectively. Thus Susan's standardized scores in Advanced Calculus and British History are

$$Z = \frac{8 - 6}{2.83} \approx .71 \qquad Z = \frac{8 - 7}{1.55} \approx .65$$

Susan did slightly better in Advanced Calculus.

If you subtract $\bar{x}$ from each score, the resulting scores will have a mean of 0; if you then divide by s, the new scores will have a standard deviation of 1 (see Problems 9 and 10). Thus standardized scores are approximately normally distributed with mean 0 and standard deviation 1; that is, they are distributed as shown below. For example, 34% of the scores can be expected to fall between 0

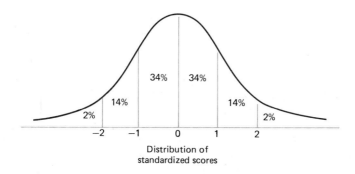

Distribution of
standardized scores

and 1, and 14% between 1 and 2. Susan's scores of .71 and .65 are well above average but certainly are not exceptional.

SUMMARY

The mean of a distribution is its balance point; it is a measure of the center of a distribution. Two distributions may have the same mean but be very different in character. For one, the values may cluster very closely around the mean; for another, they may be widely spread. To measure the spread, statisticians have introduced the standard deviation s defined by

$$s = \sqrt{\frac{(x_1 - \bar{x})^2 + (x_2 - \bar{x})^2 + \cdots + (x_n - \bar{x})^2}{n}}$$

It is always nonnegative. Roughly speaking, it measures the average distance the values lie from the mean. It is large when the values are widely dispersed and small (near 0) when they are tightly grouped.

PROBLEM SET 8.3

**TABLE 8-6
SQUARE ROOT
TABLE**

N	$\sqrt{N}$
0.5	0.71
1.0	1.00
1.5	1.22
2.0	1.41
2.5	1.58
3.0	1.73
3.5	1.87
4.0	2.00
4.5	2.12
5.0	2.24
6.0	2.45
7.0	2.65
8.0	2.83
9.0	3.00
10.0	3.16
12.0	3.46
14.0	3.74
16.0	4.00
18.0	4.24
20.0	4.47
25.0	5.00
30.0	5.48

1. Calculate the mean and standard deviation for the set {8, 6, 3, 5, 7, 6, 9, 8, 9, 9}. Use the table in the margin to approximate square roots.
2. Calculate the mean and standard deviation for the set {−3, 2, −4, 1, 0, 4, 6, −1, −5}.
3. Consider the two sets of data:
$$A = \{1, 5, 6, 9, 3, 5, 10, 5, 10, 6\}$$
$$B = \{5, 7, 6, 6, 5, 7, 6, 7, 5, 6\}$$
   (a) Make bar graphs for both sets.
   (b) Calculate the two means $\bar{x}_A$ and $\bar{x}_B$.
   (c) Calculate the two standard deviations s_A and s_B.
4. Ten students rated both Professor A and Professor B on a 1 (low) to 10 (high) rating scale. The results were the two sets of data in Problem 3. What conclusions can you draw about the two professors?
5. If the mean on a certain test is 75 with a standard deviation of 8, convert the scores 56, 70, 81, and 90 to standardized scores.
6. What is the meaning of a standardized score of 2? Of −1?
7. Suppose that sets A and B in Problem 3 are students' scores on a quiz in German and in Math, respectively. Amy got 5 on both quizzes. Compute her standardized score on both quizzes and decide whether she did better in German or in Math.
8. Rachel scored 456 on the national college entrance examination for which the mean is 500 with a standard deviation of 50. Raymond scored 25 on the BAT test which has a mean of 30 with a standard deviation of 6. If Hudson College weights these two examinations equally, which student has the best chance of being admitted to Hudson College?

9. Add 5 to each of the data in Problem 1. Compute the new mean and standard deviation. Make a conjecture about what happens to the mean and standard deviation when you add the same number to each data point.

10. Multiply each of the data in Problem 1 by 3. Compute the new mean and standard deviation. Make a conjecture about what happens to the standard deviation when you multiply each data point by a constant.

11. If a variable is normally distributed, approximately what percentage of its values will lie more than 2 standard deviations away from the mean? What percentage will lie more than 1 standard deviation above the mean?

12. Which of the following is more likely to be approximately normally distributed: the weights of all babies measured at birth, or the last digits of all the numbers in the Manhattan telephone directory? Explain.

13.* Consider the word lengths of all the words in the first paragraph of this section.
   (a) Make a frequency chart.
   (b) Calculate the mean.
   (c) Calculate the standard deviation.

14.* Roll a pair of dice 50 times, keeping track of the total number of spots showing on each roll.
   (a) Make a frequency chart.
   (b) Compute the mean.
   (c) Compute the standard deviation.

15.* If all the numbers in a set are equal, what is the standard deviation? Show that, if the standard deviation is 0, all the numbers in the set are equal.

16.* If the mean, median, mode, and standard deviation of two sets of five numbers are the same, are the two sets identical?

TABLE 8-7 BUS TIME DATA		
X	Y	Z
10.1	13.3	13.2
12.9	12.5	12.3
15.1	13.3	14.1
12.4	14.3	13.4
13.1	14.3	12.3
13.4	12.2	11.4
10.5	14.7	11.2
13.4	10.5	11.2
14.1	11.5	21.2
12.4	12.4	12.4
13.7	13.1	12.4
11.2	11.4	10.3
11.1	12.2	14.2
12.5	12.5	9.6
13.1	12.5	12.3
11.2	14.4	11.2
13.1	14.3	15.2

Table 8-7 contains data from the Metropolitan Transit Commission of Minneapolis and St. Paul. It gives the elapsed time between two specific bus stops for 17 different buses on three different days (X: April 22, Y: April 23, Z: April 28)

c 17. For Table 8-7 calculate the mean and standard deviation for X. If your calculator has mean and standard deviation buttons, use them. However, be aware that some calculators use $n - 1$ rather than n in the definition of the standard deviation which may lead to slight variations in the answer.

c 18. For Table 8-7, calculate the mean and standard deviation for Y.

c 19. For Table 8-7, calculate the mean and standard deviation for Z.

"IS THAT CLEAR?"

8.4 / Sigma Notation

Many people have the idea that mathematicians like nothing better than long, involved calculations. It is a popular myth, but very misleading. The thought of calculating the standard deviation for a set of 100 scores is at least as distasteful to mathematics teachers as it is to students. What mathematicians do enjoy is looking for shortcuts, for laborsaving devices, and for elegant ways of doing things. Professor Witquick's blackboard may look complicated. Actually he has just shown his class a new way to calculate the standard deviation, which cuts down on the labor involved.

There is another point to make about mathematicians. They like to use symbols, symbols that represent big ideas in a compact way. If the information on Professor Witquick's blackboard were written out in words, it would take several blackboards and would actually be much harder to comprehend. One of the professor's symbols has become so common that you occasionally see it printed in newspapers and popular magazines. We think that liberally educated people ought to be familiar with his use of the Greek letter Σ.

WRITING SUMS COMPACTLY

"Sum" begins with "S"; and the Greek letter for "S" is Σ, pronounced "sigma." Think of Σ as standing for sum. In particular,

$$\sum_{i=1}^{n} x_i = x_1 + x_2 + x_3 + \cdots + x_n$$

> $\sum_{i=1}^{n} x_i$ is read, "sum of x sub i as i runs from 1 to n."

Here Σ means add up; the symbol $i = 1$ below the sigma tells where to start adding; the n at the top tells where to stop. Try to follow these examples:

$$\sum_{i=1}^{5} a_i = a_1 + a_2 + a_3 + a_4 + a_5$$

$$\sum_{i=3}^{9} b_i = b_3 + b_4 + b_5 + b_6 + b_7 + b_8 + b_9$$

$$\sum_{i=1}^{5} i^2 = 1^2 + 2^2 + 3^2 + 4^2 + 5^2$$

$$\sum_{i=1}^{50} (2i - 1) = 1 + 3 + 5 + 7 + \cdots + 99$$

If you were able to follow them, you should be able to write each of these expressions in sigma notation:

$$c_1 + c_2 + c_3 + \cdots + c_{100}$$
$$1 + 2 + 3 + \cdots + 1000$$
$$1 + \tfrac{1}{2} + \tfrac{1}{3} + \tfrac{1}{4} + \tfrac{1}{5} + \tfrac{1}{6} + \tfrac{1}{7} + \tfrac{1}{8}$$

The answers appear in the summary.

PROPERTIES OF SUMS

Not only does sigma notation allow us to write sums in a very compact way; it also allows us to state certain properties of sums in an elegant fashion. For example, the three familiar facts:

$$\underbrace{c + c + c + \cdots + c}_{n \text{ terms}} = nc$$

$$ca_1 + ca_2 + ca_3 + \cdots + ca_n = c(a_1 + a_2 + a_3 + \cdots + a_n)$$

$$(a_1 + b_1) + (a_2 + b_2) + \cdots + (a_n + b_n)$$
$$= (a_1 + a_2 + \cdots + a_n) + (b_1 + b_2 + \cdots + b_n)$$

become, when written in sigma notation,

$$\sum_{i=1}^{n} c = nc$$

$$\sum_{i=1}^{n} ca_i = c \sum_{i=1}^{n} a_i$$

$$\sum_{i=1}^{n} (a_i + b_i) = \sum_{i=1}^{n} a_i + \sum_{i=1}^{n} b_i$$

SIGMA NOTATION IN STATISTICS

Statisticians take great delight in sigmas; their books are full of them. This is because most statistical measures require that we add many terms. Here are the formulas we studied earlier for means and standard deviations, now written as statisticians prefer to write them:

$$\bar{x} = \frac{1}{n} \sum_{i=1}^{n} x_i \qquad\qquad s = \sqrt{\frac{1}{n} \sum_{i=1}^{n} (x_i - \bar{x})^2}$$

We have already hinted that s is messy to calculate; if you worked out the problems at the end of Section 8.3, you don't need to be told. This is particularly true when $\bar{x}$ is not an integer, which is usually the case. There is another formula for s, which is much nicer, and the properties of sigma mentioned above are very useful in deriving it. To avoid the square root we initially consider s^2 rather than s:

$$s^2 = \frac{1}{n} \sum_{i=1}^{n} (x_i - \bar{x})^2$$

$$= \frac{1}{n} \sum_{i=1}^{n} (x_i^2 - 2\bar{x}x_i + \bar{x}^2)$$

$$= \frac{1}{n} \left[\sum_{i=1}^{n} x_i^2 + \sum_{i=1}^{n} (-2\bar{x}x_i) + \sum_{i=1}^{n} \bar{x}^2 \right]$$

$$= \frac{1}{n} \left(\sum_{i=1}^{n} x_i^2 - 2\bar{x} \sum_{i=1}^{n} x_i + \sum_{i=1}^{n} \bar{x}^2 \right)$$

$$= \frac{1}{n} \left(\sum_{i=1}^{n} x_i^2 - 2\bar{x}n \frac{1}{n} \sum_{i=1}^{n} x_i + n\bar{x}^2 \right)$$

$$= \frac{1}{n} \left(\sum_{i=1}^{n} x_i^2 - 2\bar{x}n\bar{x} + n\bar{x}^2 \right)$$

$$= \frac{1}{n} \sum_{i=1}^{n} x_i^2 - \bar{x}^2$$

Thus

$$s = \sqrt{\left(\frac{1}{n}\sum_{i=1}^{n} x_i^2\right) - \bar{x}^2}$$

a formula we claim is generally much easier to use than the original one.

Let's illustrate it first for Professor Witquick's first quiz in Section 8.3

Score	Score Squared	
$x_1 = 10$	$x_1^2 = 100$	$\bar{x} = \frac{36}{6} = 6$
$x_2 = 8$	$x_2^2 = 64$	
$x_3 = 7$	$x_3^2 = 49$	$s = \sqrt{\frac{264}{6} - (6)^2}$
$x_4 = 5$	$x_4^2 = 25$	$= \sqrt{44 - 36}$
$x_5 = 5$	$x_5^2 = 25$	≈ 2.83
$x_6 = \underline{1}$	$x_6^2 = \underline{1}$	
TOTAL 36	TOTAL 264	

Naturally, the result is the same as the one calculated in Section 8.3.

But the real merit of our new formula shows up when we have a large number of values and the mean is not integral. In the table below we have worked out an example in which there are 20 values.

	TABLE 8-8	
x_i	x_i^2	Calculation of $\bar{x}$ and s
10	100	
10	100	
9	81	$\bar{x} = 135/20 = 6.75$
9	81	
9	81	$s = \sqrt{(1027/20) - (6.75)^2}$
8	64	
8	64	$= \sqrt{51.35 - 45.56}$
8	64	
8	64	$= \sqrt{5.79}$
7	49	≈ 2.41
7	49	
7	49	
7	49	
6	36	
6	36	
5	25	
4	16	
3	9	
3	9	
$\underline{1}$	$\underline{1}$	
135	1027	

SUMMARY

Two new ideas were introduced in this section. The first was simply a matter of notation, the sigma notation for sums. One could argue that the choice of notation is a trivial subject, hardly worthy of comment. But mathematicians have learned that the use of a carefully chosen, compact symbol can help them handle a complicated idea with ease and efficiency. This is particularly true of the symbol Σ which is used to denote sums. Its use is illustrated by

$$c_1 + c_2 + c_3 + \cdots + c_{100} = \sum_{i=1}^{100} c_i$$

$$1 + 2 + 3 + \cdots + 1000 = \sum_{i=1}^{1000} i$$

$$1 + \frac{1}{2} + \frac{1}{3} + \cdots + \frac{1}{8} = \sum_{k=1}^{8} \frac{1}{k}$$

The following problems give you an opportunity to confirm and improve your understanding of this symbol.

The second idea was a new formula for the standard deviation, which is generally easier to use in practice. It is conveniently written using the sigma symbol and was derived with its aid:

$$s = \sqrt{\left(\frac{1}{n} \sum_{i=1}^{n} x_i^2\right) - \bar{x}^2}$$

PROBLEM SET 8.4

1. Calculate:

   (a) $\sum_{i=1}^{8} 2i$ (b) $\sum_{i=1}^{8} (2i + 1)$ (c) $\sum_{i=1}^{5} i^2$ (d) $\sum_{i=1}^{4} \frac{1}{i}$

2. Calculate:

   (a) $\sum_{i=1}^{5} 3i$ (b) $\sum_{i=1}^{5} (3i - 2)$ (c) $\sum_{i=1}^{5} (i - 1)^2$ (d) $\sum_{i=2}^{4} \frac{1}{i - 1}$

3. Rewrite using sigma notation.
   (a) $1 + 2 + 3 + 4 + \cdots + 10$
   (b) $1 + 4 + 9 + 16 + \cdots + 100$
   (c) $y_1 + y_2 + y_3 + \cdots + y_{37}$
   (d) $(y_1 - 3)^2 + (y_2 - 3)^2 + \cdots + (y_n - 3)^2$

4. Rewrite using sigma notation.
   (a) $2 + 4 + 6 + 8 + \cdots + 30$
   (b) $1 + \frac{1}{2} + \frac{1}{3} + \cdots + \frac{1}{75}$
   (c) $w_3 + w_4 + w_5 + \cdots + w_{400}$
   (d) $x_1 y_1 + x_2 y_2 + x_3 y_3 + \cdots + x_n y_n$

5. Let $\sum_{i=1}^{100} x_i = 431$. Calculate:

(a) $\bar{x}$

(b) $\sum_{i=1}^{100} 3x_i$

(c) $\sum_{i=1}^{100} (3x_i + 1)$

6. Let $\sum_{i=1}^{10} x_i = 20; \sum_{i=1}^{10} x_i^2 = 90$. Calculate:

(a) $\bar{x}$
(b) s

(c) $\sum_{i=1}^{10} (x_i^2 + 2x_i)$

7. Let $\sum_{i=1}^{100} x_i = 200; \sum_{i=1}^{100} x_i^2 = 4000$. Calculate $\bar{x}$ and s.

8. Let $\sum_{i=1}^{64} y_i = -32$ and $\sum_{i=1}^{64} y_i^2 = 1280$. Calculate $\bar{y}$ and s.

9. Redo Problem 1 in Section 8.3 using our new formula for s.
10. Find the mean and standard deviation of the set $\{25, 27, 30, 25, 32, 24, 27, 30, 28, 26\}$.
11* Show, using sigma notation, that if $y_i = x_i + c$ where c is constant, $\bar{y} = \bar{x} + c$ and $s_y = s_x$.
12* Show, using sigma notation, that if $y_i = cx_i$, where c is a positive constant, $\bar{y} = c\bar{x}$ and $s_y = cs_x$.

13. Show, using sigma notation, that $\sum_{i=1}^{n} (x_i - \bar{x}) = 0$.

14. Show that, if $z_i = x_i + y_i$, then $\bar{z} = \bar{x} + \bar{y}$.

8.5 / Correlation

TABLE 8-9 A RANDOM SAMPLE OF 15 PODUNK UNIVERSITY STUDENTS		
Geometry Grade, x	Calculus Grade, y	SMOB Score, z
4	3	48
4	4	60
4	3	78
4	4	62
4	3	70
3	4	80
3	3	60
3	3	72
3	2	64
3	2	62
2	2	52
2	1	56
2	2	60
1	0	58
1	1	40

Professor Witquick faces a problem which is very familiar to all scientists. What relationship (if any) exists between two variables? The great achievements in the physical sciences over the last 300 years are largely due to the discovery of almost exact relationships between important variables. For example, a physicist knows that $v = 32t$ and $s = 16t^2$ describe how velocity v and distance s are related to time t when an object falls under the influence of gravity.

The situation is much more complicated in the social sciences. Rarely do we find exact relationships. Yet there is great interest in establishing even the slightest tendency of one variable to depend on another and to measure the strength of that tendency. In particular, Professor Witquick would like to measure the dependence of y (grade in college calculus) on x (grade in high school geometry), and in turn the dependence of y on z (the SMOB score).

Here is a good way to begin. Take a random sample of the students who took first-semester calculus at Podunk University last year. Record the high school geometry grade, college calculus grade, and SMOB score for each student as in the table at the right. Then make a comparison. But how?

SCATTER DIAGRAMS

By now we are conditioned to the notion that we should try to display data pictorially. The appropriate picture here is a **scatter diagram.** This is simply a plot of the (x, y) and (z, y) data on a standard coordinate plane (see diagrams below).

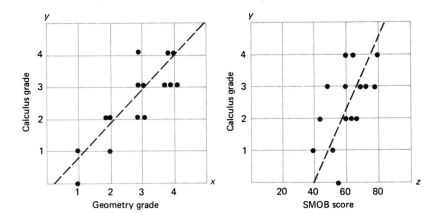

In each case, there is some indication of a trend, a trend that the dotted lines on the scatter diagrams are meant to suggest. Admittedly, neither scatter diagram comes close to the ideal situation in which the data points lie almost exactly on a straight line. That is too much to expect, for then we could predict with near certainty what the college calculus grade would be. Maybe some people can see from the scatter diagrams that high school geometry grades are a better predictor than SMOB scores, but we confess that this is far from obvious to us. We need stronger evidence than we have as yet.

THE PEARSON CORRELATION COEFFICIENT

To measure the tendency toward a linear (i.e., straight line) relationship, Karl Pearson (1857–1936) introduced the sample correlation coefficient r. If (x_1, y_1), (x_2, y_2), ..., (x_n, y_n) is a sample of paired values for the variables x and y, the **sample correlation coefficient** between x and y is defined by

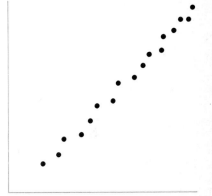

The ideal situation

$$r_{xy} = \frac{1}{n s_x s_y} \sum_{i=1}^{n} (x_i - \bar{x})(y_i - \bar{y})$$

Here $\bar{x}$, $\bar{y}$, s_x, and s_y are the respective means and standard deviations for the x and y data.

There are several reasons for Pearson's choice of this measure.

First, $\Sigma (x_i - \bar{x})(y_i - \bar{y})$ is chosen because it is large and positive when the data closely approximate a line that slopes upward

Example A,
r near 1

Example B,
r near -1

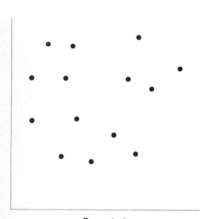

Example C,
r near 0

(example A), large but negative when the data approximate a line sloping downward (example B), and near 0 when the data spread out randomly (example C). Note, for instance, that when data follow a pattern like that in example A, x_i and y_i tend to be bigger or smaller than $\bar{x}$ and $\bar{y}$, respectively, at the same time. This means that the factors $x_i - \bar{x}$ and $y_i - \bar{y}$ tend to be positive or negative at the same time, thus making the product $(x_i - \bar{x})(y_i - \bar{y})$ consistently positive. This in turn makes the sum $\Sigma(x_i - \bar{x})(y_i - \bar{y})$ large and positive. See if you can reason out examples B and C in a similar fashion.

Second, the constant $1/ns_x s_y$ that stands in front of Σ is put there to standardize all correlation coefficients, that is, to put them on the same scale. It has the effect of making a correlation coefficient take a value between -1 and 1 (see Problem 9). A value of r near 1 indicates a strong tendency for the data to lie along a line with a positive slope, that is along a line on which y increases as x increases (example A). A value of r near -1 indicates a strong tendency for the data to lie along a line with a negative slope, that is a line on which y decreases as x increases (example B). Values of r close to 0 indicate that x and y are not linearly related.

A SIMPLER FORMULA FOR r

The formula as given is complicated to use. A somewhat better formula for computational work can be derived using the properties of Σ noted in Section 8.4. We include the derivation for algebraic experts and especially for wise but skeptical souls who won't believe anything unless they see it proved.

$$r = \frac{1}{ns_x s_y} \sum_{i=1}^{n} (x_i - \bar{x})(y_i - \bar{y})$$

$$= \frac{n \sum_{i=1}^{n} (x_i y_i - \bar{x} y_i - \bar{y} x_i + \bar{x}\bar{y})}{n \sqrt{\frac{1}{n}\sum_{i=1}^{n} x_i^2 - \bar{x}^2} \; n \sqrt{\frac{1}{n}\sum_{i=1}^{n} y_i^2 - \bar{y}^2}}$$

$$= \frac{n\left(\sum_{i=1}^{n} x_i y_i - \bar{x}\sum_{i=1}^{n} y_i - \bar{y}\sum_{i=1}^{n} x_i + \sum_{i=1}^{n}\bar{x}\bar{y}\right)}{\sqrt{n\sum_{i=1}^{n} x_i^2 - n^2\bar{x}^2}\sqrt{n\sum_{i=1}^{n} y_i^2 - n^2\bar{y}^2}}$$

$$= \frac{n\left(\sum_{i=1}^{n} x_i y_i - \bar{x}n\bar{y} - \bar{y}n\bar{x} + n\bar{x}\bar{y}\right)}{\sqrt{n\sum_{i=1}^{n} x_i^2 - (n\bar{x})^2}\sqrt{n\sum_{i=1}^{n} y_i^2 - (n\bar{y})^2}}$$

$$= \frac{n \sum\limits_{i=1}^{n} x_i y_i - n\bar{x}n\bar{y}}{\sqrt{n \sum\limits_{i=1}^{n} x_i^2 - (n\bar{x})^2} \sqrt{n \sum\limits_{i=1}^{n} y_i^2 - (n\bar{y})^2}}$$

Thus

$$r = \frac{n \sum\limits_{i=1}^{n} x_i y_i - \sum\limits_{i=1}^{n} x_i \sum\limits_{i=1}^{n} y_i}{\sqrt{n \sum\limits_{i=1}^{n} x_i^2 - \left(\sum\limits_{i=1}^{n} x_i\right)^2} \sqrt{n \sum\limits_{i=1}^{n} y_i^2 - \left(\sum\limits_{i=1}^{n} y_i\right)^2}}$$

With this formula for r in hand, we illustrate a calculation for the (x, y) data from Professor Witquick's sample of 15 students.

TABLE 8-10

x_i	x_i^2	y_i	y_i^2	$x_i y_i$	Calculation of r
4	16	3	9	12	
4	16	4	16	16	$r = \dfrac{(15)(121) - (43)(37)}{\sqrt{(15)(139) - (43)^2} \sqrt{(15)(111) - (37)^2}}$
4	16	3	9	12	
4	16	4	16	16	
4	16	3	9	12	$= \dfrac{224}{\sqrt{236} \sqrt{296}}$
3	9	4	16	12	
3	9	3	9	9	$\approx .85$
3	9	3	9	9	
3	9	2	4	6	
3	9	2	4	6	
2	4	2	4	4	
2	4	1	1	2	
2	4	2	4	4	
1	1	0	0	0	
1	1	1	1	1	
43	139	37	111	121	

The calculation yields $r_{xy} = .85$. An identical exercise gives $r_{zy} = .51$.

THE ANSWER TO THE ORIGINAL QUESTION

Well, which is the better predictor of college calculus grades? High school geometry grades or SMOB scores? The calculations we have made (namely, $r_{xy} = .85$ and $r_{zy} = .51$) strongly support the view that high school geometry grades did a better job last year. As indicated earlier, we always hope for a value of r near 1 or near -1. A value of $r = .85$ is really quite good. One rarely finds correlation coefficients better than this arising from problems in the social sciences.

But now some words of caution. First, our data were completely artificial; we just made them up. But even if they were real data, we should not make exorbitant claims. Our sample size of 15 was rather small. Perhaps it was not representative of the total population of students who took calculus at Podunk University, and certainly we must be careful in claiming that it represents students who will take calculus in the future. It may be well to do the same experiment over, using a much larger sample. Then, too, we should avoid claiming that what is true at Podunk University is necessarily true at other colleges. Nevertheless, the hypothetical data give support to the conclusion that at Podunk University high school geometry grades are a better predictor of college calculus grades than SMOB scores. And, in fact, they suggest that geometry grades are a very good predictor.

SUMMARY

We have introduced the correlation coefficient as a measure of the tendency for two variable data to lie along a straight line. The number r always lies between -1 and 1. A value of r near 1 or -1 indicates a strong linear relationship, near 0 none at all.

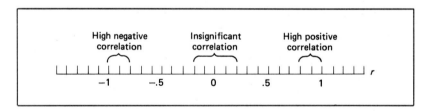

We emphasize that r is designed to measure the tendency toward a straight-line relationship. The sample data from two variables may lie along some other curve, indicating a more complicated but definite relationship between these variables, yet the value of r could be very small (see Problem 10).

TABLE 8-11 DATA SET A		TABLE 8-12 DATA SET B	
x	**y**	**x**	**y**
2	4	10	2
4	7	9	3
3	6	9	4
4	6	8	4
1	4	6	4
5	6	5	6
6	7	5	7
0	2	3	8
8	9	3	9
7	9	2	13

PROBLEM SET 8.5

1. Make a scatter diagram for data set A. Then calculate $\bar{x}$, $\bar{y}$, s_x, s_y, and r.
2. Make a scatter diagram for data set B. Then calculate $\bar{x}$, $\bar{y}$, s_x, s_y, and r.

3. How would you interpret each of the following values for a correlation coefficient (assuming a sample size of at least 25)?
   (a) $r = -.9$ (b) $r = -.41$
   (c) $r = .12$ (d) $r = .79$

4. Make a scatter diagram for the following xy data: $\{(1, 0); (1, 4); (5, 0); (5, 4)\}$. What do you estimate for a value of r? Now calculate r.

5. Would you expect a high positive, low positive, low negative, or high negative correlation between the following variables?
   (a) The weight of a car and its gas consumption in miles per gallon.
   (b) Median salary and gross national product.
   (c) Athletic ability and mathematical talent.
   (d) Average winter temperature and the cost of heating a home.
   (e) The weight of a diamond ring and its cost.
   (f) The amount of rain in the Midwest and the cost of beef in New York City.
   (g) Intelligence and hat size for males.
   (h) Weight and height for people.
   (i) College grade-point average and median salary 20 years after graduation.

6. If $n = 20$, $\sum_{i=1}^{n} x_i y_i = 160$, $\sum_{i=1}^{n} x_i = 30$, $\sum_{i=1}^{n} y_i = 40$, $\sum_{i=1}^{n} x_i^2 = 180$, and

$\sum_{i=1}^{n} y_i^2 = 200$, find r approximately.

7.* Obtain a set of two-variable data as follows. Roll a pair of dice (marked so that you can distinguish between them) 25 times. Let x be the number of spots on one die, and y the number of spots on the other. Guess the correlation coefficient. Now actually calculate r.

8.* Suppose $y_i = 3x_i + 1$ for $i = 1, 2, \ldots, n$. What does this imply about the scatter diagram for the xy data? Show, using properties of Σ, that r is 1 in this case.

9.* A famous inequality, due to A. Cauchy, states that, for any sets of numbers $a_1, a_2, \ldots, a_n$ and $b_1, b_2, \ldots, b_n$,

$$\left| \sum_{i=1}^{n} a_i b_i \right| \leq \sqrt{\sum_{i=1}^{n} a_i^2} \sqrt{\sum_{i=1}^{n} b_i^2}$$

Use this inequality to show that $|r| \leq 1$. Hint: Let $a_i = x_i - \bar{x}$ and $b_i = y_i - \bar{y}$.

10.* Suppose that all the xy data happen to lie scattered around a circle after they are plotted. You should expect to find a small value of r, but certainly x and y seem to be related very strongly. Explain this paradox.

c 11. Calculate r_{xy}, the correlation coefficient between X and Y, for the data of Table 8-7 of Problem Set 8.3. Interpret the result.

c 12. Calculate r_{xz} for the data of Table 8-7. Interpret the result.

c 13. Calculate the correlation coefficient between N and $\sqrt{N}$ using the data of Table 8-6. Interpret the result.

Chapter 9
Geometric Paths

A mathematician, like a painter or a poet, is a maker of patterns. If his patterns are more permanent than theirs, it is because they are made with ideas . . . . The mathematician's patterns, like the painter's or poet's, must be beautiful; the ideas, like the colours or the words, must fit together in a harmonious way. Beauty is the first test; there is no permanent place in the world for ugly mathematics.

G. H. HARDY

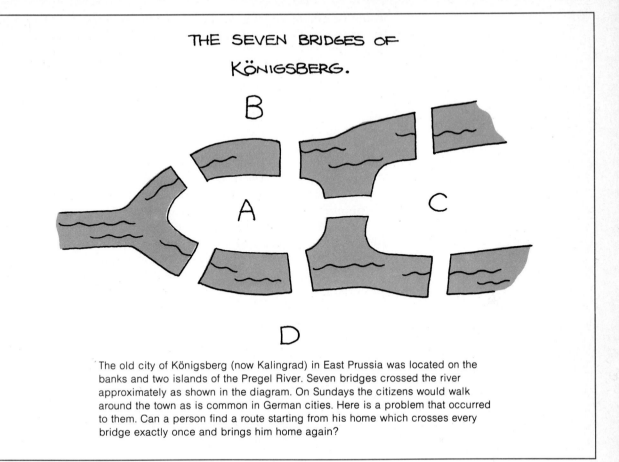

THE SEVEN BRIDGES OF
KÖNIGSBERG.

B

A

C

D

The old city of Königsberg (now Kalingrad) in East Prussia was located on the banks and two islands of the Pregel River. Seven bridges crossed the river approximately as shown in the diagram. On Sundays the citizens would walk around the town as is common in German cities. Here is a problem that occurred to them. Can a person find a route starting from his home which crosses every bridge exactly once and brings him home again?

9.1 / Networks

The problem of the seven bridges somehow came to the attention of the great Swiss mathematician, Leonhard Euler (1707–1783), who was residing in Russia as mathematician at the court of Catherine the Great. With customary acumen, he solved not only this problem but all others of the same type. His celebrated solution, here to be described, was presented to the Russian Academy at St. Petersburg in 1735.

It is clear to anyone who tries to attack this problem with the head instead of the feet that the size of the islands, the length of the bridges, and the fact that these walks occurred on Sundays are irrelevant details (recall Section 1.1) that should be stripped away. Euler saw that, if each land mass were represented by a point and each bridge by an arc joining two points, the problem

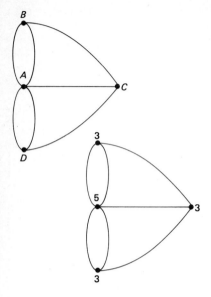

would be one of moving around the network shown in the margin traversing each arc exactly once, and returning to the starting point.

Before reading further, try your hand at traversing this network with a pencil, being sure not to lift it off the paper or retrace any arc. Try starting from different places. You will probably decide that it can't be done—but are you sure?

The next move seems simple enough, but it is the key to all problems of this type. It nicely illustrates the value of finding succinct notation that calls attention to the significant features of a problem. Simply attach to each of the four points (or **nodes,** as points are often called in network theory) a number that counts the number of arcs meeting at that node.

Now, any pass through a node contributes two to that number, one for the entrance and the other for the exit. Thus every node in a traversible network must have an even number attached to it. This includes the initial node, since the problem demands that the walker end where he or she started. The Königsberg bridges network doesn't have a ghost of a chance; all four of its nodes are odd. No one will ever find the required route across the seven bridges.

EULER CIRCUITS

The seeds of a beautiful theory have already been planted. To see it in full bloom, we require precise terminology. The building blocks of a network are a set of points, from now on called **vertices** (or nodes), and a set of arcs, curved or straight, called **edges.** We require that each edge have two distinct vertices (its end points) and that each vertex be the end point of at least one edge. Two edges can meet only at a vertex. A **network** is a finite, connected collection of vertices and edges. Here "connected" means that between any two vertices of the network there is a **path,** that is, a continuous sequence of edges, joining them. All this is meant to say that a network is just what you think it ought to be; the examples in the margin help clarify some special points. Finally, the **degree** of a vertex is the number of edges that meet at that vertex.

The general problem Euler attacked and solved was this: Is there a path around a network that traverses each edge exactly once and ends where it started? It is a matter of simple justice to call such a path an **Euler circuit.** Here is Euler's grand theorem on networks:

Theorem. A network has an Euler circuit if and only if every vertex is of even degree.

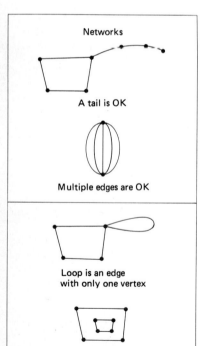

Networks

A tail is OK

Multiple edges are OK

Loop is an edge
with only one vertex

Not connected

Not networks

Note that the theorem says two things. It says that, if a network has one or more odd vertices, there is no Euler circuit. We have already seen, in the case of the seven-bridges problem, an argument for this that suffices just as well in the general case. The theorem also says that a network having all even vertices has an Euler circuit. You may wish to enhance your appreciation for Euler's deceptively easy arguments by trying to prove this second part for yourself. Then read on to see how Euler leads us through such a network.

Take any network whatever that has all even vertices. Pick a vertex A, start a path there, and walk as far as you can, always leaving a vertex on an edge not previously used. Eventually, you will reach a vertex with no possible exit. But this dead-end vertex must be A, since at any other vertex there is an exit if there was an entrance, the degree being even.

But alas, the path P so obtained may leave out part of the network (see diagram). If so, there is a vertex B on path P where some unused edges meet. Start from B and obtain a path Q using previously unused edges and returning to B just as you did to A above. Now adjoin path Q to the original path P by following it until you get to B; then traverse path Q around its circuit back to B, finally completing path P back to A. If this combined path traverses the network, you are finished. If not, repeat the process. Eventually you will get an Euler circuit.

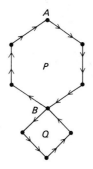

HAMILTONIAN CIRCUITS

Horace and his good friend Hugo have discovered a strange coincidence. Horace is road inspector for the county highway department; Hugo is a salesman for a greeting card company. Both of them cover exactly the same territory; it is shown in the marginal diagram. Each Monday, Horace is required to make a complete inspection of the road system, checking for potholes and washouts. Each Tuesday Hugo calls at the drugstores in each of the towns shown. Naturally both men want to do their jobs as efficiently as possible.

Horace's job is easy. He needs an Euler circuit through the network. After checking that all the vertices are even, he draws a circuit on his map (see margin, where successive steps are numbered).

Hugo's problem is quite different. He doesn't care a whit about the roads of the network. He simply wants a path that visits every vertex exactly once, returning him to his starting point. Such a path is called a **Hamiltonian circuit,** after another fine mathematician, William Rowan Hamilton. Fortunately there is one (see margin). Both men can do their jobs efficiently.

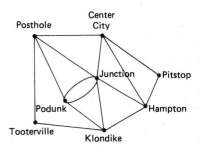

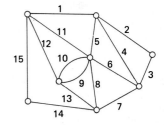

An Euler circuit

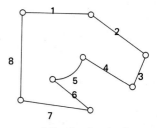

A Hamiltonian circuit

W. R. Hamilton (1805–1865)
Known as the Royal Astronomer of
Ireland, Hamilton was a great
mathematician but an even greater
physicist. Among his minor accomplishments was the invention of a game
which required construction of the
circuit now called by his name.

But now you can guess what the big question is. Is there a simple procedure for deciding whether a given network has a Hamiltonian circuit? No one knows, though many great mathematicians have looked long and hard. It is one of the famous unsolved problems of mathematics. A solution would be of considerable interest even to chemists who use networks to describe the structure of molecules — but this is a connection we don't have space to discuss.

SPIDERWEBS AND SUBWAY SYSTEMS

Networks occur naturally all around us. One of the most interesting is the beautiful orbed web of the familiar garden spider. It is both an engineering marvel and a work of art. Here is an intriguing question for both people and spiders: What is the fewest number of continuous runs required for a spider to spin the web shown below, assuming it does not want to double any strands?

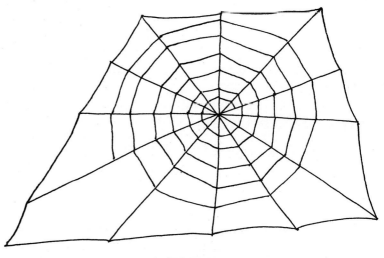

A SPIDERWEB

Surely the answer has something to do with the degrees of the vertices. A quick check reveals 14 odd vertices — 12 around the rim and 1 at each end of the spiral. As Euler noted, odd vertices correspond to starting and ending points on noncircular paths. It therefore takes at least seven runs to spin the web; and seven will do. See if you can find them. Once again Euler's method gives us the complete solution for all problems of this kind.

Theorem. If k is greater than or equal to 1, a network with $2k$ odd vertices contains a family of k distinct paths which together traverse each edge exactly once. Moreover, no family of fewer paths can do the job.

Washington, D.C., is building a subway system which looks something like the diagram below. Assuming the city wants the minimum number of lines and will not allow two different lines to use the same track, how many lines will be required? There are 10 odd vertices, so it will take at least 5 lines. Readers should be cautioned that the city fathers may have chosen different assumptions and may even have revised the network by the time this book is published. After all, there are other factors to consider (e.g., cost, congestion, and congresspeople).

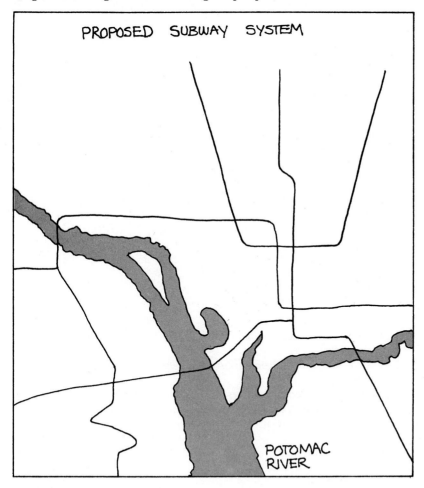

PROPOSED SUBWAY SYSTEM

POTOMAC RIVER

SUMMARY

Networks are all around us, though it sometimes takes a person like Euler to notice them. Two problems about networks are: (1)

Is there a circuit that traverses every edge exactly once? (2) Is there a circuit that visits every vertex exactly once? To the first we gave a simple answer — yes if all the vertices are even, no otherwise. To the second, we simply say — try your luck. No one knows the complete answer.

PROBLEM SET 9.1

1. Determine which of the following has an Euler circuit. If there is one, draw it.

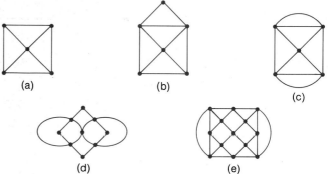

2. Follow the directions in Problem 1.

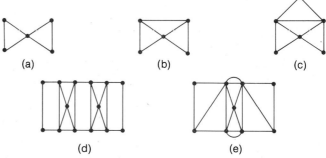

3. Try to draw a Hamiltonian circuit for each of the networks in Problem 1.
4. Try to draw a Hamiltonian circuit for each of the networks in Problem 2.
5. As in the Königsberg bridges problem, decide whether or not there is an Euler circuit for the following network of islands and bridges. If there is one, draw it.

6. Can you add one more bridge to the Königsberg bridges network so that it will have an Euler circuit? If not, how about two bridges?

7. Call a path through a network that traverses each edge exactly once an **Euler path.** It differs from an Euler circuit in that it doesn't necessarily begin and end at the same vertex. Try to draw an Euler path for each of the following.

(a)

(b)

(c)

8. Try to draw Euler paths for the following networks.

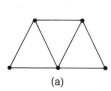

(a)

(b)

(c)

9. Based on your experience with Problems 7 and 8 and your reading of this section, suggest a theorem that tells us when there will be an Euler path through a network. Hint: Study the degrees of the vertices.

10. Use the theorem you suggested in Problem 9 to decide whether or not the following have Euler paths.

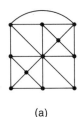

(a)

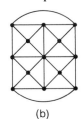

(b)

11. Each of the following represents a house plan. Is it possible to find an Euler path (not necessarily a circuit) that takes a person through each door exactly once? If so, draw such a path. Hint: Superimpose a network as in part (a).

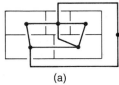

(a)

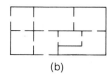

(b)

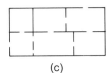

(c)

12. Follow the directions in Problem 11.

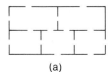

(a)

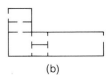

(b)

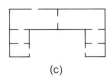

(c)

13. Find a Hamiltonian circuit for each of the following.

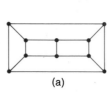

(a)

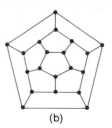

(b)

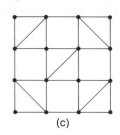

(c)

14. Assume that each of the networks in Problem 13 represents a spider web. How many runs are required to spin each web?

15. The diagram in the margin represents the subway system for a large city. How many separate lines are required when no two use the same track?

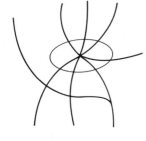

16.* Draw five fairly complicated networks. Label the degree of each vertex. Note for each network whether the total number of odd vertices is an even or an odd number. Make a conjecture. See if you can give an argument establishing your conjecture: Hint: Remove one edge after another from a network until you are down to one edge.

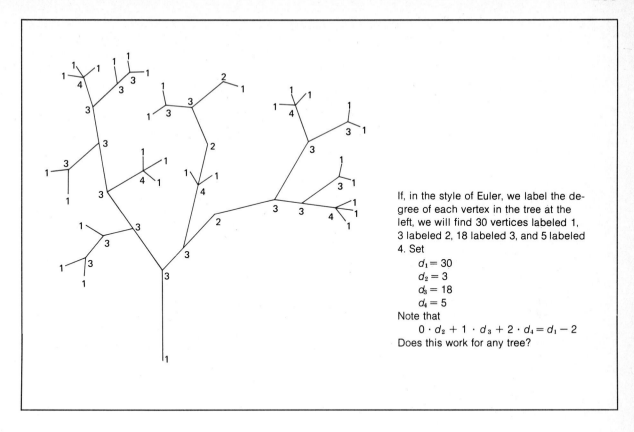

If, in the style of Euler, we label the degree of each vertex in the tree at the left, we will find 30 vertices labeled 1, 3 labeled 2, 18 labeled 3, and 5 labeled 4. Set

$$d_1 = 30$$
$$d_2 = 3$$
$$d_3 = 18$$
$$d_4 = 5$$

Note that

$$0 \cdot d_2 + 1 \cdot d_3 + 2 \cdot d_4 = d_1 - 2$$

Does this work for any tree?

9.2 / Trees

The reader will be forgiven for thinking the authors have no appreciation for what is beautiful about a tree. In our defense, we claim to have the imagination required to see in the diagram above much more than the skeleton of a fig tree shorn of that which makes for decency. We see, for example, rivulets flowing into creeks, thence into streams, and finally into a mighty river. Or, starting from the bottom, we can see a huge natural gas network coming from the oil fields of Texas and branching out to all the little towns of a midwestern state. We can even interpret it as a series of decisions to be made, each junction displaying the possible choices (with the three vertices of degree 2 indicating places where options have been cut off).

In any case, we have pictured a mathematical tree. For mathematicians, a **tree** is a (connected) network with no circuits or loops of any kind. Can something so simple be worth studying? We think so.

A fool sees not the same tree that a wise man sees.
William Blake

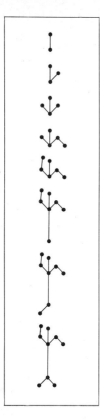

SOME RELATIONSHIPS THAT MUST HOLD

Here is the first question to ask: Is there a relationship between the number of vertices and the number of edges in a tree? Check the examples in the margin and make a conjecture. Then try to demonstrate that you are right.

Here is a simple way to look at it: Start with the simplest possible tree, the one-edge, two-vertex tree shown at the top. Note that vertices lead edges by one. Now start building from there to any complicated tree that you wish, one edge at a time. Each time you add an edge, you add a vertex. Thus you maintain the relationship; vertices always stay one ahead of edges. We have shown that

> A tree with n edges has $n + 1$ vertices.

Recall the definition of degree of a vertex; it's just the number of edges meeting there. Let d_1 be the number of first-degree vertices, d_2 the number of second-degree vertices, etc.; and let n be the total number of vertices altogether. For example, the last stick figure in the margin has $d_1 = 5$, $d_2 = 2$, $d_3 = 1$, $d_4 = 1$, and $n = 9$. Is there any relationship among these numbers? One that's trivial and not very interesting is

$$d_1 + d_2 + d_3 + \cdots = n$$

We have something far deeper in mind, something showing that mathematical trees, like the more familiar kind, have a balance and proportion which makes them quite lovely also. We claim that

$$1d_3 + 2d_4 + 3d_5 + \cdots = d_1 - 2$$

Try it on our stick figure:

$$1 \cdot 1 + 2 \cdot 1 = 5 - 2$$

Try it on a dozen examples of your making. Then try to find a proof (we outline our own in Problem 14).

Here is an application to river systems: Such systems have no two-way junctions and only rarely exhibit junctions of degree greater than 3. Thus most river systems obey the law $d_3 = d_1 - 2$. Now d_1, the number of first-degree vertices, is really just the number of sources plus the one exit. *Thus, for river systems, the number of triple junctions is equal to the number of sources minus 1.* For example, a system with 40 sources has 39 triple junctions.

And this our life,
Exempt from public haunt,
Finds tongues in trees,
Books in the running brooks,
Sermons in stones,
And good in everything.
William Shakespeare

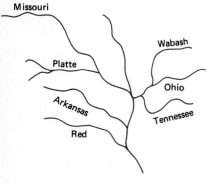

Mississippi River system

MINIMIZING THE COST OF A TREE

Imagine that the 11 dots below represent buildings on a new college campus, which are to be connected by underground electric cables. The building code states that junctions can occur only within buildings. The contractor wishes to build the network so as to minimize its cost, that is, its total length. Clearly the network should not contain any circuits, for one link of a circuit could be left out and the buildings would still be connected. The contractor wants an **economy tree.** Here are the rules for building one:

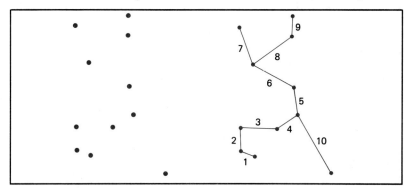

1. First build the shortest edge.
2. Thereafter connect the vertex closest to the part already constructed.
3. Break any ties in an arbitrary manner.

For our example, the result of following the rules gives the tree shown above (with the steps numbered).

There remains the essential point: to show that these rules produce a tree of minimum length. That of course is the rub; we know of no easy demonstration (see Problem 15).

SPIRALS, MEANDERS, AND EXPLOSIONS

Sometimes there are factors other than total length to be considered in building a tree. If edges represent steam heating lines, one should make sure no building is too far from the central heating plant and that no one line services too many buildings. For similar reasons, no branch of an oak tree can be too long; otherwise the trunk will not be able to force nutrients out to the farthest twig.

Suppose, for example, that a heating plant is located at H in the accompanying diagrams, where for simplicity the 25 buildings are at the intersection points of a regular grid. The total lengths of the spiral and the meander are 24 units, the least possible (they are economy trees); but we can guarantee that the dorm resi-

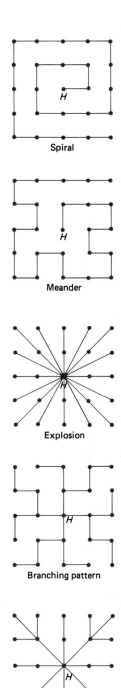

Spiral

Meander

Explosion

Branching pattern

Compromise

dents at the end of the line will complain on a cold February morning.

The explosive pattern in the middle excels in directness, but at the expense of adding length; it measures 37 units overall. Now the contractor will complain. To find some middle ground takes us back to Section 1.3; experiment, guess, demonstrate. The branching pattern is an economy tree (of length 24) and is quite good on directness. But the compromise shown at the bottom achieves almost the best of both worlds. It measures 27.3 units, and yet each building is within 2.8 units of H.

SUMMARY

We have discovered two important laws that trees obey. Vertices always exceed edges by exactly one, and their degrees inevitably conform to the formula

$$1 \cdot d_3 + 2 \cdot d_4 + 3 \cdot d_5 + \cdots = d_1 - 2$$

Moreover, for a given set of vertices, we gave a clearly prescribed set of rules that will yield a tree of shortest length (an economy tree).

PROBLEM SET 9.2

1. Label the degree of each vertex in tree A. Then calculate d_1, d_2, d_3, ..., and verify the formula displayed in the summary above.
2. Follow the instructions in Problem 1 for tree B.
3. Count the number of vertices in tree A. Without counting, calculate the number of edges. Hint: We derived a formula for this.
4. Count the number of edges in tree B. Without counting, calculate the number of vertices.
5. Draw (if possible) a tree with $d_1 = 17$, $d_2 = 5$, $d_3 = 3$, $d_4 = 3$, and $d_5 = 2$.
6. Draw (if possible) a tree with $d_1 = 20$, $d_2 = 1$, $d_3 = 2$, $d_4 = 4$, $d_5 = 4$, and $d_6 = 1$.
7. Assuming that all junctions in a certain river system are triple junctions and that there are 49 of them, how many sources does the system have?
8. Tree C represents the pairings in a tournament. Note that there are no vertices of degree greater than 3, so $d_3 = d_1 - 2$. A vertex of degree 3 corresponds to a game. Derive a rule for the number of games if there are n contestants. Compare with Section 1.4.
9. Make two copies of set D, shown in the margin (do this by placing a sheet of onionskin paper over the figure and marking the dots).
   (a) Use one copy to draw an economy tree.
   (b) Use the other copy to draw a good distribution system for steam pipes, assuming H is the heating plant and that connections can occur only at dots.

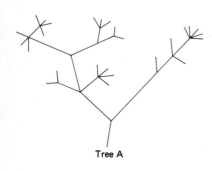

Tree A

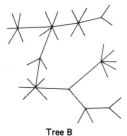

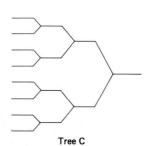

Tree B

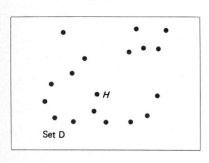

Tree C

Set D

10. Follow the directions in Problem 9 for set E.
11. Show that every tree (with at least one edge) has at least two vertices of degree 1 (i.e., $d_1 \geq 2$).
12.* Draw all possible different trees connecting the points in set F. (What should "different" mean?)
13.* Draw all possible different trees connecting the points in set G.
14.* Show that the formula $1 \cdot d_3 + 2 \cdot d_4 + 3 \cdot d_5 + \cdots = d_1 - 2$ holds by convincing yourself of the following.
   (a) We can build any tree one step at a time, each step (after the first) consisting of adding one edge and one vertex.
   (b) The formula holds for the simplest tree, one edge with two vertices.
   (c) Each step does two things. It replaces a vertex of degree j by a vertex of degree $j + 1$, and it adds a vertex of degree 1. The result is simply to add 1 to both sides of the formula.

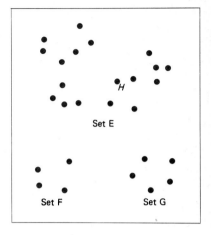

Set E

Set F

Set G

15.* Show that the rules given in the text for producing an economy tree actually work by filling in the details of the following argument.
   (a) Let E be a tree constructed according to the rules and suppose there is a tree of shorter length reaching all the given vertices. Let F be a shortest such tree.
   (b) Let $E_1, E_2, E_3, \ldots, E_n$ be the edges of E listed in the order in which they were adjoined to make E. There is a first edge, say E_i, in E that is not in F.
   (c) Adjoin E_i to F. The result, $F \cup E_i$, must have a circuit, since there are two paths connecting the vertices joined by E_i. This circuit must have an edge, say E_j, not in E.
   (d) Create a new tree G from F by removing E_j and replacing it with E_i:
   $$G = (F - E_j) \cup E_i$$
   (e) If L denotes length,
   $$L(G) = L(F) - L(E_j) + L(E_i)$$
   Since $L(F)$ is as small as is possible for a tree, $L(E_i) \geq L(E_j)$.
   (f) But E_i was (according to the rules of construction) the shortest edge that could be connected to $E_1, E_2, \ldots, E_{i-1}$ without making a circuit. Since E_j can also be joined to $E_1, E_2, \ldots, E_{i-1}$ without making a circuit, $L(E_j) \geq L(E_i)$ and so $L(E_j) = L(E_i)$. Thus G and F have the same length.
   (g) Note that G has one more edge, namely, E_i, in common with E than did F. Now repeat the above operations on G. We eventually obtain a tree coinciding with E, which has the same cost as F, contrary to our supposition.
16.* A, B, C, and D are the four corners of a square of side length 1 unit. A tree of minimum length is to be constructed connecting these four points but for this problem we will allow the introduction of new vertices. For example, we will allow the tree shown in the margin which has length $2\sqrt{2} \approx 2.8284$. It can be shown that the shortest tree connecting A, B, C, and D (allowing new vertices) has length $1 + \sqrt{3} \approx 2.7321$. Try to find the construction that leads to this answer.

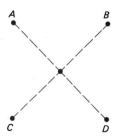

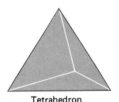

Tetrahedron

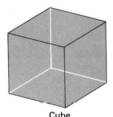

Cube

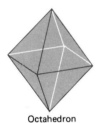

Octahedron

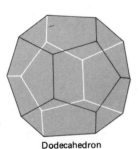

Dodecahedron

Icosahedron

The lithograph above serves to introduce our readers to the remarkable work of the late Dutch artist, M. C. Escher. Though he had no formal training in mathematics, Escher's engravings and lithographs reflect strong mathematical themes. The interlocking alligators appearing in the open book form a mosaic, the subject of Section 9.4. Here we are interested in the beautiful symmetric block with pentagonal faces all alike where the alligator reaches the high point of his existence and snorts with triumph. It is one of the Platonic solids. How many such regular solids are there?

9.3 / The Platonic Solids

We are about to study a topic which at first glance may seem to have little to do with networks; but appearances are deceptive, as we shall soon see. The Greeks discovered the five regular solids shown in the margin and attached a certain mystical significance to them. Why, they asked, could they find only five of these solids with identical regular polygons as faces and the same configuration at the corners. Both they and those who followed found ingenious philosophical and religious reasons for this strange fact of nature. Kepler, the great sixteenth-century astronomer, thought the five Platonic bodies provided an explanation for the orbits of the six planets then known (see next page). We leave such speculations to the philosophers; our aim is to demonstrate in a rigorous mathematical way that only five regular solids exist.

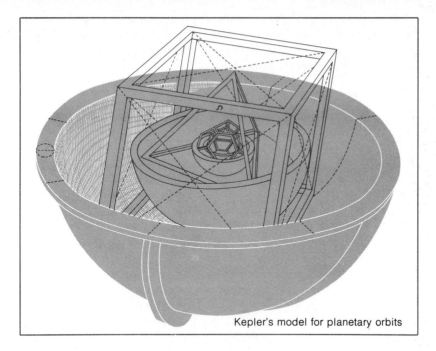

Kepler's model for planetary orbits

Triangular prism

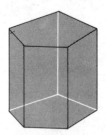

Pentagonal prism

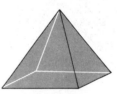

Square pyramid

Pentagonal pyramid

EULER'S FORMULA FOR POLYHEDRA

A **polygon** is a planar figure without holes, which is bounded by line segments (e.g., a triangle, a quadrilateral, etc.). A **polyhedron** is a solid figure, all of whose faces are polygons. Examples are the five Platonic solids, but there are hosts of others. A few are shown in the margin. We are primarily interested in convex polyhedra, that is, those without holes or dents. (In technical language, a figure is convex if any two of its points can be joined by a line segment within the figure.) Each convex polyhedron has a certain number of faces F, vertices V, and edges E. Is there a relationship between these three numbers that is always valid? Let's make a table for the solids we have pictured and see if any pattern appears.

TABLE 9-1			
Polyhedron	F	V	E
Triangular prism	5	6	9
Pentagonal prism	7	10	15
Square pyramid	5	5	8
Pentagonal pyramid	6	6	10
Roofed box	9	9	16
Tetrahedron	4	4	6
Cube	6	8	12
Octahedron	8	6	12
Dodecahedron	12	20	30
Icosahedron	20	12	30

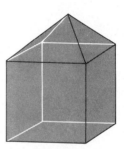

Roofed box

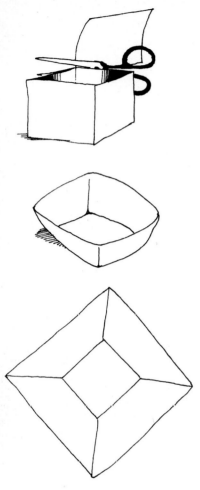

We dare say that most readers will see little regularity in the numbers they observe in the table. But Euler saw a relationship (some people claim that Descartes had noticed it even earlier); today we call it **Euler's Formula:**

$$F + V = E + 2$$

We urge the reader to check it for the convex polyhedra listed in the table and then on any others that can be imagined.

We have accumulated some evidence, but we'd prefer a proof. Take any convex polyhedron and suppose that its insides have been removed, leaving only its outer surface which we imagine to be made of thin, flexible rubber. Cut out one face and then stretch out the part that's left so that it lies flat in the plane (see marginal illustration for the case of a cube). This transforms the surface of the polyhedron into a network in the plane and in the spirit of Section 1.4, transforms our problem as well. Imagine the region around the outside of the network as corresponding to the removed face. Then the network will have exactly the same number of faces, vertices, and edges as the polyhedron. If we can prove Euler's Formula for a network in the plane, we will get it free for convex polyhedra.

EULER'S FORMULA FOR NETWORKS

Consider an arbitrary network in the plane with F faces (remember that the region outside the network is counted as a face), V vertices, and E edges. We claim

$$F + V = E + 2$$

To prove this, note that any network can be built starting with a single edge, which certainly satisfies Euler's Formula ($F = 1$, $V = 2$, $E = 1$), and then adding one edge at a time. At each subsequent stage, one of two things happens. The new edge either joins two existing vertices (thereby adding 1 to each of F and E but leaving V fixed), or it joins a new vertex to an old one (thereby adding 1 to each of V and E but leaving F fixed). In either case, Euler's Formula is preserved; we have simply added 1 to each side.

A case in which flat-ery gets you somewhere

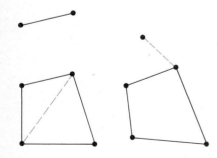

REGULAR POLYHEDRA

A polygon is regular if it has equal sides and equal angles. Examples are equilateral triangles, squares, regular pentagons, etc. A convex polyhedron is called **regular** if

1. Each face is a regular p-gon (a polygon with p edges).

2. Each vertex is surrounded by the same number, say q, of these faces (equivalently, q edges meet at each vertex).

Regular convex polyhedra are also called Platonic solids. There are only five of them. Let's see why.

Notice that each edge is shared by two faces. Thus pF counts each edge twice; i.e.,

$$pF = 2E \quad \text{or} \quad \frac{F}{E} = \frac{2}{p}$$

However, q edges meet at each vertex. So qV counts the number of edge ends, which again is twice the number of edges.

$$qV = 2E \quad \text{or} \quad \frac{V}{E} = \frac{2}{q}$$

Now take Euler's Formula, divide by E, substitute the above results, and do a bit of algebra, and another remarkable formula appears.

$$F + V = E + 2$$

$$\frac{F}{E} + \frac{V}{E} = 1 + \frac{2}{E}$$

$$\frac{2}{p} + \frac{2}{q} = 1 + \frac{2}{E}$$

$$\frac{2}{p} + \frac{2}{q} > 1$$

$$\frac{1}{p} + \frac{1}{q} > \frac{1}{2}$$

What integers p and q can possibly satisfy this last formula? Keep in mind that both p and q must be at least 3. Observe that, if $p \geq 6$, then $q < 3$, which we have just ruled out. We quickly conclude that both p and q are between 3 and 5. On checking we find exactly five solutions, and each of them corresponds to a Platonic solid. The possible values for p and q and then for E, F, and V are shown below.

TABLE 9-2					
Solid	p	q	E	F	V
Tetrahedron	3	3	6	4	4
Cube	4	3	12	6	8
Octahedron	3	4	12	8	6
Dodecahedron	5	3	30	12	20
Icosahedron	3	5	30	20	12

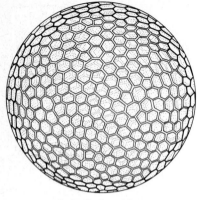

Aulonia hexagona

Actually we have proved considerably more than we have claimed. Nowhere did we use the assumption that the faces are regular, but only that they have the same number of edges. We have proved in fact that any convex polyhedron with p-gons as faces, q of them meeting at each vertex, fits into one of five classes. A class is determined by one of the five possible values for p and q. The skeleton of a radiolarian, called *Aulonia hexagona*, appears to defy this law. It seems to be made of a huge number of hexagons ($p = 6$, $q = 3$). But a close check reveals that some of the faces are pentagons, this being the only way that Euler's Formula can survive.

SUMMARY

Certainly the most important result in this section is Euler's Formula, $F + V = E + 2$, which relates the number of faces, vertices, and edges of a convex polyhedron or of a planar network. It is so simple that anyone can understand it; yet its consequences are far-reaching. We saw one in this section: There are only five Platonic solids. We will see another in Section 9.4.

PROBLEM SET 9.3

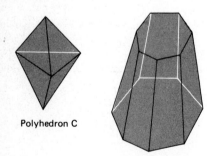

Network A

Network B

1. Calculate F, V, and E for network A shown in the margin. Verify Euler's Formula. Remember that the region outside the network counts as a face.
2. Follow the directions in Problem 1 for network B.
3. Calculate F, V, and E for polyhedron C and verify Euler's Formula. Note that polyhedron C, which is formed by pasting two regular tetrahedra together, has six equilateral triangles as faces; yet it is not a regular solid. Why?
4. Calculate F, V, and E for polyhedron D and verify Euler's Formula.
5. If a convex polyhedron has five faces and six vertices, how many edges does it have? Sketch such a polyhedron.
6. Is there a convex polyhedron with 8 vertices, 5 faces, and 13 edges? Justify your answer.
7. Show that a convex polyhedron cannot have the same number of vertices as edges.
8. Show that, if a convex polyhedron has an even number of vertices and faces, it also has an even number of edges.
9. Which of the five regular polyhedra have Euler circuits (see Section 8.1)? Hint: Draw the corresponding networks in the plane. For example, the dodecahedron network is shown at right.

Polyhedron C

Polyhedron D

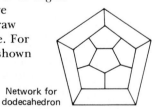

Network for dodecahedron

10. Which of the five regular polyhedra have Hamiltonian circuits (see Section 8.1)? Follow the hint in Problem 9.

11. Calculate F, V, and E for the nonconvex polyhedron E. Does Euler's Formula hold?

12. Does Euler's Formula hold for polyhedron F which has a hole through it?

13.* Give an argument showing that a nonconvex polyhedron with dents but no holes still satisfies Euler's Formula (see Problem 11).

14.* Draw several polyhedra, each having one hole (see Problem 12). Make sure all their faces are polygons. Calculate $F + V - E$ for each of them. Make a conjecture.

15.* Draw several polyhedra, each having two holes. Calculate $F + V - E$ and make a conjecture.

16.* Conjecture a formula for $F + V - E$ for a polyhedron with n holes.

17.* Without cutting it into more than one piece, determine how to fold and cut a piece of construction paper to make models of a tetrahedron and an octahedron (see below for a cube). Make these models.

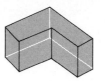

Polyhedron E

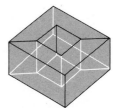

Polyhedron F

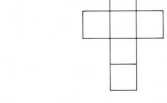

18.* Follow the directions in Problem 17 for a dodecahedron and an icosahedron.

19.* Show that every convex polyhedron satisfies $V \geq 2 + \frac{1}{2}F$.

20.* Show that every convex polyhedron satisfies $F \geq 2 + \frac{1}{2}V$.

21.* Imagine a cube made of clear plastic sheets. Suppose that a hole is drilled at the center of each face and stiff wires are inserted through these holes. The outline of another regular solid, an octahedron, is obtained (see marginal illustration). A similar phenomenon occurs if we start with any of the other regular solids. Figure out exactly what happens in each case. Hint: The centers of the faces of the original solid become the vertices of the new solid.

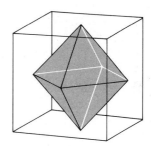

22.* By truncating (slicing off) the corners of a cube we obtain a semi-regular polyhedron—a truncated cube. A convex polyhedron is semiregular if its faces are regular polygons of two or more types and if each vertex of the solid is surrounded by these polygons in the same way. There are exactly 14 semiregular polyhedra. See how many you can find.

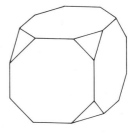

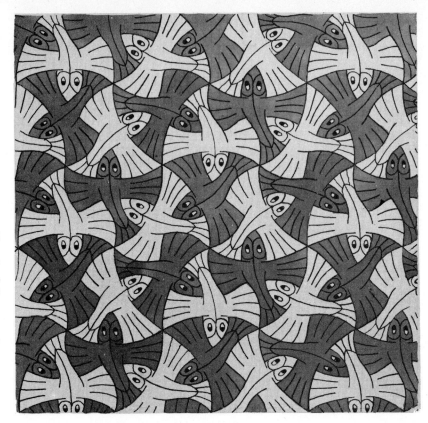

This is the richest source of inspiration I have ever struck; nor has it dried up yet. The symmetry drawings . . . . show how a surface can be regularly divided into, or filled up with, similar-shaped figures which are contiguous to one another, without leaving any open spaces. The Moors were past masters at this. They decorated walls and floors, particularly in the Alhambra in Spain, by placing congruent, multi-coloured pieces of majolica together without leaving any spaces between.
M. C. Escher

9.4 / Mosaics

Take a jigsaw puzzle piece (for example, one shaped like the bird above) and reproduce it *ad infinitum.* Then try to fit the resulting identical pieces together so they fill the whole plane without gaps or overlap. If you succeed, you will have made a pattern called a **mosaic** (or tessellation), and the basic piece you started with is said to **tile** the plane. The great Dutch artist, M. C. Escher, created many beautiful mosaics using a bird or an animal as the basic piece. Lest you think it is easy, try to create a mosaic of your own. You will soon discover that not just any old piece will work. In fact, this is the question we want to consider: What shape pieces will tile the plane?

REGULAR MOSAICS

Among all mosaics, certainly the simplest and in some ways the most elegant are those having regular polygons as faces. Three come immediately to mind: There is the familiar hexagonal pattern seen in honeycombs and on many bathroom floors; there is the pattern of squares that appears on chessboards and maps of city streets; there is the pattern of equilateral triangles occasionally found in American Indian arts and crafts. But there are no others. This will take some explaining.

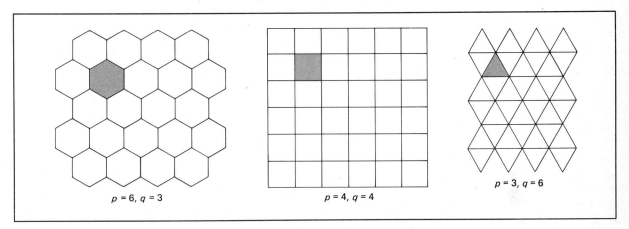

$p = 6, q = 3$ $p = 4, q = 4$ $p = 3, q = 6$

Recall that a regular polygon is a polygon with equal sides and equal angles. Now call a mosaic a **regular mosaic** if

1. Each face is a regular p-gon (a polygon with p sides).
2. Each vertex is surrounded by the same number, say q, of these faces (equivalently, q edges meet at each vertex).
3. Two faces abut only along a common edge.

Condition 3 may seem superficial. It is included to eliminate certain modifications of regular mosaics, which can be obtained by sliding a whole row of polygons one way or the other along a line (see margin). With this understanding, we assert that there are only three regular mosaics.

We are going to resurrect a fact from high school geometry. Almost everyone remembers that the angles of a triangle add up to 180°; some may remember that the angles of a quadrilateral (a four-sided polygon) add to 360° or twice 180°. The general result is that the angles of a p-gon add up to $(p - 2)180$ degrees, a result you can see by decomposing the polygon into triangles as indicated in the margin. Now if a p-gon is regular, so that all its angles are equal, each of them will measure

$$\frac{(p - 2)180°}{p}$$

Now in a regular mosaic, q of these angles surround each vertex. This implies that each angle measures 360/q degrees. If

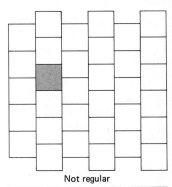

Not regular

Polygon	Sum of angles
	180°
	360°
	540°

we equate these two expressions and do a bit of algebra, an interesting formula appears (compare it with a similar formula in Section 9.3):

$$\frac{(p-2)\,180}{p} = \frac{(2)\,180}{q}$$

$$\frac{p-2}{p} = \frac{2}{q}$$

$$1 - \frac{2}{p} = \frac{2}{q}$$

$$1 = \frac{2}{p} + \frac{2}{q}$$

$$\boxed{\frac{1}{2} = \frac{1}{p} + \frac{1}{q}}$$

My house is constructed according to the laws of a most severe architecture; and Euclid himself could learn from studying the geometry of my cells.
The bee in Arabian Nights

Take a look at the last formula which we remember holds for any regular mosaic. What positive integers p and q can satisfy this equation? It takes only the least bit of experimenting to discover that there are only three solutions:

$$\begin{array}{ccc} p = 6 & p = 4 & p = 3 \\ q = 3 & q = 4 & q = 6 \end{array}$$

These correspond to the three regular mosaics we have already discussed. That's all there are.

For untold ages, bees have employed a hexagonal mosaic as the basic pattern in the construction of their honeycombs. This may be dictated by another fact about this particular arrangement of cells. Among all divisions of the plane into parts of equal area, this is the one for which the network of edges has the least length. The bee has learned to economize on the use of wax for cell partitions. However, it would take us too far afield to try to demonstrate this.

QUASIREGULAR MOSAICS

Let's relax our requirements. Call a mosaic **quasiregular** if

1. Each face is an identically shaped convex p-gon (convex means without dents or holes).
2. Two faces abut only along a common edge.

This allows for great variety but maybe not as much as you would expect. We now show that no convex polygon of more than six sides works.

First note that any triangle works. Simply take two copies of

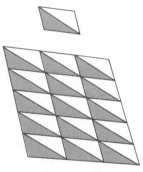

Triangular tiling

it, rotate one through 180° and fit them together as a parallelogram. The resulting parallelogram clearly tiles the plane. A similar argument applies to any convex quadrilateral. Fit two copies together, one flipped over, and the resulting hexagon tiles the plane (see Problem 11).

Things get more complicated and more interesting when we move to pentagons. Not all pentagons work; we already know that a regular pentagon fails. Several authors have at one time claimed to have found all convex pentagons that will tile the plane only to have some one else find a new example. In 1980, it was still not known whether the list was complete. For an interesting discussion of this problem, see Martin Gardner, Mathematical Games, *Scientific American,* July 1975, pp. 112–117, and December 1975, pp. 116–119, or D. Schattschneider, Tiling the Plane with Congruent Pentagons, *Math. Magazine* 51, 1978, pp. 29–44.

Finally, we mention that there are three different types of hexagons that can tile the plane. An example of a pentagonal and a hexagonal tiling appears below.

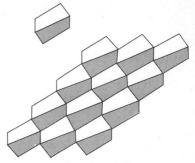

Quadrilateral tiling

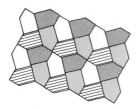

Pentagonal tiling

Hexagonal tiling

Why is it that a convex polygon that tiles the plane can have at most six sides? Once again, it is Euler's Formula that dictates this fact. Consider any mosaic formed by using copies of a convex p-gon. Look at some bounded portion of it, say, the part inside a big circle. Note that each edge is shared by two faces; so

$$pF = 2E \qquad \text{or} \qquad \frac{F}{E} = \frac{2}{p}$$

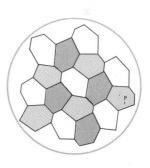

Furthermore, at least three edges enter each vertex; that is, there are at least $3V$ edge ends. Since each edge has two ends, we see that

$$3V \le 2E \qquad \text{or} \qquad \frac{V}{E} \le \frac{2}{3}$$

Oh, we admit to having cheated slightly. Our analysis is not quite correct for the part around the outer boundary. But what happens on the boundary becomes less and less significant as the circle gets bigger and bigger; and the shaded formulas get better and better. Now proceed as we did in Section 9.3. Take Euler's Formula, divide by E, and substitute the above results:

$$E + 2 = F + V$$

$$1 + \frac{2}{E} = \frac{F}{E} + \frac{V}{E}$$

$$1 + \frac{2}{E} \leq \frac{2}{p} + \frac{2}{3}$$

$$\frac{1}{3} + \frac{2}{E} \leq \frac{2}{p}$$

Note that, as the circle gets bigger, E gets larger and $2/E$ fades to 0. We are left with the inequality

$$\frac{1}{3} \leq \frac{2}{p}$$

which implies that $p \leq 6$.

SUMMARY

The repetitive pattern one obtains when tiling a floor with identically shaped pieces is called a mosaic. While great variety is possible, the laws of mathematics, especially Euler's Formula, $F + V - E + 2$, impose surprising restrictions. There are, for example, only three possible regular mosaics. And no mosaic can be composed of convex polygons with more than six edges.

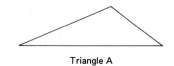

Triangle A

PROBLEM SET 9.4

1. Make a quasiregular mosaic using triangle A.
2. Make a quasiregular mosaic using quadrilateral B.
3. Make a quasiregular mosaic using quadrilateral C.

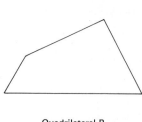

Quadrilateral B

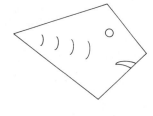

Quadrilateral C

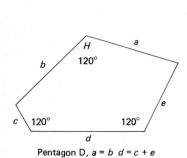

Pentagon D, $a = b$ $d = c + e$

4. Make a quasiregular mosaic using pentagon D. Hint: Fit three pentagons around point H and then reproduce the resulting figure.

5. Make a quasiregular mosaic using pentagon E. Hint: Fit six pentagons around point *H* and then reproduce the resulting figure.

6. Make a quasiregular mosaic using hexagon F. Hint: Fit two hexagons together and then reproduce the resulting figure.

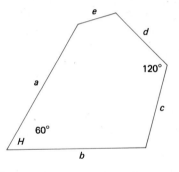

Pentagon E, *a* = *b* *c* = *d*

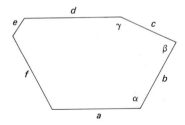

Hexagon F, $\alpha + \beta + \gamma = 360°$
a = *d*

7. Two identical squares joined to have a common edge form a domino. Three squares joined together form a tromino, and they come in 2 shapes. Pentominoes, composed of five squares, come in 12 different shapes, 4 of which are shown below. We have illustrated a tiling of the plane using one of them. Show how the plane can be tiled with each of the others. Suggestion: Use graph paper.

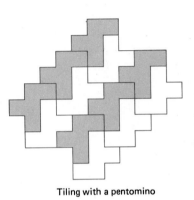

Tiling with a pentomino

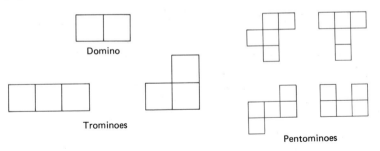

Domino

Trominoes

Pentominoes

8. Find some hexaminoes (six squares) that can tile the plane.
9. Show how to tile the plane with the heptomino shown at right.
10. Show how to tile the plane with the heptomino shown at far right.
11* Show that any convex quadrilateral can tile the plane. Hint: Fit four copies together around a vertex as shown below. Note that the angles add up to 360°. This can be done at each vertex.

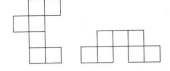

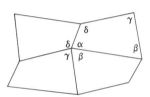

12* Show that the plane can be tiled with the nonconvex quadrilateral shown in the margin. Does any nonconvex quadrilateral work?

A nonconvex quadrilateral

13* A mosaic is called semiregular if it is composed of regular polygons of two or more types so that each vertex is surrounded by these polygons in the same way. It can be shown that there are only eight semiregular mosaics. One is shown below. See how many others you can find.

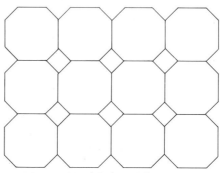

A semiregular mosaic

14* Many beautiful mosaics (including the one by Escher shown at the beginning of this section) can be created using an equilateral triangle as the basic starting piece. The steps are illustrated in the margin.

a. Draw a curve along one side of the triangle.
b. Reproduce this curve along a second side but in a reflected position.
c. Draw a curve along one-half of the third side.
d. Reflect this curve in the midpoint of the third side.

Make a mosaic of your own following this procedure.

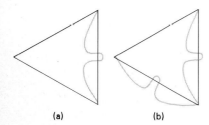

(a)   (b)

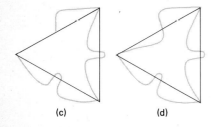

(c)   (d)

9.5 / Map Coloring

What do you say? Is there a map that requires five colors in order that regions with a common boundary will be differently colored? This innocent-sounding question, posed by an obscure student, defied the best efforts of the world's best mathematicians for well over 100 years. As late as 1975, the complete solution seemed farther away than it must have to DeMorgan in 1852. Oh, we admit that progress had been made. For instance, it had been shown that five colors would always suffice, and that any map with less than 39 countries (or compartments or regions) could be colored with four colors. But the answer to the question, Can any map whatever be colored with four or less colors? seemed as elusive as ever. In late 1976, K. Appel and W. Haken reported [Every Planar Map is Four Colorable, *Bulletin Amer. Mathematical Society*, Vol. 82 (1976), 711–712] that they had demonstrated an affirmative answer by making extensive use of a large computer.

Three observations are in order. (1) A good teacher listens to a good question. (2) There are easily stated mathematical problems that even the best mathematicians find terribly difficult. (3) It pays to study a hard problem even if you don't solve it. It is to the last-mentioned point that we turn our attention.

A conversation between Huck Finn and Tom Sawyer in their flying boat:
"We're right over Illinois yet. And you can see for yourself that Indiana ain't in sight. . . . Illinois is green, Indiana is pink. You show me any pink down there, if you can. No sir; it's green."
"Indiana pink? Why, what a lie!"
"It ain't no lie; I've seen it on the map, and it's pink."
Mark Twain

A student of mine asked me today to give him a reason for a fact which I did not know was a fact, and do not yet. He says that if a figure be anyhow divided, and the compartments differently coloured, so that figures with any portion of common boundary line are differently coloured — four colours may be wanted, but no more. . . .

What do you say? And has it, if true, been noticed? My pupil says he guessed it in colouring a map of England. The more I think of it, the more evident it seems.
Augustus DeMorgan to W. R. Hamilton, October 23, 1852

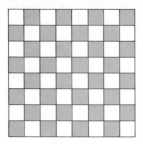

Chessboard

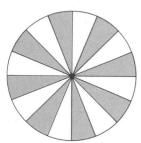

Dart board 1

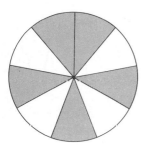

Dart board 2

TWO-COLOR MAPS

If the four-color problem is too hard for us, maybe we can solve a simpler one. What kind of maps can be colored with one color? Clearly, maps with just one country. That was easy enough. Well, what kinds of maps can be colored with just two colors? Think of some two-color maps with which you are familiar, such as a chessboard or a dart board of the kind pictured. What is special about them?

The dart board is a good example to think about. Not every dart board can be colored with two colors. Try as you will, you can never make two colors do for the second dart board. That's because the colors must alternate as you go around a vertex; and that's possible only if that vertex is of even degree.

We must stop and point out a temporary assumption being made. Each of our maps (the chessboard and the two dart boards) are to be thought of as islands imbedded in an infinite ocean requiring no color. With this in mind, call any vertex not on the shore of the island a dry vertex. Then one conclusion is already obvious. If the regions of an island can be colored with two colors, then each dry vertex is of even degree.

The truly remarkable fact is that the converse also holds. If each dry vertex is of even degree, then the regions of the island can be colored using two colors.

Any proof of the latter statement requires some ingenuity. Ours is borrowed from Sherman Stein, *Mathematics, the Man-made Universe,* 3rd ed., Freeman, San Francisco, 1976, pp. 336–340. Consider an arbitrary island with dry vertices of even degree, for example, one like the island shown below. Stick a tack somewhere in the center of each region, making sure that the direct line between any two tacks misses all the vertices. Pick two of the tacks; label one London and the other Bath.

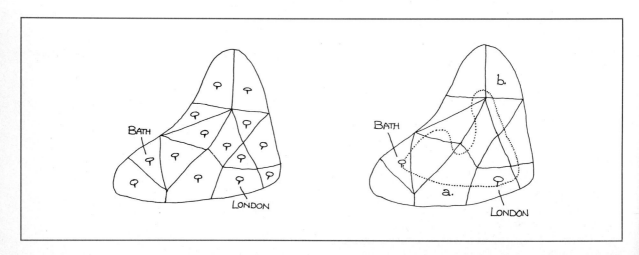

Now any traveler from London to Bath has her choice of many different paths. She can take the direct route we have labeled *a*; she can take a meandering path such as *b*. The astute traveler notices a very strange fact. If the direct path has an odd number of border crossings (as it does in our example), so does every other path. If it has an even number of border crossings, so does every other path. Why is this so?

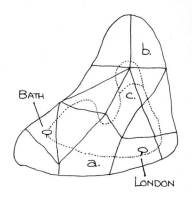

Imagine any meandering path from London to Bath to be an elastic string which is steadily contracting toward the direct path. As it passes through a vertex (moves from position *b* to position *c*), it loses some border crossings and perhaps picks up some brand new ones. The number lost plus the number gained is the degree of the vertex, an even number. Now, when the sum of two numbers is even, so is their difference. Thus the change in the number of crossings is even.

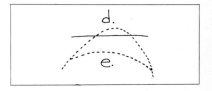

The same is true when the string frees itself of a border, as in going from position *d* to position *e*. The change is even. It follows that the *parity* (oddness or evenness) of the path is unchanged as it shrinks. We conclude that any path from London to Bath has an odd or even number of border crossings according as the most direct path does so. And keep in mind that Bath could have been located in any of the regions.

That was the tricky part. Now we need only to give our imaginary traveler some red and blue paint and some instructions. Tell her to paint London and its region red. Tell her to change colors whenever she crosses a border. Tell her to keep going until she has painted every region. Our argument shows that no matter what path she takes she can't possibly get mixed up.

That takes care of the island. What if we want to color the ocean too? We won't get by with two colors in the example just studied but, if all the wet vertices are of even degree, we will have no trouble. We just treat the ocean like another country and forget the distinction between dry and wet vertices. Here is the general two-color theorem:

Theorem 1. Suppose that a network partitions the plane (or the surface of a sphere) into two or more regions. The resulting map can be colored in two colors if and only if each vertex is of even degree.

We have glossed over one little point. What is a region (or country)? We wish to rule out certain kinds of countries, for example, those shaped like a figure eight or even those like the United States which consists of three separate parts. To include such countries complicates the coloring problem still further. To avoid these difficulties, we insist on talking about regions that can be thought of as having a single loop of string as their boundary, a loop that neither meets nor crosses itself.

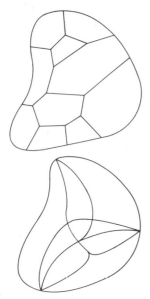

THREE-COLOR MAPS

What kind of maps can be colored with just three colors? No one knows a very good answer to this question. But there are two partial results worth stating.

Theorem 2. If the plane (or surface of a sphere) is partitioned into regions, each with an even number of edges, and if each vertex is of degree 3, the resulting map can be colored with exactly three colors.

Theorem 3. If the plane (or surface of a sphere) is partitioned into at least five regions, each sharing its borders with exactly three neighboring regions, the resulting map can be colored in three colors.

COLORING SPHERES AND DOUGHNUTS

We have already hinted that coloring spheres and coloring planes are equivalent problems. To see why, take a sphere and place it on a plane so that it rests on its south pole. Poke a long straight hatpin through its north pole and on through point P until you hit the plane at P'. This process, since it can be done for any point P, transfers a map on the sphere to a map on the plane. Conversely, it can be used to transfer a map of the plane back onto the sphere. It follows that, if you can color one, you can color the other.

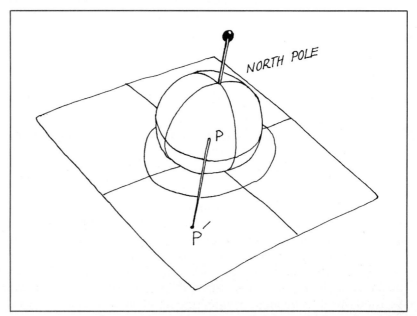

The surface of an inner tube (or doughnut) is an entirely different matter. Take a rectangular sheet of rubber and color it with seven colors as indicated. Then roll it up into a cylinder (like the paper on a cigarette). Finally, bring the two ends around and stick them together, forming a surface like the inner tube in a bicycle tire. Note that each of the seven regions touches the other six. It takes seven colors to color them. Now the clincher. Mathematicians have shown that any map whatever on a doughnut can be colored with seven colors.

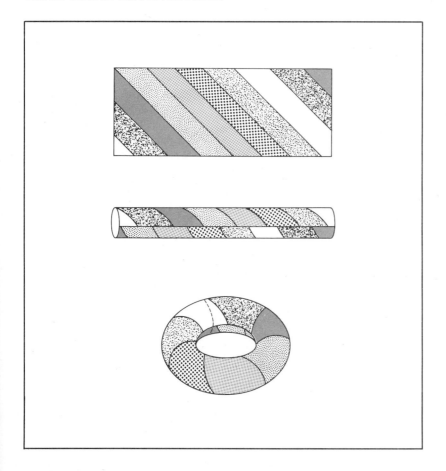

SUMMARY

In 1852, a student posed a simple-sounding question which has given mathematicians fits for over 100 years. Can every map on a sphere be colored with four or fewer colors? As late as 1975, no one knew the answer to this question, though an affirmative answer was announced by two mathematicians in 1976. We showed that a map can be colored with two colors if and only if all its vertices are of even degree. And to top that, we stated that the general coloring problem for a doughnut is solved. It may take up to seven colors, but never more than that, to color a doughnut.

PROBLEM SET 9.5

1. Color each of the following islands using the smallest number of colors.

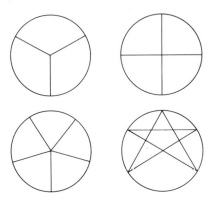

2. Color each of the following islands using the smallest number of colors.

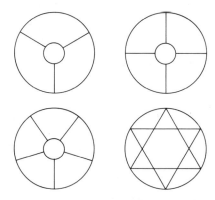

3. Color map A with the fewest number of colors.
4. Color map B with the fewest number of colors.

 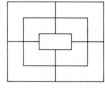

Map A   Map B

5. Draw a map of your own that illustrates Theorem 2.

6. Draw a map of your own that illustrates Theorem 3.

7. A map covering a sphere has 50 regions, each with five edges. How many vertices does it have? Hint: Euler's Formula should help.

8. Fifty dots are placed on a sphere. Homer and Horace are each asked to draw maps so that each region has three edges, using the dots as vertices. They inevitably wind up with the same number of regions. Why?

9. What is the minimum number of colors required to color a
   (a) Tetrahedron?
   (b) Cube?
   (c) Octahedron?
   (d) Dodecahedron?
   (e) Icosahedron?
   See Section 9.3 for pictures of these solids.

10.* What is the least number of colors required for a map of the United States if no adjoining states get the same color? Color a map of the United States to make sure.

11.* Show that any map on a sphere must have at least one region with five or fewer edges.

12.* In how many essentially different ways can a cube be colored with three given colors?

Part III
Reasoning and Modeling

REASONING AND MODELING

A scientist has been compared to a person trying to understand the mechanism of a watch from observations he can make without removing the case. He naturally looks around for some kind of model that can be used to describe what makes the thing tick. The difference is of course that a watch is a human creation, and so the possibility of duplicating it actually exists. But most scientists do not expect to duplicate the world of their experience, so they rely on being able to form models.

The models of which we speak are not made with plastic and balsa wood. Rather, they consist of ideas. They are assumptions, sometimes (misleadingly) called laws, which together with their consequences seem to explain some phenomena we observe in the world. The entire enterprise, dare we admit it, is an activity of the mind in which one begins not with something proved beyond doubt, but with something tentatively set forth as a hypothesis. Proof enters the picture only when we ask what certain conclusions must follow if the hypotheses are accepted. Then, armed with these conclusions, we go back to the world of our experience, asking whether or not the conclusions proved are consistent with the facts observed. If they are, the assumptions together with their implications become our model.

It is this view of science that explains not only why mathematics has played a central role in the natural sciences, but why it is playing an increasingly important role in the social sciences. When reduced to its logical foundations, mathematics is a subject that clearly states its assumptions (axioms) and, without asking whether they are true or false, sets about deducing the consequences (theorems). This is the axiomatic method. It lies at the heart of modeling.

Albert Einstein (1879–1955)

If asked to name a famous scientist, many if not most people may well respond with Einstein's name. Born in Germany to a family owing its circumstances to modest beginnings and the happy-go-lucky character of his father, Albert showed little early evidence of what he was to become. A teacher, when asked what profession Albert should adopt, reportedly answered, "It doesn't matter; he'll never make a success of anything." Unable to enter a German university because he never finished his education at the gymnasium (high school) he went to Zürich to attend the Swiss Federal Institute of Technology. Thus began the chain of events in which wars, his Jewish background and Zionist convictions, and his professional affiliations combined to make him in one sense an international citizen and, in another, a man without a country. On graduation from the institute, he needed the influence of friends to obtain a post as Technical Expert, Third Class (he applied for Second Class, but lacked qualifications) in the Swiss Patent Office. It was in this position, not from a post in a major research center, that he used the tools of mathematical reasoning, not the instruments of a scientific laboratory, to fashion a model that was to change the way in which we understand our universe.

Chapter 10
Methods of Proof

Considering that among all those who have previously sought truth in the sciences, mathematicians alone have been able to find some demonstrations, some certain and evident reasons, I had no doubt that I should begin where they did, although I expected no advantage except to accustom my mind to work with truths and not to be satisfied with bad reasoning.

RENÉ DESCARTES

10.1 / Evidence but Not Proof

We seldom think about the processes by which we come to believe something, and some people may be happier not thinking about them. We believe, however, that it is useful to examine a few of these processes so that we can distinguish between evidence for believing something and an argument that constitutes a "proof."

AUTHORITY

We believe many things, whether or not we recognize it, because we learned them from a source we regard as authoritive. To be sure, we are selective in deciding what constitutes an authoritive source, and this decision varies with topic and time. A child believes his parents when they tell him about their family tree, but he is likely to check with his friends any information his parents give on how new twigs come into being.

As we grow older, our sources of information increase. We choose and then listen to certain people we believe (or fervently hope) to be competent: a dentist, a lawyer, a mechanic, perhaps (if we may say so) a teacher. We turn to the written word and, after making the proper disclaimers about not believing everything in print, we decide which book on gardening, which history book, and which newspaper or magazine we will quote as the gospel truth. It is surprising, sometimes even alarming, to realize how many things we believe on the authority of a spoken or written word.

No proposition should ever be considered as proved, however, just because some authority says it is so. Never! This is easy to illustrate in mathematics, and we cite an example. Pierre de Fermat, a mathematician with a towering and deserved reputation, once wrote that he could prove the following.

Fermat's Last Theorem. For $n > 2$, there are no three positive integers x, y, and z such that

$$x^n + y^n = z^n.$$

(The reader will recall, or easily verify, that for $n = 2$, there are integer solutions. For example, $x = 3$, $y = 4$, $z = 5$ is a solution since $3^2 + 4^2 = 5^2$; try $x = 5$, $y = 12$, $z = 13$.)

Fermat did not, however, prove the theorem; neither have succeeding generations of very clever people; so mathematicians do not accept this theorem as proved. Of course, they don't claim it's false either. It stands today as it has stood for more than 300 years, waiting to be proved or disproved.

INDUCTION

It is no problem in elementary science classes to convince youngsters that dark clouds carry water. They already believe it; they have probably been baptized into faith by being sprinkled. This process of drawing conclusions on the basis of many observations is called **induction.** We expect, consciously or unconsciously, that what has happened before will happen again. If it doesn't, we feel betrayed.

Once again, however, we find in mathematics some forceful reminders that induction does not constitute proof.

Mathematicians tenaciously cling to the principle that no number of examples suffices to prove an assertion. Again we draw attention to Fermat's Last Theorem. For years mathematicians have struggled to prove or disprove this result. Following the advice offered in Section 1.3, they have worked on special cases, considering particular values of n. Their combined efforts, augmented in recent years by computers, have established that the theorem is true for all *n* through 600, and for every prime *n* less than 125,000. Yet, lacking a general proof for arbitrary *n*, it is universally held by mathematicians that Fermat's Last Theorem is still unproved.

EXPERIMENTATION

Sometimes, as in the controlled procedures of the laboratory or in the informal testing of a new product in our home, we deliberately

---

How Many Are Enough?

An assertion is not to be believed simply because one observation seems to confirm it. Neither is it to be believed on the basis of two or three examples. Well, okay. How about 1500 examples? 600,000 examples? Consider the following. Let us call an integer $k > 1$ an even type if, when factored into primes, the number of prime factors is even. (A prime is a whole number greater than 1 with no factors other than itself and 1.) And call it an odd type if the number of prime factors is odd. Examples: $4 = 2 \cdot 2$, $6 = 3 \cdot 2$, and $9 = 3 \cdot 3$ are even types; $8 = 2 \cdot 2 \cdot 2$, $12 = 2 \cdot 2 \cdot 3$, and $18 = 3 \cdot 3 \cdot 2$ are odd types; any prime number is an odd type.

Considering all the numbers from 2 through 12, we find

Odd types: 2, 3, 5, 7, 8, 11, 12
Even types: 4, 6, 9, 10

Counting the odd types and the even types, we find that through 12 the odds lead the evens seven to four. The mathematician George Polya found in 1919 that, for all *k* up to 1500, the odds led the evens. He then made the following conjecture.

For any integer *k*, the number of integers from 2 through *k* of odd type exceeds the number of integers from 2 through *k* of even type. Polya knew it was true for $k = 1, 2, \ldots, 1500$. It was later proved for all *k* through 600,000. But in 1960, R. S. Lehman showed that, for $k = 906,180,359$, from 2 through *k* there are exactly as many integers of even type as there are of odd type.

This is a remarkable example. An assertion, known to be true in 600,000 cases (and probably more), fails to be true in general. Mark it! No particular number of examples proves an assertion.

Is this peculiar to mathematics? Are there areas of human endeavor in which a given number of examples constitutes final proof? What things do you believe on the basis of a certain number of examples?

put a proposition to the test. At other times, as when we hit a nonfunctioning television set — only to discover that it then works — we come upon learning experiences quite by chance. But however it happens, much of what we believe is the result of something we have tried.

Learning from an experiment is of course just a form of induction in which we substitute for many observations a few test runs we hope can be repeated. As such, the experiment shares the limits of any inductive process as a means of proof. But there is an additional problem with experimentation, which is once again conveniently illustrated in mathematics. Consider the Theorem of Pythagoras, which says that, in every right triangle,

$$x^2 + y^2 = z^2$$

Suppose we wish to verify this experimentally in the case of the triangle pictured. The sides, measured in centimeters, are

$$x = 2.7 \qquad y = 1.8 \qquad z = 3.2$$

Now $x^2 + y^2 = (2.7)^2 + (1.8)^2 = 10.53$ and $z^2 = 10.24$. Does $x^2 + y^2 = z^2$? How much deviation do you allow for experimental and/or roundoff error? This is of course a problem encountered in all quantitative sciences. In mathematics, the answer is easy. Mathematical propositions are not established by physical experiments.

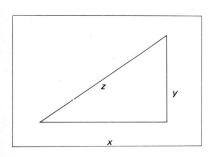

Can We Forget about Small Errors?

In 1840, a young French astronomer named Urban Jean Leverrier showed that the observed path of the planet Mercury deviated from the path predicted by Newton's theory. The deviation amounted to about 42 inches each century! The layman may wonder, in view of the astronomical distances involved, how such a deviation could be detected, much less worried about. Nevertheless, astronomers did worry about it. Several possible explanations were offered, all of them so complicated as to seem unlikely. Finally, in 1915, Einstein demonstrated that his new theory offered an explanation. This demonstration was in fact one of the few actual pieces of experimental evidence Einstein was able to cite as support for his theory of relativity.

SUMMARY

We have examined several of the rationales we commonly offer as evidence for believing something: authority, induction, and experimentation. We have illustrated each of these with examples of things people commonly learn in these ways, and we have shown, using examples from mathematics, that none of these methods can be used as proofs.

Ideas from mathematics that were used included Fermat's Last Theorem, Polya's classification of numbers into even and odd types, and the Theorem of Pythagoras.

SELF-TEST

Choose the completion that seems best to you. If none of them seem quite correct to you, be prepared to tell why and to supply an ending of your own in class discussion.

1. Experimentation
   (a) has no place in mathematics.
   (b) never really proves anything, in mathematics or anywhere else, because it is really a form of induction.
   (c) is the only certain way to prove anything.
2. Polya's classification of the integers into even and odd types was mentioned
   (a) because it illustrates how much a great mathematician can see in so familiar an idea as the classification of integers into evens and odds.
   (b) because it cautions us not to believe that something is sure to happen every time just because it happened the first 600,000 times we tried it.
   (c) because it illustrates that even in mathematics, people tend to accept a thing as true if a well-known mathematician says he thinks it is true.
3. Fermat's Last Theorem
   (a) would be considered proved by almost anyone (except for a professional mathematician) since it has been shown to be true for every one of the more than 600 cases that have been investigated.
   (b) cannot be verified experimentally because, as is the case with the Theorem of Pythagoras, one cannot distinguish between a genuine error and an error that is introduced by rounding off decimals.
   (c) is an example of a theorem that cannot ever be proved for certain because it asserts that something is true for all numbers n, and we cannot possibly know that something will be true for all n.

PROBLEM SET 10.1

1. Make a list of four things you believe because
   (a) You read or heard them from a source you consider reliable (authority).
   (b) You have seen them so many times that you have come to expect them (induction).
   (c) You have personally put them to a test (experimentation).

2. Think of at least one example of something you once believed on the basis of authority, which you no longer believe. Do the same for induction and experimentation. Then identify the reasons that caused you to change your mind (different authority, a chance to make observations under new circumstances, contradictory experiences, etc.).

3. Do you believe or disbelieve the following statements? On what grounds?
   (a) George Washington and his troops camped at Valley Forge.
   (b) The moon gives off no light of its own.
   (c) When water freezes, it expands.
   (d) Everyone should be vaccinated against smallpox.
   (e) Air travel is safer than automobile travel.
   (f) Each January, there are some days when the temperature in St. Paul, Minnesota, drops below 0°F.

4. Do you believe or disbelieve the following statements? On what grounds?
   (a) Abraham Lincoln debated Stephen Douglas.
   (b) An American has stood on the moon.
   (c) When chilled, iron contracts.
   (d) Children should drink lots of milk.
   (e) Americans spend more money on commercially prepared dog food than they contribute to cancer research.
   (f) High humidity increases discomfort on a hot day.

5. Verify Polya's conjecture about numbers of odd type and even type for $k = 27$.

6. Let $k \geq 5$ be fixed. For each even number n less than k, there is another number m less than n such that m and n are of different type. Explain.

Mathematicians distinguish sharply between a conjecture (for which experimental evidence and induction are very much in vogue) and a proof (for which these methods are out of style). In Problems 7 through 10, you are given some examples from which you are to make a conjecture. How can your conjectures be proved or disproved?

7. (a)
$$\begin{aligned} 1 &= 1^2 \\ 1 + 2 + 1 &= 2^2 \\ 1 + 2 + 3 + 2 + 1 &= 3^2 \end{aligned}$$

   (b) We have, in the following array of numbers, circled the prime numbers

⑦	8	9	10	⑪	12
⑬	14	15	16	⑰	18
⑲	20	21	22	㉓	24

8. (a)
$$\begin{aligned} (1)(1) &= 1 \\ (11)(11) &= 121 \\ (111)(111) &= 12321 \end{aligned}$$

   (b) $2^2 - 2 = 2$, $3^2 - 3 = 6$, $4^2 - 4 = 12$, . . . are all divisible by 2.
   $2^3 - 2 = 6$, $3^3 - 3 = 24$, $4^3 - 4 = 60$, . . . are all divisible by 3.

9. (a) Draw three line segments Aa, Bb, and Cc so that they intersect at a common point. Then draw triangles ABC and abc, extending the sides as necessary to form the intersections of AB with ab, AC with ac, and BC with bc.

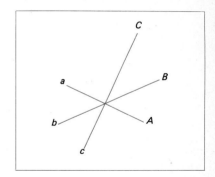

(b) The external tangents to two circles are the two tangent lines that do not cut the line segment connecting the centers. Draw three circles. Taking them in pairs, draw the three sets of external tangents, noting the three points of intersection thus determined.

10. (a) Locate points A, B, and C on one line, and points a, b, and c on another. Consider the intersections of Ab with aB, Ac with ab, AC with ac, and BC with bc.

(b) Draw a circle and circumscribe an arbitrary quadrilateral, labeling the points of tangency in clockwise order A, B, C, and D. Draw segments AC and BD. Draw the two diagonals of the quadrilateral.

11. Measure the sides of the right triangle below. How close can you come to verifying the Theorem of Pythagoras?

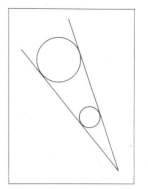

12. Measure the angles in the above right triangle. How close do you come to verifying the theorem that says that the sum of the interior angles is 180°?

13.* Draw a circle of radius 3 centimeters on onionskin paper. Lay it over a grid of 1-millimeter squares. By counting the squares included in the circle, see how close you come to verifying that the area $A = \pi r^2$. (Use $\pi = 3.1416$.)

14.* Lay a piece of string along the circumference of the circle drawn in Problem 13. Try to verify that the circumference $C = 2\pi r$.

15.* Summarize the reasons for rejecting authority, induction, and experience or experimentation as methods of proof in mathematics.

16.* In the history of some subject other than mathematics that you are now studying, look for examples of ideas once believed but no longer accepted. On what grounds were they believed? Why were they set aside?

The black and white hat game is played by seating contestants in a circle so they can see each other. A referee places on each head a hat which may be black or white. No player sees his own hat. The rule is that a player who sees a black hat on any opponent must raise his hand. The first player to deduce (not guess) the color of his own hat wins.

At a certain party, Simple, Simon, and Professor Witquick were all given black hats. Of course all three raised their hands immediately. After a few moments of silence, the professor confidently announced that his hat was black. How did he deduce this?

10.2 / Deduction

If you believe that Jack, because of his conviction that the national government should own the railroads, is a communist, and if you also believe that all communists are atheists, it follows that you believe that Jack is an atheist. This process, in which we infer a further proposition on the basis of some principle or principles already accepted, is called **deduction.** It tells us, on the basis of what we say we believe, what else we must believe. And the process works in reverse, telling us what not to believe. Having accepted the proposition that rabbits can be produced (and probably will be produced) only by other rabbits, we smile when the magician produces a rabbit out of the air, ignore the apparent evidence, and assume that our powers of observation have somehow betrayed us.

As a means for deciding whether or not to accept a proposition, the method of deduction depends on two things. In the first place, we must have some previously accepted proposition with which to start. In the second place, we must be able to reason from premise to conclusion in a way that we (to say nothing of others) believe to be reliable.

In winning the black and white hat game described above, Professor Witquick used both features of deductive thinking. The professor reasoned that, if his hat were white, then Simon, seeing Simple's hand up, would know that his (Simon's) own hat was black. He realized from the silence that Simon was momentarily confused by seeing two black hats and so concluded that his own hat was black. Thus, on the basis of an assumption (Simon wants to win), confidence in Simon's ability to think (if he saw a white hat on me, he'd figure out that his was black), and the fact of Simon's silence, the professor was able to deduce the color of his own hat.

The truth of the conclusion of a deductive argument depends of course on the truth of the proposition accepted to start with. If the initial proposition is wrong, the best reasoning in the world cannot give us confidence in the conclusion. But if we believe the initial proposition to be true, and if the reasoning is sound, we are compelled to believe the conclusion.

The nature of the initial propositions is usually emphasized by referring to them as the assumptions, the premises, or the hypotheses. In assessing the soundness of an argument, we do not allow ourselves to be drawn into a discussion of the truth of the assumptions. Neither therefore are we in a position to discuss the truth of the conclusion. We ask rather if the conclusion is valid. Is it the inescapable consequence of the assumptions?

> Books on game theory sometimes point out that, in planning a strategy, you must assume that your opponent will do his best, and that his best includes intelligent reaction to your moves.

VENN DIAGRAMS

One picture, it is said, is worth a thousand words. And it is true that a picture called a **Venn Diagram** is sometimes of help in analyzing an argument. The technique calls for drawing ovals to represent the sets under consideration and then asking whether or not the hypotheses, the assumed part of the argument, force you to draw a picture that corresponds to the desired conclusion. Consider the following argument.

Hypotheses:	All tubs float.
	Some tubs are gray.
	All battleships are gray.
	She is a tub.
Conclusions:	She is a battleship.
	She is gray.
	She floats.
	All battleships float.

We have drawn, in Fig. A, a picture in which all the hypotheses are satisfied. That is, tubs are completely contained in the set of things that float, and at least some tubs are gray things. Similarly, battleships are completely contained in the set of gray things, and she is indicated as being a member of the set of tubs. At the same time, as Fig. A is drawn, all the conclusions are satisfied. She is pictured in the set of battleships, in the set of gray things, and in the set of things that float; and the set of battleships lies entirely in the set of things that float.

The situation is quite different in Fig. B, however. We again have a picture in which all the hypotheses are satisfied, but this time only one of the conclusions is satisfied.

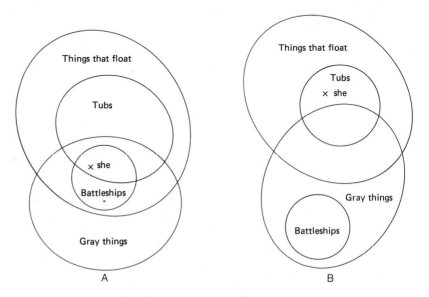

The one conclusion that is satisfied, "She floats," is inescapable. Logic drives us to it. This is what is meant by a valid conclusion; it is a conclusion from which we cannot escape if we accept the hypotheses. In terms of Venn Diagrams, we are driven to it in the sense that any picture that satisfies the hypotheses must also satisfy a valid conclusion.

This example gives us one more opportunity to comment on the distinction between validity and truth. It may be true that all battleships float, but we are not driven to this conclusion on the basis of the stated hypotheses. We did not reject as invalid the statement, "All battleships float," because we were thinking about those that have sunk. We rejected the statement because it was possible to satisfy all the hypotheses without satisfying that conclusion.

When the hypotheses and conclusions are stated as clearly as they are in the example above, it is generally easy to draw a Venn Diagram showing which conclusions are valid and which are not. When one is confronted with a paragraph in a written argument (editorial, political commentary, etc.) or an oral presentation, however, it is more difficult to determine the validity of a conclusion, because one must first state clearly what seem to be the assumptions (hypotheses) being made and what conclusions are being drawn. Consider the following argument.

> Inflation is very damaging to the worker who does not hold capital assets that increase in value with the inflation. Since inflation can only be curbed by balancing the federal budget, it is in the best interests of the working class to balance the federal budget.

What are the hypotheses? Does the writer assume that no members of the working class hold capital assets, or are we talking in the first sentence only about the subset of workers who happen not to hold capital assets? Since such questions can be asked, we must remember that, when we list the hypotheses and conclusions, we are probably listing just one of several reasonable interpretations (one of the reasons why it is difficult to assess the position held even after you read or hear a political candidate, for example). Suppose we decide the following is what is meant by the paragraph above.

Hypotheses: Inflation is very damaging to those who do not hold capital assets.

Some workers do not hold capital assets.

Inflation can be curbed only by balancing the federal budget.

Conclusion: It is in the best interest of the working class to balance the federal budget.

Our Venn Diagram can now be drawn as indicated. We see that the conclusion is not valid. We also see that it can be made valid by changing our interpretation of the second hypotheses to:

No workers hold capital assets.

or by changing the conclusion to:

> It is in the best interests of some workers to balance the federal budget.

The first change has the effect of significantly decreasing the number of people willing to accept our hypothesis; the second change leads to a conclusion not nearly so dramatic. A writer wishing to make a strong case that appeals to a large number of people is best served by the rather imprecise prose so familiar to us all.

SUMMARY

> By deductive reasoning, we are enabled only to reveal to ourselves implications already included in our assumptions.
> *Paul Samuelson*

Deduction is a method of arguing in which we show on the basis of a compelling argument that, if we accept certain propositions (the hypotheses of the argument), we are forced to accept some other proposition (the conclusion). Since no judgment is made about the truth or falsity of the hypotheses, none can be made about the conclusion. We only ask whether the argument is valid and, if it is, we say the conclusion has been proved.

SELF-TEST

Choose the completion that seems best to you. If none of them seem quite correct to you, be prepared to tell why and to supply an ending of your own in class discussion.

1. If a statement has been proved by deduction, then
    (a) it certainly is true.
    (b) it will be true if and only if the hypothesis on which it is based is true.
    (c) it is properly said to be valid.
2. A statement obtained by valid deduction
    (a) must be true.
    (b) may be true.
    (c) will be false if it is based on a false hypothesis.
3. Suppose we have drawn a Venn Diagram in which all of the hypotheses are satisfied.
    (a) If the conclusion is also satisfied, then we may be sure that the conclusion is valid.
    (b) If the conclusion is not satisfied, then we may be sure that the conclusion is not valid.

PROBLEM SET 10.2

In Problems 1–8, decide which (if any) of the conclusions follow from the hypotheses by a valid argument.

1. *Hypotheses:*   Some x are y.
                   All y are z.
                   Some y are w.
      Conclusions: (a) Some x are not z.
                   (b) Some x are z.
                   (c) Some x are w.

2. *Hypotheses:*   All x are y.
                   All x are z.
                   Some w are y.
      Conclusions: (a) Some z are y.
                   (b) Some y are z but not x.
                   (c) Some x are not w.
                   (d) Some x are w.

3. *Hypotheses:*   Some humans are hairy animals.
                   All dogs are hairy animals.
                   All dogs should be kept on a leash.
      Conclusions: (a) Some dogs are not humans.
                   (b) Some dogs are human.
                   (c) Some hairy animals that are not dogs should be kept on a leash.
                   (d) Some animals that should be kept on a leash are hairy.

4. *Hypotheses:*   Everyone who works hard is well educated.
                   Some who work hard are rich.
                   Some professors work hard.
      Conclusions: (a) Some professors are not well educated.
                   (b) Some professors are rich.
                   (c) Some professors are well educated.

5. *Hypotheses:*   Full professors are old creatures.
                   No goats receive social security checks.
                   Some old creatures receive social security checks.
                   Some full professors are good teachers.
      Conclusions: (a) Some good teachers are old creatures.
                   (b) Some good teachers receive social security checks.
                   (c) No full professors are old goats.
                   (d) Some professors are old goats.

6. *Hypotheses:*   Some hot things are very colorful.
                   All red things are very colorful.
                   Red things make us cautious.
                   Some dogs are red.
      Conclusions: (a) Some dogs are colorful.
                   (b) Anything that makes us cautious is colorful.
                   (c) There are hot dogs.
                   (d) There are red hot dogs.
                   (e) Some hot things are red.

7. *Hypotheses:*   All capable people are cantankerous.

                  No conceited people are capable.

                  Some clever people are conceited.

                  Some clever people are capable.

   Conclusions: (a) There are no clever people who are not cantankerous.

                  (b) All clever conceited people are cantankerous.

                  (c) No cantankerous people are conceited.

8. *Hypotheses:*   All timid people are followers.

                  All timid people are quiet.

                  All followers contribute to the success of the leader.

                  Some timid people are irresponsible.

   Conclusions: (a) All quiet people contribute to the success of the leader.

                  (b) Some people who contribute to the success of the leader are irresponsible.

                  (c) Some followers are quiet.

9. Professor Witquick classifies people as conversationalists as follows:

     Some informed people are provocative.

     Some, but not all, overbearing people are provocative.

     All overbearing people, by definition, at least have the virtue of being informed.

     Scintillating people are always provocative.

   Can we conclude that the Professor believes the following?

   (a) Some informed people are not provocative.

   (b) Scintillating people are sometimes overbearing.

   (c) People may be scintillating without being informed.

10. John Q. Public classifies politicians as follows:

     Some are crooks who are getting rich at public expense.

     Many are do-gooders, but all politicians are incompetent.

     There are a few sincere ones who are not getting rich at public expense.

   From these opinions, is it clear that John Q. Public also believes the following?

   (a) The do-gooders are all crooks.

   (b) The sincere politicians are all competent.

   (c) Some crooks are competent.

In Problems 11 through 14, we cite a quotation which is, for the purpose of the problem, accepted. You are then to decide whether or not the statements that follow are logical deductions from the given information.

11. "Silence is the best tactic for him who distrusts himself" (F. de La Rochefoucauld).

   (a) Hugo distrusts himself, so he is silent.

   (b) Hugo is silent, so Horace concludes that Hugo distrusts himself.

   (c) Hugo distrusts himself, so silence is his best tactic.

12. "We always like those who admire us" (F. de la Rochefoucald).

   (a) Professor Witquick admires Librarian Hardback, so we can be sure that Witquick likes Hardback.

(b) Professor Witquick admires Librarian Hardback, so we can be sure that Hardback likes Witquick.

(c) Professor Witquick likes Hardback, so we can be sure that Witquick admires Hardback.

(d) Professor Witquick likes Hardback, so we can be sure that Hardback admires Witquick.

13. "A really busy person never knows how much he weighs" (Edgar Watson Howe).

(a) A person who never knows how much he weighs is really busy.

(b) A person who is not really busy sometimes knows how much he weighs.

(c) A person who is not really busy always knows how much he weighs.

(d) Since Professor Witquick is a really busy person, there are times when he does not know what he weighs.

14. "The greatest minds are capable of the greatest vices as well as the greatest virtues" (René Descartes).

(a) Professor Witquick has a great mind; his vices are therefore likely to be great.

(b) Those who exhibit the greatest virtue are likely also to exhibit the greatest vice.

(c) Those capable of the greatest vices have the greatest minds.

(d) One not capable of the greatest vices and the greatest virtues is not numbered among those having great minds.

15. For each of the following paragraphs, identify the hypotheses and the conclusion.

(a) Since my home state is certain to collect income taxes on any money made in the state, I was careful to conduct all my business in another state. That way, I won't have to pay income tax in my state.

(b) Dogs can't pull a sled that is too heavy. If you insist on taking all that gear, the sled will be too heavy. The dogs won't be able to pull it.

(c) A sensitive person understands that he or she cannot continually refuse offers of help. Homer accepts any offer of help that comes his way. He must be a very sensitive person.

16. For each of the following paragraphs, identify the hypotheses and the conclusion.

(a) Every school in the state is trying to terminate several positions on its professional staff. I have decided to accept a position in another state so that I can avoid the anxiety of wondering if my position will be terminated.

(b) If a person combines hard work with ability in this business, financial reward is certain. Jones has received great financial reward from the business. He must have combined hard work with ability.

(c) There is no way a person with ability can fail if he works hard. Smith has failed, in spite of hard work. Poor Smith just doesn't have the ability.

17. Suppose that you, as a student government leader, employ some-one to survey students standing in line outside the dining hall. You are told the next day that, of the 141 students interviewed, 72 were satisfied with the food service; 97 were satisfied with their dormitory room, 61 were satisfied with both, and 39 were dis-satisfied with both. Deduction should serve to warn you not to trust the work of the one who took the survey. Why?

18. Hugo asked Horace if he had any change, and when Horace reported that he had $1.15 in change, Hugo asked for change for his $1.00 bill. Horace responded that he could neither change a $1.00 bill nor a half-dollar. What coins did Horace have?

19. Can you give examples of valid arguments in which you would say
    (a) The hypotheses are false and the conclusion is false?
    (b) The hypotheses are false and the conclusion is true?
    (c) The hypotheses are true and the conclusion is false?
    (d) The hypotheses are true and the conclusion is true?

20. Can you give examples of arguments that are not valid in which you would say
    (a) The hypotheses are false and the conclusion is false?
    (b) The hypotheses are false and the conclusion is true?
    (c) The hypotheses are true and the conclusion is false?
    (d) The hypotheses are true and the conclusion is true?

21.* Analyze the black and white hat game if the professor plays with three others. In what situations can he determine the color of his hat? Are there any situations in which he cannot determine it?

Mouse is a syllable.
A mouse eats cheese.
So some syllables eat cheese.

10.3 / Difficulties in Deductive Thinking

Since we have claimed that deduction is the only way to prove something conclusively, it is perhaps time to acknowledge that such a proof is only as good as the logic on which it depends. Let us look therefore at some of the common errors that can mar a deductive argument.

In the first place, we often get into trouble because of the language we use. Words have multiple meanings, a fact often exploited in jokes and quips. More seriously, most of us have been in arguments in which the meaning of a term was shifted in midstream. There are also the more subtle problems that come up when we confuse the name of a thing (such as "mouse") and the thing itself (a little rodent that eats cheese), or when we find it necessary to use language to make statements that are supposed to apply to the language itself, such as:

Every statement on this line is false.

More common than troubles with language are errors often made in connection with the simple statement that A implies B, written $A \Rightarrow B$ or expressed in the form, "If A, then B."

The statement, "If B, then A," is the **converse** of "If A, then B." When expressed so succinctly, there is little difficulty in getting people to see that a statement can be true while its converse is false. But in the heat of an argument or even in ordinary discourse,

The converse, which the child not only supplies, but believes is, "If I do not buy you ice cream, then I will buy you some candy."

this distinction is often blurred over. Tell a child, "If I buy you some candy, then I will not buy you ice cream." Then drive past the ice cream shop. No logic compels you to buy candy, but the child may.

Or, to illustrate the same idea with an example closer to the reader's heart, consider the college president who is quoted in the student paper, "If we provide more scholarship aid for those needing it, then we will have to have a general raise in tuition." Suppose this is followed several weeks later with the announcement of a tuition increase. What will be the "logical" expectation of the student body?

A second problem with $A \Rightarrow B$ is that there are those who draw conclusions from "not A." A father tells his son, "If it is a nice day tomorrow, then I want you to help me with some yard work." The next day turns out to be, by mutual agreement, not nice. If the father decides to go ahead with the yard work anyhow, he may have to listen to the complaint, "You said I'd have to work if it was a *nice* day." The father of course had said nothing at all about what he would do if it were not a nice day. It's a point of logic. It may be the only point he gets in the ensuing discussion.

Finally, let us turn to what is potentially the most damaging criticism of deduction as a method of proof. We have pointed it out before. Deduction is absolutely dependent on the idea that we are able to reason from hypotheses to conclusion in a reliable manner. We bank on the idea that what appears to be a compelling argument to one person is just as compelling to another, that we all share something so universal that we call it "common sense." The great thinkers have commented on this thread of good sense running through mankind, and it does seem that most people, even those with appetites almost insatiable in other respects, feel that they are abundantly supplied with common sense.

A **paradox** is an argument which, at each step, looks correct. Yet it leads to an absurd conclusion. One might expect that encounters with a few paradoxes would cause us to have doubts about how much confidence we should place in our sense of reason. In fact, however, the reaction of most people to a paradox seems only to confirm that our confidence in our power of reason is deep-seated indeed.

We see this in a reaction typical of many people when they first encounter Zeno's Paradox. They follow the argument a step or two (or three) until they see where it is leading. Then they break into a grin and try to offer some objection to the argument that seemed so logical to them a moment before. Finally, (after failing to identify any real error in the argument), they exhibit the confidence in their good sense that we referred to above. Amused, and perhaps a little perplexed, they shrug off the arguments of the learned Zeno

---

Of a wordy and illogical colleague in the nation's first Congress, Thomas Jefferson wrote, "To refute was easy indeed, but to silence impossible."

The power of forming a good judgement and of distinguishing the true from the false which is properly speaking what is called Good Sense or Reason, is by nature equal in all men.
René Descartes

Everyone complains of his memory and no one complains of his judgement.
F. de La Rochefoucauld

and walk away no less certain that Achilles will surely pass the tortoise.

The hopelessness of conversation between two people not sharing a common sense of logic is illustrated very cleverly in the story, "What the Tortoise Said to Achilles" (J. R. Newman, *The World of Mathematics,* Simon and Schuster, 1956, pp. 2402–2405) by Lewis Carroll. Lewis Carroll, the children's story teller (*Alice in Wonderland,* etc.) was in fact Charles Dodgson, a professor of mathematics with a lively interest in logic.

ZENO'S PARADOX

Achilles and a tortoise are to have a race. Achilles, being twice as fast, gives the tortoise a head start of 1 mile. The race is begun, and in due time, Achilles reaches the spot T_1 where the tortoise had been when the race started. The tortoise of course has moved on to a point T_2, $\frac{1}{2}$ mile up the road.

The race is not over and in due time Achilles reaches T_2. By then the tortoise is at T_3, $\frac{1}{4}$ mile ahead of Achilles. Again we observe that the race isn't over, that Achilles will eventually reach T_3. And so on. Is it now clear that Achilles will never catch the tortoise?

It is fortunate of course that our thinking processes are similar, for if it were not so, we would have great difficulty in communicating with each other. What, after all, remains to be said to someone who completely understands what you said, but fails to see that it makes any sense. Even the technique of raising one's voice under such circumstances seems not to be terribly effective.

SUMMARY

We have identified several difficulties commonly encountered in deductive arguments: words can carry a multiplicity of meanings, statements can appear self-contradictory, the converse of a true statement is sometimes mistakenly taken to be true, and we are tempted to draw conclusions from $A \Rightarrow B$ when we know that A is false.

We have also noted that deduction rests squarely on the idea of common sense. The great thinkers have generally supported this notion, and most people trust their ability to think in spite of paradoxes in which we seem to get misled.

A VISUAL PARADOX

We have described in this section an example which raises questions about our ability to reason correctly. Perhaps it is not entirely inappropriate to remind the reader that one can't always trust one's eyes either.

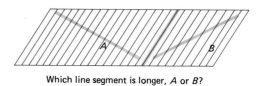

Which line segment is longer, *A* or *B*?

SELF–TEST

Choose the completion that seems best to you. If none of them seem quite correct to you, be prepared to tell why and to supply an ending of your own in class discussion.

1. This section begins with a "logical" syllogism which concludes that some syllables eat cheese. This absurdity occurs because
   (a) we have played on the fact that words often have multiple meanings.
   (b) we have constructed a paradox.
   (c) we are trying to use language to make a statement about the language itself.
   (d) we are confusing a symbol with the actual thing represented by the symbol.
2. A valid argument that reasons from a true hypothesis to a false conclusion
   (a) illustrates one of the pitfalls of deductive thinking.
   (b) is a contradiction in terms; it simply can't happen.
   (c) is called a paradox.
3. Zeno's paradox is used in this section to illustrate
   (a) that arguments that seem correct at each step can lead to conclusions that are not correct.
   (b) that people can be made to lose confidence in their "common sense."
   (c) that the people of ancient Greece reasoned in ways that seem wrong to us today.

PROBLEM SET 10.3

1. State the converse of each of the following ideas of Henry Thoreau, rewritten here in if-then style.
   - (a) If a government imprisons any person unjustly, then the true place for a just man is in prison.
   - (b) If a man is thinking or working, then he is alone.
   - (c) If one is truly rich, there are many things he can afford to leave alone.
   - (d) If a man does not keep pace with his companions, it is because he hears a different drummer.
   - (e) If a man must earn his living by the sweat of his brow, he sweats easier than I do.

2. State the converse of each of the following proverbs of Benjamin Franklin, rewritten here in if-then style.
   - (a) If there's marriage without love, then there will be love without marriage.
   - (b) If two are dead, then three can keep a secret.
   - (c) If the well runs dry, we learn the worth of water.
   - (d) If you are willing to give up essential liberty to obtain a little temporary safety, then you deserve neither liberty nor safety.
   - (e) If you would not be forgotten as soon as you're dead and rotten, then either write things worth reading or do things worth the writing.

*The **contrapositive** of the statement, "If A, then B," is, "If not B, then not A." A statement and its contrapositive are both true or both false.*

3. State the contrapositive of each of the statements in Problem 1.
4. State the contrapositive of each of the statements in Problem 2.
5. In each instance, identify the error in the reasoning.
   - (a) When the Republicans are in office, the country always heads for a depression. My doctor says I am showing signs of depression. If the Republicans win, I will be in a depression.
   - (b) You said that, if I didn't turn in this paper, I'd flunk. But I did turn it in, so I don't see how you can flunk me.
   - (c) It is clear that, if the courts are too lenient, crime in the streets will increase. Moreover, statistics make it evident that crime in the streets is increasing. The conclusion is inescapable; the courts must be too lenient.
6. In each instance, identify the error in the reasoning.
   - (a) Jake (the Brake) Stout is no mean linebacker. Unless a linebacker is mean, he cannot succeed at his position. Jake is not a success as a linebacker.
   - (b) He said that, if it rains, he will be in a bad mood. I'm happy the sun is shining so that he'll be in a good mood.
   - (c) Senator Hornblower said that, unless the Senate adopted her plan, the country would be in a terrible mess. Considering the mess we're in, I take it that the Senate did not adopt her plan.

There are numerous popular puzzles in which logical analysis seems to defy good sense. Problems 7 through 9 are illustrations.

7. Two exhibitors are selling identical wooden gizmos at a sidewalk art sale. One is selling them at three for $1, and the other at two for $1. Both have 30 items left late in the afternoon, and they decide to leave for the day, asking a third party if he will sell them at five for $2. If they had each sold their items separately, they plainly would have netted $25; so they were surprised when their friend gave them $24, reporting that all 60 had been sold at the requested five for $2. What happened to the other $1?

8. Three men register for a hotel room and are told that the charge for the room is $30. They pay $10 apiece and have gone to their room when the clerk realizes that she has made an error; the charge should have been only $25. She sends a bellhop up with the $5, but the bellhop, anticipating the difficulty of dividing $5 among three men, decides they'll be pleased to get back $3 which they can easily split up. For this thoughtfulness, he rewards himself with the $2. Then the men will have paid just $27 for the room, and the bellhop will have $2. Your problem: Shouldn't there be another dollar somewhere?

9. A farmer decides that, on his death, his oldest son should have one-half of his possessions, the next son should have one-third, and the youngest should have one-ninth. He does die of course, and when he goes he has 17 horses. With the market for butchered horses depressed at the time, the sons are in a dilemma as to how to split up the legacy. A neighbor with good horse sense comes to the rescue, giving them a horse of his own. They then take, respectively, 9, 6, and 2 horses. To everyone's surprise, they have a horse left over to return to the neighbor. How come?

10. Here is another question to which an incorrect answer seems plausible. The large wheel rolls along the lower track, so the distance AA' is the circumference of the large circle. The small wheel, physically fastened to the large one, moves along the upper track. Does BB' equal the circumference of the small circle?

11. Some writers include as paradoxes any question to which an incorrect answer seems immediately obvious. Some questions of this sort were considered in Chapter 1. Here are a few more.
   (a) Suppose a 5-pound ball and a 10-pound ball are dropped simultaneously from a building rooftop. Will the 10-pound ball reach the ground faster?
   (b) Two identical coins are laid adjacent to each other, and the one on the left is rolled along the half-circumference of the

other. Will the head that started right side up again be right side up or upside down?

12.* Zeno's Paradox about Achilles is nicely explained by making use of what we learned in Section 4.3 about the sum of a geometric series. Suppose Achilles runs 2 mph. Then in going from A to T_1, he uses up $\frac{1}{2}$ hour; in going from T_1 to T_2, he uses up $(\frac{1}{2})(\frac{1}{2}) = (\frac{1}{2})^2$ hour; etc. He will therefore reach T_n in the total elapsed time of

$$\tfrac{1}{2} + (\tfrac{1}{2})^2 + \cdots + (\tfrac{1}{2})^n$$

That he will catch the tortoise follows from the fact that an infinite number of terms can have a finite sum.

(a) Use the formula in Section 4.3 for the sum of the terms indicated above.

(b) Compute this for $n = 4, 5, 6$.

(c) What happens as n gets very large? How long, then, before Achilles catches the tortoise?

(d) Does your answer to part (c) correspond to your commonsense answer to the question, If Achilles, running at 2 mph, gives the tortoise, running at 1 mph, a 1-mile head start, how long will it be until Achilles catches the tortoise?

There are some deep, difficult paradoxes arising from the way we use language, especially when we talk about classes of all objects of a certain kind. The reader who finds the following examples engaging should consult the article, "Paradox," by W. V. Quine in the April 1962 issue of Scientific American.

13. Think about the following statements. Are they true or false?

(a) Every rule has an exception.

(b) Never say never.

(c) Every generalization is false.

14. (a) The sheriff of a western town, angered at the bearded youths hanging around town, instructs the town's only barber that he is to shave all those, and only those, who don't shave themselves. What happens to the barber's beard?

(b) The same hapless sheriff decides to rid the town of long-haired men. He summarily cuts hair he considers too long. He asks people whom he picks up one question and decides whether or not they are telling him the truth. Those he judges to be telling the truth get a very short haircut; those whom he believes to be lying are shaved bald. Hearing that this was his policy, one young man—logically inclined—answered the question, "Do you know what is going to happen to you?" by saying, "I am going to be shaved bald." What should the sheriff do?

**Galileo
(1564–1642)**

For just as in nature itself there
is no middle ground between
truth and falsehood, so in
rigorous proofs one must either
establish his point beyond
doubt, or else beg the question
inexcusably. There is no
chance of keeping one's feet
by invoking limitations, dis-
tinctions, verbal distortions,
or other mental acrobatics.
One must with a few words
and at the first assault become
Caesar or nothing at all.
Galileo

10.4 / Deduction in Mathematics

A mathematical theorem must be true without exception. If there
is an exception, someone is sure to find a counterexample that
makes it embarrassingly obvious that the theorem does not always
hold. In this case, there is no way to hedge; the theorem is false.
The terms used in mathematics do not carry the multiple meanings that create (or sometimes, we suspect, allow) the ambiguity
often encountered in other disciplines. And while the paradoxes
encountered in Section 10.3 do not shake the general faith we have
in common sense, mathematicians are very much aware that they
often deal with complicated relationships in which one cannot rely
on visual pictures or intuition to correct lapses in step-by-step
arguments. For this reason, they have given considerable thought
to the kinds of arguments they allow in their proofs.

It is often alleged that studying the methods of reasoning
allowed in mathematics has beneficial effects on one's reasoning
in general. While we hesitate to venture an opinion as to whether
this is so, we do believe that an introduction to these techniques
is essential to understanding what mathematics is about. We therefore turn now to a discussion of some of these methods.

In every man there is an eye
of the soul which, when by other
pursuits is lost and dimmed, is
by these [arithmetic, geometry]
purified and re-illuminated; and
is more precious far than ten
thousand bodily eyes, for by it
alone truth is seen.
Plato

THE ARBITRARILY CHOSEN ELEMENT

A theorem of elementary plane geometry asserts that all triangles inscribed in a semicircle are right triangles. Suppose the following proof is offered.

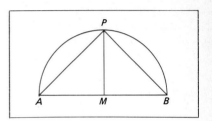

> Erect at M the perpendicular bisector of AB, letting P be the point of intersection with the semicircle. Since $AM = MP$, $\triangle APM$ is an isosceles right triangle; thus $APM = 45°$. Similarly, $MPB = 45°$. This means that $APB = 90°$; the inscribed $\triangle APB$ is a right triangle.

Someone will surely point out that this proof works for only a very special inscribed triangle. If the theorem is to be proved for all inscribed triangles, we must begin by considering an *arbitrary* inscribed triangle. The proof must begin as follows.

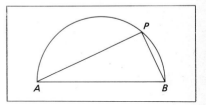

> Let P be a point chosen arbitrarily on the semicircle; draw the inscribed $\triangle APB$.

The argument that follows must then show that triangle APB is a right triangle. See Problem 18 for the complete argument. The principle we are trying to illustrate is the following.

> *To establish an assertion about all members of a certain set, choose an arbitrary member; that is, choose a member having no properties not shared by all members of the set under consideration. Show the assertion to be true for this arbitrary member.*

A professor who taught a large section of an introductory course was annoyed at the seeming indifference of students to his lectures. He went so far as to claim that not one student in the class brought anything to write with so as to be able to take notes. When challenged, he offered the following proof. He arbitrarily selected a student seated about halfway back in the lecture hall and asked this unfortunate if he had a pencil or a pen with him. The student admitted that he had neither. Since the student had been chosen arbitrarily, the professor claimed that he had proved his point. What is wrong with his argument?

The idea is of course that the argument given for the arbitrary element can be given for any other element just as well.

CONTRADICTION

One commonly hears the phrase, "Suppose for the sake of the argument that. . . ." What is to follow can be anticipated. The

The first person he met was
Rabbit. "Hello, Rabbit," he said.
"Is that you?"

    "Let's pretend it isn't," said
Rabbit, "and see what happens."
A. A. Milne

speaker intends to show that this supposition leads to something undesirable if not impossible, hoping then that some interior linkage in the thinking apparatus of the listener will cause him or her to reject the supposition.

This is the idea of the form of argumentation known as *reductio ad absurdum* (reduction to the absurd) or proof by contradiction. The principle is as follows.

> *To prove that A is the case, assume the negation of A (not A). Show that the assumption of not A, logically developed, leads to something absurd.*

We illustrate this by proving a very simple fact about even and odd numbers.

Let us recall several definitions. A positive integer is even if it has a factor of 2; that is, n is even if it can be written in the form $n = 2k$ for some integer k. And n is odd if it is 1 more than some even number; that is, n is odd if it can be written in the form $n = 2k + 1$. Now we are ready to prove a theorem by contradiction.

Theorem.   If n^2 is even, n is even.

$$
\begin{array}{r}
2k + 1 \\
2k + 1 \\
\hline
2k + 1 \\
4k^2 + 2k \\
\hline
4k^2 + 4k + 1
\end{array}
$$

We prove this by assuming that it is false; that is, we assume that n^2 is even but n is odd. Then

$$n = 2k + 1$$

from which it follows that

$$n^2 = 4k^2 + 4k + 1$$
$$= 2(2k^2 + 2k) + 1$$

It is now clear, however, that n^2 is odd, contradicting the given fact that it is even. This shows that n cannot be odd; it must be even.

THE PIGEONHOLE PRINCIPLE

Suppose a mailperson has five letters, all to be delivered to a building having just four apartments, hence four mailboxes. What conclusion can you draw? (We once put this question to a student who, after a period of nervous silence, responded, "It's been several years since I've had any mathematics.") We hope that, independent of the reader's training in mathematics, it is obvious that at least one mailbox will get more than one letter.

The principle illustrated by this example is called the **pigeonhole principle:**

> *If n objects are to be placed in m slots with $n > m$, at least one slot will get more than one object.*

A similar principle is also referred to as the pigeonhole principle:

If n objects are to be placed in n slots with no more than one object in each slot, each slot will get exactly one object.

There is a well-known problem often used to illustrate the pigeonhole principle. The idea is to show that, in any sufficiently large city, at least two people have exactly the same number of hairs on their head. We quote one solution to this problem (M. Kac and S. M. Ulam, *Mathematics and Logic*, Praeger, New York, 1968, p. 11) and then comment on it.

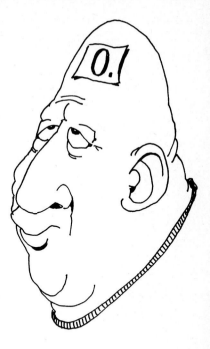

The Solution to a Hairy Problem

In the case of New York City all one needs to know is that the number of hairs on any head is less than the city's population of roughly 8,000,000. (A person would collapse under the weight of 8,000,000 hairs.) If each person is tagged by his specific number of hairs, at least two people must be tagged by the same number (i.e., have the identical number of hairs).

Before including the solution to this problem, we decided to check the assertion that a person could not stand up under 8 million hairs. Accordingly, we went to the chemistry department and had the chemists weigh a hair about $2\frac{1}{2}$ inches in length. The announced result was 0.0004 gram. A little computation (the chemists did it for us) showed that 8 million such hairs would weigh about 7 pounds. Few people would collapse under such a burden, but it is also true that few people are likely to buy a 7-pound hat; and in several class discussions we have gotten general agreement that even the hairiest people on campus probably do not carry 7 pounds of hair. In any case, the solution gives us an opportunity to identify the several stages of an attempt to solve a problem.

Observation. Some hair was weighed. Some computations were made. Varying hair styles were observed.

Premises. No human being has as many as 8 million hairs on his or her head. New York City has a population of more than 8 million.

Mathematical Argument. Consider the residents of New York City. Put persons with no hairs on their heads in group 0, one hair on their heads in group 1, those with two hairs in group 2, etc. By hypothesis, we will need at most 8 million groups to accommodate everyone. But there are more than 8 million people to be put into these groups. By the pigeonhole principle, at least one group must contain more than one person.

Conclusion. At least two people in New York City have the same number of hairs.

METHODS
1. Use of an arbitrary
   element.
2. Contradiction.
3. Pigeonhole principle.

ILLUSTRATIONS
1. Triangles inscribed in
   semicircles.
2. If n^2 is even, n is even.
3. Hair counts in New York.

Now notice something. If you wish to dispute our conclusions, you will need to question the initial assumptions (premises). You may convince someone (in fact, you may be right) that, if one takes careful account of the short hairs on the nape of the neck, the average weight of a hair will be much less than 0.0004 gram. Or you may question whether or not New York really has 8 million people; there has in fact been a decrease since Kac and Ulam's book was published. All kinds of questions can be raised. But our point is that they all have to do with the premises. Once they are granted, the conclusion is inescapable. A mathematician need not worry that his or her part of the job will be spoiled by future investigation, or even the news that the analytical balance was out of whack on the day of the observations. A mathematician's satisfaction derives from having showed that those who accept the premises must accept the conclusion.

SUMMARY

Certain methods of deductive argument have proved themselves very useful in mathematics. We have described three of them in this section, summarizing each in an italicized statement. In our discussion, we have given (or at least started) three proofs the reader should not only understand, but also see as illustrations of the corresponding methods.

SELF-TEST

Choose the completion that seems best to you. If none of them seem quite correct to you, be prepared to tell why and to supply an ending of your own in class discussion.

   1. No one has ever seen more than a finite number of primes, but we know that there are infinitely many of them.
      (a) This statement is nonsense.
      (b) This statement was proved by Euclid, using contradiction.
      (c) We owe our ability to make a statement like this to the computer age, because computers are able to examine larger sets of numbers that we will ever live to examine ourselves.
   2. We quoted Galileo's assertion that in a rigorous proof, one must either establish his point beyond doubt, or else beg the question inexcusably. In this quote, we see
      (a) that Galileo failed to grasp the idea that in the final analysis, nothing can be proved.
      (b) an example of the arrogance that ultimately brought Galileo into conflict with the church.
      (c) that Galileo identified a rigorous proof with what we have called a deductive argument.

3. In our proof that there are two individuals in New York City with the same number of hairs on their heads,
    (a) we see that the role of experiment is to help us decide whether the pigeon hole principle is an appropriate argument to use.
    (b) we see that an experiment is useful in helping us decide on the truth of the hypothesis, but is of no help at all in establishing the validity of the argument.
    (c) we see that one can give perfectly good mathematical proofs of statements that no one would really believe.

PROBLEM SET 10.4

1. Using the notion of an arbitrarily chosen element, show that the square of an even number (a number that can always be written in the form $2n$) is again an even number.
2. Show that the square of an odd number (a number that can always be written in the form $2n + 1$) is again an odd number.
3. Show that the square of a number not divisible by 3 (a number of the form $3n + 1$ or $3n + 2$) is not divisible by 3.
4. Show that the square of a number divisible by 3 is again divisible by 3.
5. Show that the product of an odd number and an even number is even.
6. Show that the product of two odd numbers is odd.
7. Show that the square of an odd number can be written in the form $8n + 1$.
8. Show that the product of three consecutive positive integers is divisible by 6.
9. It is a fact that one cannot construct the trisectors of an arbitrary angle using only a straightedge and compass. Can you construct an angle equal to one-sixth of an arbitrary angle? Hint: Assume you can; then use contradiction.
10. In a certain city, while you can get from A to B by bus, you cannot get from A to C by bus. Prove that you cannot get from B to C by bus.
11. Prove that, if n^2 is odd, n is odd. Hint: Assume n is not odd; note Problem 1.
12. Prove that, if n^2 is divisible by 3, then n is divisible by 3. Hint: Assume n is not divisible by 3; note Problem 3.
13. Assume there are more maple trees in the world than there are leaves on any single maple tree. Show that it follows that there must be at least two maple trees having exactly the same number of leaves.
14. Assume there are more chickens in the world than there are feathers on any one chicken. Without any chicken plucking, show that there must be at least two chickens having the same number of feathers.

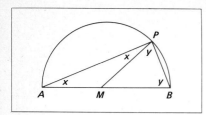

15. If all the sand on the shore of Lake Michigan were scooped into thimbles, would two thimbles necessarily contain exactly the same number of grains of sand?

16. Prove that if 5 points are placed anywhere within or on the boundary of an equilateral triangle of side 1, then at least two of the points must be within $\frac{1}{2}$ of each other.

17. Let P be an arbitrary polyhedron. Show that at least two faces have the same number of edges.

18. Complete the proof that a triangle inscribed in a semicircle is a right triangle, using the following hints. Draw the radius MP, forming two isosceles triangles, $\triangle AMP$ and $\triangle PMB$. Recall that the base angles of isosceles triangles are equal. Recall also that the sum of the interior angles of any triangle is 180°.

Chapter 11
From Rules to Models

I think that everything that can be an object of scientific thought at all, as soon as it is ripe for the formation of a theory, falls into the lap of the axiomatic method and thereby indirectly of mathematics. Under the flag of the axiomatic method mathematics seems to be destined for a leading role in science.

DAVID HILBERT

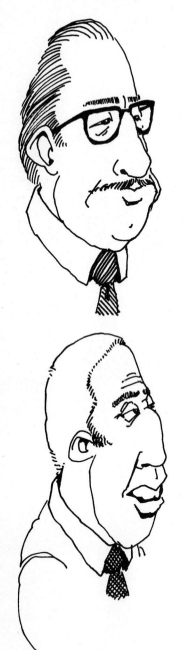

A college president, a professor, an instructor, and a janitor are named Mr. Brown, Mr. Green, Mr. White, and Mr. Black, but not respectively. Four students with the same names will be designated here as Brown, Green, White, and Black. The student with the same name as the professor belongs to Black's fraternity. Mr. Green's daughter-in-law lives in Philadelphia. The father of one of the students always confuses White and Green in class, but is not absent-minded. The janitor's wife has never seen Mr. Black. Mr. White is the instructor's father-in-law and has no grandchildren. The president's oldest son is seven. What are the names of the president, professor, instructor, and janitor?
Litton's *Problematical Recreations* compiled by James F. Hurley Van Nostrand Reinhold, New York, 1971

11.1 / The Consequences of Given Rules

A good first step in a problem like the one above is to rewrite the given information in a series of clear, crisp sentences, each stating one bit of information in an unambiguous way. One student began the problem above by listing the given information as follows:

1. The student with the same name as the professor belongs to Mr. Black's fraternity (who, from the common use of language, we therefore take to be someone other than Mr. Black).
2. Mr. Green has a daughter-in-law living in Philadelphia.

3. One of the teachers, not named Mr. White or Mr. Green, is the father of one of the students (who therefore has the same name as that teacher).
4. Mr. White is the instructor's father-in-law and has no grandchildren.
5. The president's oldest son is 7 (and so, we presume, is unmarried).

She then identified the people as follows:

6. *The professor is Mr. Brown.* The father of one of the students, named either Mr. Black or Mr. Brown, teaches (item 3); so he is either the professor or the instructor. He must be the professor, since the instructor has no children (item 4). The professor is not named Mr. Black (item 1), so he must be Mr. Brown.

7. *The instructor is Mr. Black.* The instructor has no children (item 4), so he cannot be Mr. Green (item 2). He is not Mr. White (item 4) and he is not Mr. Brown (item 6).

8. *The president is Mr. White.* We know that neither Mr. Brown (item 6) nor Mr. Black (item 7) is the president. The president does not have a married son (item 5), so Mr. Green is not the president (item 2). That leaves Mr. White.

The reader will note, among other things, that we are making use of the pigeonhole principle in this argument. The final identification follows directly from this principle:

9. *The janitor is Mr. Green.*

Several instructive observations can be made about this student's solution. In the first place, we call attention to the parts of items 1 and 5 that are enclosed in parentheses. It can be argued that one or both of these bits of information go beyond what is given in the statement of the problem. The only comeback to this is that she at least placed the argument where it belongs; that is, it is clear that the argument concerns what information is given to start with. Her solution makes use of the information as stated in items 1 and 5. Anyone who refuses to accept this information must either find another solution or argue that the problem as stated cannot be solved.

Second, we note that she never used certain information that was given. It is always a good exercise to review unused information. Did we use it unconsciously? If we argued differently so as to use it, could we have avoided some of the questionable assumptions we put in parentheses? Or is the information actually irrelevant to solving the problem?

Finally, we note that several times she made use of the idea that a father and his son have the same last name. This is generally true of course, but not always. Hence it should be stated as one of the things taken to be understood at the beginning. We have more to say later about making certain that all initial assumptions have been identified.

> In any argument in which inferences are to be drawn, it is essential that everyone be clear as to what information is being accepted to start with. In the case of questionable information, it is of course desirable to resolve the problem without using it. If this is impossible, the next best procedure is to accept the questionable information, clearly point out that you have done so, and proceed to give the solution.

The approach we have observed in solving the puzzle above can be of use in practical situations. The warmth and good fellowship with which many a social organization hopes to settle necessary business sooner or later gives way to the formulation of a set of by-laws. The well-known public admiration for bureaucracy notwithstanding, most of the grand goals of political oratory emerge as rules of a regulatory commission. It is tricky business to put into writing all the rules to be followed, even if by chance there is a will to do so. And all too often we discover only by sad experience the natural consequences of the rules we adopt.

Torn between a desire to describe realistic situations and a desire to use relatively simple situations, we consider two fictitious situations.

THE PROBLEM OF DORMITORY GOVERNANCE

When State University changed some of its policies so that the Gertrude Smith Residence Hall for Women became the Smith Residence for People, the governing council of that dormitory decided to propose simultaneous changes in the way the committees of the council were set up. Several criticisms were made of the old system. It was noted that the old by-laws called for committees (such as the one to mete out punishment for violating the 10:00 P.M. weeknight curfew) that had not functioned for years. It was recalled that one fall, when apathy ran rampant, only one section of the dorm remembered to elect its representative, so that later in the year, when an issue arose, the dorm had a one-woman governing council. Some criticism was directed at two women who managed to dominate three key committees, and some was directed at two other women on the council who refused to serve on any committees. Finally, it was pointed out that one of the committees had a member who was not even an elected representative.

At the next meeting of the governing council, a member proposed that the new governing council be required to abide by the following rules.

Dormitory Governing Council Rules

Rule 1: Committees shall be composed of elected representatives.

Rule 2: The council shall not function unless at least two representatives have been elected.

Rule 3: Any two elected representatives must serve together on at least one and no more than one committee.

Rule 4: For each committee, there must be at least one representative not on that committee.

Rule 5: If A is a committee and x is a representative not on committee A, then x must serve on one and only one committee that has no members in common with committee A.

She defended her proposal in the following ways. In the first place, they were simple to check. Second, they were general enough for each year's governing council to set up the committees needed that year. Third, she pointed out that many of the situations criticized under the old arrangement couldn't happen with her system because they were precluded either explicitly or implicitly. She then proceeded to point out some of the features implicitly built into her proposal.

Consequences of the Rules

Consequence 1: Every representative must serve on at least two committees.

In the best manner of a good mathematics student, she argued as follows. Consider an arbitrary representative x. There is a second representative y (rule 2), and members x and y serve on a common committee, say A (rule 3). There must be a representative z not in A (rule 4), and since x and z have a common committee (rule 3), say B, we have x in both A and B.

Note that in the course of this argument she also showed the following.

Consequence 2: There must be at least three elected representatives.

She next reminded the group of the criticism that some committees were still part of the official structure even though no members had been assigned to serve on these committees for

years. That, she claimed, could not happen if her suggested rules
were enforced.

Consequence 3:   There are no "empty" committees; that is,
                    every committee has at least one member.

Again her argument was along the lines a mathematician might
use, this time employing proof by contradiction. She supposed
there was an empty committee, called *E*. Starting as in the argument for consequence 1, she pointed out that we have

$$x \text{ and } y \text{ in } A$$
$$x \text{ and } z \text{ in } B$$

This, however, violates rule 5, because *x* is not in the empty committee *E,* hence should belong to only one committee having no
members in common with *E*. But *x* belongs to both *A* and *B*. This
is a contradiction.

Not content with having proved that every committee has at
least one member, she claimed that even more was true:

Consequence 4:   Every committee must contain at least two
                    members.

Again she supplied an argument to support her claim. You may wish to try supplying a convincing argument for the last assertion. (It will be worthwhile for you to try this not trivial problem, even if frustration drives you to peek ahead to Section 11.2 where an argument is given.) There are of course other questions about the rules that may come to the mind of the governing council. Several of them are suggested in the problems at the end of this section.

THE BUS COMPANIES

In the small, developing Imagin nation, the government agency of public transportation is charged with the regulation of bus companies operating between cities. This agency is not to regulate companies operating within a city; to fall under its jurisdiction, a company has to have a route between at least two cities. For the purposes of the government agency, a route is simply a group of cities served by a particular bus run; thus a company may list:

Route A: Pitstop, Center City, Posthole
Route B: Pitstop, Klondike

If a company claims to serve a certain two cities, it is required that these two cities appear on a common route, but that only one route list both cities. In an effort to force bus companies to provide service to smaller cities off the main highways, it is further required that, for any route listed by a company, it must serve a city not on that route. Finally, in an attempt to control the size of the companies the agency imposes the following restriction. If *R* is a route and *x* is a city served by a company, but not on route *R*, the company can have only one route through *x* that has no cities in common with *R*.

Regulations for Intercity Bus Companies

Regulation 1:　A route is a collection of cities through which a bus passes on its run.

Regulation 2:　A company must serve at least two cities.

Regulation 3:　Any two cities served by a company must be on one and only one common route.

Regulation 4:　Given any route, a company must serve at least one city not on that route.

Regulation 5:　If *R* is a route and *x* is a city served by a company but not on *R*, there must be one and only one route through *x* having no cities in common with *R*.

The perceptive reader notes that, rhetoric aside, these regulations are a restatement of the rules of the dormitory governing council discussed above. There should be little trouble, once this observation has been made, in establishing the following operation guidelines.

Guideline 1:　Every city served by a company must be on at least two routes.

Guideline 2:　A company must serve at least three cities.

Guideline 3:　Every route must include at least two cities.

In fact, every question asked about the dormitory regulations gives rise to a question about the bus regulations. And, more to the point, every answer to a question about one problem answers the corresponding question about the other. The economy thus achieved is a feature to which we return later.

SUMMARY

Any collection of interrelated statements (clues to a puzzle, rules for a game, or regulations governing an organization) is likely to have unexpected implications. To discover them, we should be sure that we begin with a clear understanding of what information is given; and when the information is ambiguous, we should be clear as to how we have decided to interpret it. We then must discipline ourselves to use only the given information, nothing else, unless—stymied—we decide to use a plausible idea, in which case we clearly identify our added assumptions. Finally, we should review information not used, asking whether it has been used unconsciously, whether it can be used to shorten the argument, or whether it really is extraneous.

A GOOD ARGUMENT

1. Get a clear understanding of the given information.
2. Use nothing that is not given.
3. If you decide to make further assumptions, say so.
4. Review any given information that you didn't use.

SELF-TEST

Choose the completion that seems best to you. If none of them seem quite correct to you, be prepared to tell why and to supply an ending of your own in class discussion.

1. In the problem of dormitory governance, we proved that every committee has at least one member.
   (a) This is actually obvious from rule 1.
   (b) In light of the fact that consequence 4 can be proved (stating that every committee must contain at least two members), it was a waste of time to prove that every committee has at least one member.
   (c) If we found a residence hall somewhere and noted that because of resignations, some committees had no members, then we could conclude that the rules proposed in our example were not the rules being used to govern that residence.
2. The logic used in solving puzzles
   (a) differs from the logic used in mathematics because mathematics deals with actual situations in the real world.
   (b) is no different than the logic used in mathematics.
   (c) is the key to finding the solution. If one can follow logic essential to the solution, then one can surely find the solution for himself.
3. In titling this section *The Consequences of Given Rules,* we meant to emphasize
   (a) that rules have consequences which, though perhaps not immediately evident, will be exactly the same for everyone who abides by those rules.
   (b) that everyone who studies the rules will discover the same set of consequences.
   (c) that while rules may look the same, different people can, in applying these rules, come to different and even contradictory conclusions.

PROBLEM SET 11.1

1. Following the proofs used in discussing the dormitory governance problem, prove guidelines 1 and 2 of the regulations for intercity bus companies. Write out all the details, being sure that you understand each step.
2. As in Problem 1, write out the proof of guideline 3.
3. In the argument for consequence 3 of the dormitory governing rules, we began by assuming that there could be an empty committee. Is this not already in violation of Rule 1?
4. We have heard the following complaint about the proof of consequence 1. While x has been shown to be on two committees, y has been shown to be on only one committee. Yet the assertion says that *every* representative must serve on two committees. Have we really proved what we set out to prove? Why?

Problems 5–9 ask some further questions about the dormitory problem. They are of course questions about the bus regulations as well.

5. Prove consequence 4.
6. Although the stated requirements are that at least two representatives must be elected, it was easily shown (consequence 2) that at least three had to be elected if all rules were to be satisfied. Show that even more is true: If all the rules are to be satisfied, at least four representatives must be elected.
7. Show that there must be at least six committees if all the rules are to be satisfied.
8. Show that all the rules will be satisfied if four representatives are elected and they form six committees.
9. Interpret Problem 8 for the bus route problem. Draw a map indicating the four cities and the six routes.
10. Here is another problem about forming committees. Since it is a new problem based on rules quite different from those used in the dormitory governance example, you cannot use the results proved for that situation. A legislative group agrees that the following rules shall be enforced with regard to any committees set up.
   a. Any two committees must have at least one common member.
   b. Any two committees shall have at most one common member.
   c. Every member of the legislature must serve on at least two committees.
   d. No member of the legislature may serve on more than two committees.
   e. There shall be four committees.
   How many members must there be in the legislature? How many members serve on each committee?
11. Four boys, Alan, Brian, Charles and Donald, and four girls, Eve, Fay, Gwen and Helen, are each in love with one of the others, and, sad to say, in no case is their love requited. Alan loves the girl who loves the boy who loves Eve. Fay is loved by the boy who is loved by the girl loved by Brian. Charles loves the girl who loves Donald. If Brian is not loved by Gwen, and the boy who is loved by Helen does not love Gwen, who loves Alan?

> J. F. Hurley, Litton's *Problematical Recreations*,
> Van Nostrand, New York, 1971, p. 66.

12. In our village we have a Mr. Carpenter, a Mr. Machinist, and a Mr. Smith. One is a carpenter, one a machinist, one a smith. None follows the vocation corresponding to his name. Each is assisted in his work by the son of one of the others. As with fathers, so with sons; none follows the trade that corresponds to his name. If Mr. Machinist is not a carpenter, what is the occupation of young Smith?

> Oswald Jacoby, *Math for Pleasure*,
> McGraw-Hill, New York, 1962 p. 91.

13. On the Island of Perfection there are four political parties—the Free Food, the Pay Later, the Perfect Parity, and the Greater

Glory. Smith, Brown, and Jones were speculating about which of them would win the forthcoming election. Smith thought it would be either the Free Food Party or the Pay Later Party. Brown felt confident that it would certainly not be the Free Food Party. And Jones expressed the opinion that neither the Pay Later Party nor the Greater Glory Party stood a chance. It turned out that only one of them was right. Which party won the election?

> E. R. Emmet, *Puzzles for Pleasure*, Emerson Books,
> Buchanan, N.Y. 1972, p. 4.

14* There are five houses, each of a different color. They are inhabited by five families who are of different nationalities, read different newspapers, keep different pets, and own different kinds of vehicles:

a. The English live in the red house.
b. The Spanish own the dog.
c. The *Examiner* is read in the green house.
d. The African reads the *Times*.
e. The green house is immediately to the right of the brown house (as you face them).
f. The owner of the antique car keeps goldfish.
g. The people in the yellow house have a sportscar.
h. The *Daily News* is read in the middle house.
i. The Norwegian lives in the first house on the left.
j. The family with the station wagon lives in the house next to the family with the cat.
k. The family with the sportscar lives next door to the family that owns the horse.
l. The Japanese drive a sedan.
m. The Norwegian lives next door to the blue house.
n. The family with the camper truck reads the *Herald*.

From the clues listed, determine who reads the *Tribune* and who owns the monkey.

11.2 / Finite Geometries

It was seen in Section 11.1 that two lists of rules, the dormitory governing rules and the regulations for intercity bus companies, were, from our point of view, the same. That is, if the words "representative" and "committee" in the first list are replaced by "city" and "route," and if appropriate changes are made in grammar, we will get the second list. The consequences of the given rules are similarly related and, more important, the arguments establishing these consequences are exactly the same. That is to say, the arguments do not depend at all on whether we use the word "representative" or "city," or the word "committee" or "route." In fact, we could "abstract" the rules stated for the dormitories or the regulations for the buses by using nonsense words as follows.

Rule 1a: Lokes are collections of pokes.

Rule 2a: There are at least two pokes.

Rule 3a: Any two pokes must be members of one and only one common loke.

Rule 4a: For each loke L, there is a poke x not in L.

Rule 5a: If L is a loke and x is a poke not in L, then x is in one and only one loke M having no members in common with L.

We can now proceed to establish the consequences of these rules.

Consequence 1a: Every poke must be in at least two lokes.

Consider an arbitrary poke x. There is a second poke y (rule 2a), and the pokes x and y are in a common loke, say A (rule 3a). There must be a poke z not in A (rule 4a) and, since x and z are in a common loke (rule 2a), say B, we have x in both A and B.

Comparison shows that we have merely copied the proof of consequence 1 from Section 11.1. In the same way we can easily prove further results.

Consequence 2a: There must be at least three pokes.

Consequence 3a: Every loke contains at least one poke.

Since we did not prove consequence 4 in Section 11.1, let us do so now.

Consequence 4a: Every loke must contain at least two pokes.

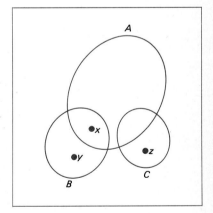

In the accepted way, we establish this by showing it to be true for an arbitrary loke that we call A. It follows from consequence 3a that A must contain some poke x, and we know from rule 4a that there is a poke y not in A. It follows in turn that there is a loke B containing x and y (from rule 3a), that there is a poke z not in B (rule 4a), and that there is a loke C that contains z and no points of B. Using dots to represent pokes, and circles to enclose pokes on the same lokes, the situation as we now have it can be pictured as in the margin. (Of course, z might be in A, but in that case we would be finished.)

Now if A and C have no common pokes, x, a poke not in C, will be a member of the two lokes A and B, neither of them having a poke in common with C. This violates rule 5a, so C and A must have some common poke, say w. This gives a second poke in A, establishing our assertion.

Except for the terminology employed, we have here an example of what is called a **finite geometry.** In finite geometries, the nonsense words are commonly called **points** and **lines;** the stated rules are called **axioms;** and the derived consequences are called **theorems.**

We worked our way into finite geometries slowly, because most students have a preconceived idea of what a point is and what a line is. They find it very difficult to let go of these ideas (thereby demonstrating, incidentally, that by 18 or 19, we are already quite set in our ways, resistant to new ideas). If we had stated consequence 4a in the form,

Every line contains at least two points

instead of in the form,

Every loke contains at least two pokes

most readers would not, in the beginning, see this as something needing proof. They would see it as an obvious statement about points and lines. Yet, in finite geometry, as in all of mathematics, the goal is to prove all consequences, showing them to be inevitable implications of the rules with which we start. It is no more acceptable to use preconceived ideas about points and lines than it would have been in the dormitory governance example to bring in personal opinions about the people who were elected representatives. The logic of any argument used to establish a theorem in finite geometry should not be affected if "point" and "line" are replaced with "poke" and "loke."

In the same way, the rules with which we begin are to be regarded as arbitrary, just as arbitrary as the bus regulations in the nation of Imagin. The first axiom, asserting that all lines are collections of points, should not be regarded as any more obvious than the assertion that all routes are collections of cities. Again, temptations to regard them as obvious can be dispelled by remembering that they can at any time be replaced by statements about pokes and lokes.

What we have said about points and lines being undefined terms and about axioms being arbitrary statements is true about all forms of geometry. Finite geometries are so named because, while employing the terminology and methods familiar from Euclidean geometry, their axioms can be satisfied by a finite collection of lines and points. Thus the axioms stated in terms of pokes and lokes at the beginning of this section are satisfied by a collection of four points and six lines, which can be represented pictorially. A series of dashes connects points that are to be thought of as lying on the same line, but it is to be noted that each line, represented by the dashes, contains just two points.

The problems at the end of this section give many examples of finite geometries, some of which have been studied extensively. The purpose of asking you to supply proofs is not to acquaint you with a lot of facts about finite geometry. Rather, writing proofs offers an unparalleled opportunity to practice saying precisely

In the beginning everything is self-evident, and it is hard to see whether one self-evident proposition follows from another or not. Obviousness is always the enemy of correctness. Hence, we must invent new and even difficult symbolism in which nothing is obvious.
Bertrand Russell

One must be able to say at all times, instead of points, straight lines, and planes—tables, chairs, and beer mugs.
David Hilbert

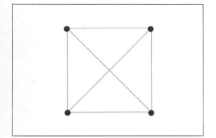

what we mean in a way that cannot be misunderstood. This is an opportunity, as the accompanying short story suggests, that should not be allowed to go to the dogs.

One's first attempt to prove a theorem often includes going up some blind alleys, including extraneous facts not necessary for the proof, etc. The proofs for which we ask, then, do not usually represent one's first effort. No author should be satisfied with his first draft; neither should a proof maker.

We are saying that a good proof should consist of attractive, well-reasoned, grammatically correct paragraphs. Every step should be clear, concise, as well as precise.

Many mathematicians have claimed that the writing of such a proof is for them an esthetically rewarding experience. In this spirit we suggest several theorems on which you can work with reasonable hope of success. But to give it a fair try, a proof must be your own—your own argument written in your own style. You can then judge for yourself whether or not you derive any satisfaction from such a project.

Gauss always strove to give his investigations the form of finished works of art. He did not rest well until he had succeeded, and hence he never published a work until it had achieved the form he wanted. He used to say that when a fine building was finished, the scaffolding should no longer be visible.
Sartorius Von Watterhouser

You know that I write slowly. This is chiefly because I am never satisfied until I have said as much as possible in a few words, and writing briefly takes far more time than writing at length.
Carl Gauss

From the Minutes of a Borough Council Meeting:

Councillor Trafford took exception to the proposed notice at the entrance of South Park: "No dogs must be brought to this Park except on a lead." He pointed out that this order would not prevent an owner from releasing his pets, or pet, from a lead when once safely inside the Park.

The Chairman (Colonel Vine): What alternative wording would you propose, Councillor?

> Mathematics must be beautiful
> . . . a mathematical proof should
> resemble a simple and clear
> cut constellation, not a scattered
> cluster in the Milky Way.
> *G. H. Hardy*

Councillor Trafford: "Dogs are not allowed in this Park without leads."

Councillor Hogg: Mr. Chairman, I object. The order should be addressed to the owners, not to the dogs.

Councillor Trafford: That is a nice point. Very well then: "Owners of dogs are not allowed in this Park unless they keep them on leads."

Councillor Hogg: Mr. Chairman, I object. Strictly speaking, this would prevent me as a dog-owner from leaving my dog in the back-garden at home and walking with Mrs. Hogg across the Park.

Councillor Trafford: Mr. Chairman, I suggest that our legalistic friend be asked to redraft the notice himself.

Councillor Hogg: Mr. Chairman, since Councillor Trafford finds it so difficult to improve on my original wording, I accept. "Nobody without his dog on a lead is allowed in this Park."

Councillor Trafford: Mr. Chairman, I object. Strictly speaking, this notice would prevent me, as a citizen, who owns no dog, from walking in the Park without first acquiring one.

Councillor Hogg (with some warmth): Very simply, then: "Dogs must be led in this Park."

Councillor Trafford: Mr. Chairman, I object: this reads as if it were a general injunction to the Borough to lead their dogs into the Park.

Councillor Hogg interposed a remark for which he was called to order; upon his withdrawing it, it was directed to be expunged from the Minutes.

The Chairman: Councillor Trafford, Councillor Hogg has had three tries; you have had only two. . . .

Councillor Trafford: "All dogs must be kept on leads in this Park."

The Chairman: I see Councillor Hogg rising quite rightly to raise another objection. May I anticipate him with another amendment: "All dogs in this Park must be kept on the lead."

This draft was put to the vote and carried unanimously, with two abstentions.

R. Graves and A. Hodges, *The Reader over Your Shoulder,*
Macmillan, New York, 1961, pp. 149–150.

SUMMARY

The strength of a logical argument does not depend on the meaning of the words used to name the objects of discussion. It does depend on the way the objects of discussion are related to one another. This dependence on interrelationships is sometimes easier to see if we can designate the objects of our discussion by names that do not carry any previous connotations; and because we work to free ourselves of the multiple, emotional connotations of normal language, mathematical argument affords us a unique opportunity to develop skills in clear communication.

SELF–TEST

Choose the completion that seems best to you. If none of them seem quite correct to you, be prepared to tell why and to supply an ending of your own in class discussion.

1. If we had used points and lines instead of representatives and committees in our discussion of dormitory governance
   (a) the proofs would have been easier; it would, for example, have been obvious that every line contains a point.
   (b) we would probably have gotten more confused, not realizing that we needed to prove statements that looked obvious.
   (c) we could at least have had the help of a picture because we know how to draw points and lines.
2. We have quoted Russell as saying "Obviousness is always the enemy of correctness."
   (a) This no doubt reflects his political views in which he scorned those who propose simplistic solutions to complex problems.
   (b) It is the spirit of this remark which led us to suggest that the theorems of finite geometry might be easier to prove if we refer to pokes and lokes instead of points and lines.
   (c) By this he means that a deductive argument, if it is to be correct by a logician's standards, must be phrased in ways that will appear to be anything but obvious to the nonlogician.
3. How would you describe a finite geometry?
   (a) It is a study in which we concentrate our attention on just a finite subset of the points in the plane.
   (b) It might be misleading to use the term geometry. Perhaps it should be called the study of finite logical systems, since it is the study of a finite collection of elements related to each other by a few rules or axioms.
   (c) We temporarily blind ourselves to the fact that lines contain an infinite number of points, thereby enabling ourselves to focus on the finite number of points essential to a particular proof.

PROBLEM SET 11.2

In Problems 1 through 3, we give the axioms for a finite geometry and then list several theorems in an order in which they can be proved. In attempting to prove a theorem, assume that all theorems listed previously in a problem have been proved. Thus inability to prove Theorem T4 in a list does not preclude trying to prove Theorem T5, etc. One problem, completed in the spirit we have described, (and illustrated with our proof of consequence 4 in this section) is a significant piece of work, worth more than hasty notes scribbled out as "solutions" to a number of problems.

1. Axioms:

   A1. There is at least one line.
   A2. Each line contains at least two points.
   A3. Any two points lie on exactly one common line.

A4. Given any line L, there exists a point not on L.
A5. Given any line L and any point x not on L, there are at least two lines through x having no point in common with L.

Theorems:

T1. There are at least 3 points.
T2. There are at least 3 lines.
T3. There are at least 5 points.
T4. There are at least 10 lines.

Draw a figure (similar to the one on page 334) that represents this geometry.

2. Axioms:

A1. If x and y are any two points, there is at least one line containing both x and y.
A2. If x and y are any two points, there is at most one line containing both x and y.
A3. If L and M are any two lines, there is at least one point that lies on both L and M.
A4. There are exactly three points on each line.
A5. If L is any line, there is at least one point not on L.
A6. There exists at least one line.

Theorems:

T1. If L and M are any two lines, there is at most one point that lies on both L and M.
T2. Two lines have exactly one point in common.
T3. If x is any point, there is at least one line that does not contain x.
T4. Every point lies on at least three lines.
T5. There are precisely seven points.

Draw a figure that represents this geometry.

3. Axioms:

A1. If x and y are distinct points, there exists at least one line containing x and y.
A2. If x and y are distinct points, there exists not more than 1 line containing x and y.
A3. Given a line L not containing a point x, there exists 1 and only 1 line M containing x and no points of L.
A4. Every line contains exactly 3 points.
A5. Given a line L, there is a point x not on L.
A6. There exists at least 1 line.

Theorems:

T1. There exist exactly 9 points.
T2. There exist exactly 12 lines.
T3. Corresponding to each line L, there are exactly 2 lines having no points in common with L.

T4. If M and N are lines neither of which have any points in common with a third line L, then M and N have no points in common.

Draw a figure that represents this geometry.

4. Let us say that in a finite geometry, two lines are parallel if they have no points in common.
   (a) In the example used at the start of this section, substitute "line" for "loke," and "point" for "poke," and use the word "parallel" to state rule 5a.
   (b) State axiom A5 in Problem 1 using the word "parallel."
   (c) State theorem T2 in Problem 2 in terms of parallel lines.
   (d) State axiom A3 and theorems T3 and T4 in Problem 3 in terms of parallel lines.

5. In seeking to write things, even nonmathematical things, in a way that is clear, it is often most helpful to use the symbols and terminology of mathematicians.
   (a) Consider the rules for the governing council stated in Section 11.1. Try to state rule 5 without resorting to the use of names like A for the committee and x for the representative.
   (b) The faculty by-laws of Macalester College include the following provision (written, it must be admitted, by a mathematician).

   In any election where a committee of n members is to be elected, the Advisory Council shall submit a slate of $2n$ nominees. Voters shall list n choices in order of preference, and in the counting of votes, a nominee listed first on a ballot shall be awarded n points, the one listed second shall receive $n - 1$ points, etc. The n nominees with the highest totals shall be declared winners, with ties settled by subsequent elections between nominees not previously elected.

   Try to state this without using n, $n - 1$, etc.

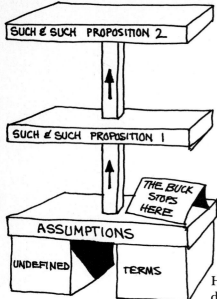

Pure mathematics consists entirely of such assertions as that, if such and such a proposition is true of anything, then such and such another proposition is true of that thing. It is essential not to discuss whether the first proposition is really true, and not to mention what the anything is of which it is supposed to be true. . . . If our hypothesis is about anything and not about some one or more particular things, then our deductions constitute mathematics. Thus mathematics may be defined as the subject in which we never know what we are talking about, nor whether what we are saying is true.
Bertrand Russell

11.3 / The Axiomatic Method

However clever we are, we don't learn a foreign language from a dictionary written in that language. A new term, whether in a foreign language or our own, must be explained to us in terms we already understand.

It's evident that the process of defining words in terms of other words must ultimately be circular. If we're to communicate with someone, we must give up the idea of defining each word or symbol. We have to depend on the other person sharing with us a common understanding of some familiar words. Stated more explicitly, some terms must be left undefined.

Recall that we felt it wise, when introducing finite geometry, to avoid using the words "point" and "line" because most people feel they know what a point is and what a line is. As we tried to show, however, such an intuitive notion is not necessary to prove the theorems. "Poke" and "loke" will do just as well. Perhaps "poke" and "loke" will do better, since there is no need to overcome an "It's obvious" feeling when asked to prove that every loke contains at least four pokes.

Any logical discussion, mathematical or not, must ultimately rest on terms left undefined. The familiar cry, "Define your terms," if pushed too far, will lead to absurdity; and it is the one making the demand who is being absurd.

[Freedom is] the right to complain about the lack of it.
Will Rogers

When I use a word, it means what I wish it to mean, neither more nor less.
Humpty Dumpty

AXIOMS

It is also clear from our work with finite geometry that nothing can be proved about points and lines (pokes and lokes) until some rules are stated which the points and lines are presumed to obey. When we discussed deductive reasoning, the only method of proof acceptable in mathematics, we emphasized that the method could not get off the ground unless there were previously accepted principles from which to work.

If we are to avoid circular reasoning, we must simply begin somewhere with statements we don't prove. These beginning statements are called **axioms.** We don't prove axioms any more than we try to prove the rules of chess. This is not because we are lazy; and it is not because they are obvious. We don't prove axioms because it can't be done. You can't prove everything.

We try to minimize the assumptions we make. In the finite geometry studied in Section 11.2, we assumed there were at least two points. We later proved that the axioms taken together implied that there were at least four points. We would have been studying the same geometry therefore if we had taken as our initial assumption the existence of at least four points. In stating axioms for a given system, however, the usual goal is to be able to derive all the theorems assuming as little as possible. However, the goal of reducing the assumptions to a bare minimum is sometimes set aside in order to make them as simple as possible to understand. For instance, the axioms given to a high school geometry class may include for clarity certain assertions that could be omitted and then derived as theorems. The effort to reduce and simplify the axioms is in the final analysis a matter of taste.

There is another consideration in setting up a system of axioms, which is decidedly not a matter of taste. We would like to be certain that it is not possible, by reasoning from the given axioms, to prove contradictory statements. If, in the axiom system discussed in Section 11.2 and in the previous paragraph, we had included an axiom stating that there were no more than three points, then (although we might not have immediately realized it), our axiom system would have been contradictory. The object is to avoid stating axioms that are contradictory.

THEOREMS

When the axioms have been stated, we can begin, in the manner of deductive reasoning, to deduce their implications. Guessing at these implications, the theorems of the subject, requires insight and imagination. A theorem is often named for the person who first suggests it, rather than for the one who first proves it.

Many mathematicians have reflected on the question of how one discovers a theorem. When asked how he made his discoveries, Newton said, "By always thinking unto them." Some mathematicians claim that once a problem is well-formulated, sleep or semisleep often plays an almost mystical role in bringing them to

Sleep on it
Old proverb

One evening, contrary to my custom, I drank black coffee and could not sleep. Ideas rose in crowds; I felt them collide until pairs interlocked so to speak, making a stable combination.
H. Poincaré

On being very abruptly awakened by an external noise a solution long searched for appeared to me at once without the slightest instant of reflection on my part.
J. Hadamard

> I believe that Newton could hold a problem in his mind for hours and days and weeks until it surrendered to him its secret. Then being a supreme mathematical technician he could dress it up, how you will, for purposes of exposition, but it was his intuition which was pre-eminently extraordinary—"so happy in his conjectures," said de Morgan, "as to seem to know more than he could possibly have any means of proving."
> *John Maynard Keynes*

a solution. The mathematician Descartes is said to have done much of his work while resting in bed in the morning, a lifelong habit developed of necessity when he was a sickly child and apparently cultivated out of a desire for privacy when he became a mature thinker. There are, then, some good arguments for spending a lot of time in bed. There are also some poor ones. This much seems clear. Those who hit upon fruitful ideas while asleep have always thought very deeply about them when wide awake.

It takes keen insight to propose a theorem, and it often takes admirable ingenuity to offer a proof. But once a theorem is proved, anyone willing to take for granted the previously proved theorems, as well as the axioms, should be able to follow the proof step by step.

SUMMARY

In any logically organized argument, there must be some terms that are not defined and some assumptions that are not proved. The trick is to be certain that one cannot deduce contradictory statements from the assumptions. Beyond this, it is desirable to keep the assumptions to a minimum. One form of mathematical insight is the ability to discover the implications (the theorems) that follow from the assumptions. These things, undefined terms, assumptions that relate them in a logically consistent way, and the implications of the assumptions, constitute what is called an axiomatic system.

SELF–TEST

Choose the completion that seems best to you. If none of them seem quite correct to you, be prepared to tell why and to supply an ending of your own in class discussion.

1. If we hope to organize any body of information into a systematic logical system,
   (a) all of our terms should be defined as carefully as possible, since

failure to use terms in the same way is a root cause of much misunderstanding.

(b) we may as well face from the beginning the impossibility of defining all of our terms.

(c) then the arguments of this section underscore the difficulty we shall have in defining our terms, but do not excuse us from trying.

2. The critical question to ask about a theorem is:
(a) Does it follow, without exception and by the accepted rules of deductive argument, from the axioms?
(b) Is it true?
(c) Who discovered it?

3. An axiom is a statement
(a) so obvious that people are willing to accept it without proof.
(b) which is simply assumed to be true.
(c) which cannot be proved to be either true or false.

PROBLEM SET 11.3

1. Illustrate the futility of trying to define everything by finding in your dictionary a series of definitions that take you in a circle. A simple way to find examples in most dictionaries is to look up a unit of foreign currency: krone, krona, guinea, etc.

2. Another problem with definitions, not discussed explicitly in the text, is that words often have a multiplicity of meanings. The problem can be illustrated:
(a) Write down all the definitions you can think of for "fair."
(b) Consider now the possible meanings of the sentence, "She is fair."
Construct other examples, beginning with a familiar word having multiple meanings.

3. In the light of this section, how might you respond to the question, What is truth?

4. In any discussion between two people on a given subject, certain of the basic words used simply have to be accepted as being commonly understood. What might these terms be in a discussion of religion? Economics? Political ideology?

In Problems 5 through 8, we give a set of hypotheses. In each case give
   (a) A conclusion that is valid in terms of the given hypotheses, but one that not everyone accepts as true.
   (b) A conclusion, the truth of which may seem so apparent to some as to lead them to claim (falsely) that the conclusion is valid.
The idea is of course to provoke discussion. Many answers are possible.

5. *Hypotheses:* Useful education should prepare a student for a job. A liberal arts education doesn't prepare a student for a job. This course is intended only for students pursuing a liberal arts education.

6. *Hypotheses:*   Poor entertainment is a waste of time.

   Entertainment that encourages us to identify with unrealistic life situations is poor entertainment.

   Television serials cause us to identify with life situations that are unrealistic.

7. *Hypotheses:*   Those who wish to improve themselves must be people who want to learn.

   People who want to learn go to school.

   Homer goes to school.

8. *Hypotheses:*   People who have a kind nature show it by being helpful to children.

   People who are helpful to children are able to exercise firm discipline.

   Children find that Mr. Jones is very helpful.

9. The following statements are probably statements that you would call true. For each statement, write down a set of hypotheses from which the desired truth follows as a valid conclusion.

   (a) You should not commit murder.

   (b) You should not belch out loud at the table.

   (c) You should exercise moderately each day.

   (d) A mother cat should care for her kittens.

10. Follow the instructions in Problem 9.

    (a) You should be willing to help a blind person across a street.

    (b) You should not steal another student's bicycle.

    (c) Smoking cigarettes is bad for your health.

    (d) To be a good citizen, you must vote.

In the same way [as an artist] the researcher . . .
depicts the world in scientific laws and concepts
Siu

11.4 / Models

We have described an axiomatic system as a collection of undefined terms, assumptions called axioms, and derived consequences
called theorems. We have stressed that theorems are to be proved
on the basis of logic and are not to depend on preconceived notions
we may have had about the undefined terms. The entire system
may seem therefore to be artificial, as certain in its conclusions as
human reason itself, but unrelated to anything in our world of
experience. We now wish to show how axiomatic systems relate to
our understanding of the world, and to correct impressions, so far
allowed to grow, about how axiomatic systems actually evolve.

MODELS FROM AXIOMATIC SYSTEMS

If an axiomatic system is to say anything about the world around
us, it is clear that we must begin by assigning meanings to the undefined terms. Moreover, this has to be done in such a way as to make
the axioms into sensible statements which appear to be true. If we
succeed in this, so that we are willing to consider the axioms true, it

> In scientific work, it is not enough to be able to solve one's problems. One must also turn these problems around and find out what problems one has solved. It is frequently the case that, in solving a problem, one has automatically given the answer to another, which one has not even considered in the same connection.
> *Norbert Wiener*

follows that the theorems will also be true. The artificial system that said nothing about the world of experience is thereby transformed into a system we think is at least a partial description of the world as it really is. We have a model.

Let us be specific. In Section 11.2 we have a set of axioms defining a finite geometry. They become sensible statements if we substitute "representative" and "committee" for the undefined terms "poke" and "loke." We may or may not find a legislative body somewhere for which these sensible statements appear to be true. If we find such a body, we will have a ready-made model for that particular situation. And if we don't find such a body, our axiomatic system will be in no way invalidated. Perhaps some other words ("city" and "route") can make the axioms into statements that appear to be true about something in the world around us. It is the potential of a given structure to serve as a model for many different real-world situations that makes the axiomatic method a versatile and attractive scientific tool.

AXIOMATIC SYSTEMS FROM MODELS

Euclid thought his axioms were self-evident statements about the world as it really is. It is a relatively recent development to think of axioms as assumptions and, even when they are recognized as assumptions, a practicing scientist is likely to think of them as assumptions that describe the world of experience. Even the most abstract-looking axiom systems are more often than not generalizations of very common mathematical structures, structures first developed as tools for handling real-world problems.

For these reasons one often finds a mathematician asking if a theorem is true. Technically, it is meaningless to ask if an axiom or a theorem is true (What is the sense of asking whether or not it is true that every loke contains three pokes?) But in practice, the working mathematician has in mind some model of the system under consideration. The question, correctly put, is, "With respect to the model that I have in mind, would the following statement be true?"

It is this interplay between an abstract axiomatic system and attempts to make it serve as a model that has enriched both mathematics and science. The concept of a "pure" mathematician constructing axiom systems without regard for the world of experience and an "applied" mathematician or scientist trying to find a model that explains his or her observations is a convenient distinction for discussion. It does not convey the way science has progressed historically, or the way in which mathematics has developed. Except for isolated instances, mathematical sciences have been carried forward by people seeking to use the mathematics already known and to develop new mathematics as it is needed to attack problems encountered in the world.

Models are tentative. It is to be stressed that our identification of undefined terms with real objects in the world, being done in such a way that the axioms *appear* to be true statements, involves a value judgment. What seems to be true to one observer may not seem true to another. What seems true at one period in history may not seem true at another. Theorems, which were certain so long as we worked in the artificial axiomatic framework where validity was the only concern, lose their certainty when taken as statements supposed to be true about the real world.

Models may have to be revised or abandoned for any number of reasons: persistent discrepancies between the model and what can be observed, predictions based on the model that don't square with experience, and new observations—perhaps made possible with the development of more sophisticated instruments. For these reasons, acceptance of any model should be tentative.

At this point an enigma presents itself which in all ages has agitated inquiring minds. How can it be that mathematics, being after all a product of human thought which is independent of experience, is so admirably appropriate to the objects of reality? Is human reason, then, without experience, merely by taking thought, able to fathom the properties of real things?

In my opinion the answer to this question is, briefly, this: as far as the propositions of mathematics refer to reality, they are not certain; and as far as they are certain, they do not refer to reality. It seems to me that complete clarity as to this state of things became common property only through that trend in mathematics which is known by the name of "axiomatics."
A. Einstein

The classic example of our continuing effort to model what we observe is the attempt to understand, explain, predict, and make use of the apparent motions of the stars and planets as viewed from the earth. Here we see models proposed, accepted, adjusted to conform to new observations, and ultimately abandoned in favor of new ones. The traditional reluctance to abandon models has in this case been compounded by religious and philosophic arguments as well.

A Classic Example

The creation of a coherent system to explain and predict the apparent motions of the planets and stars as viewed from the earth is certainly a significant triumph of human intellect. That the problem had attracted attention from the most ancient times and that the theory is still undergoing modification in this century give an indication of the enormous time and energy that have gone into its study.

Early views of a fixed and flat earth covered by a spherical celestial dome were studied by the Greeks, who devised a real model in the fourth century B.C. which accounted at least approximately for the rough observations then available. The earth was viewed as fixed with a sphere containing the fixed stars rotating about it. The "seven wanderers" (the sun, moon, and five planets) moved in between. The Greeks' concern was to construct combinations of uniform circular motions centered in the earth by which the movements of the seven wanderers among the stars could be represented. Each body was moved by a set of interconnecting rotating spherical shells. This system was adopted by Aristotle, who introduced 55 shells to account for observed motions. This real model based on geometry was capable of reproducing the apparent motions, at least to a degree consistent with the accuracy of the contemporary observations. However, since it kept each planet a fixed distance from the earth, it could not account for the varying brightness of the planets as they moved.

This system was modified by Ptolemy, the last great astronomer at the famous observatory at Alexandria, in the second century A.D. In its simplest form the Ptolemaic system can be described as follows: Each planet moved in a small circle (epicycle) in the period of its actual motion through the sky, while simultaneously the center of this circle moved around the earth on a larger circle. The basic model was capable of repeated modification to account for new observations, and such modifications in fact took place. The result was that by the thirteenth century the model was extremely complicated, 40–60 epicycles for each planet, without commensurate effectiveness.

By the beginning of the sixteenth century there was widespread dissatisfaction with the Ptolemaic system. Difficulties resulting from more numerous and more refined observations forced repeated and increasingly elaborate revision of the epicycles on which the Ptolemaic system was based. As early as the third century B.C. certain Greek philosophers had proposed the idea of a moving earth, and as the difficulties with the Ptolemaic point of view increased, this alternative appeared more and more attractive. Thus in the first part of the sixteenth century the Polish astronomer Copernicus proposed a heliocentric (sun-centered) theory in which the earth, among the other planets, revolved about the sun. However, he retained the assumption of uniform circular motion—an assumption with a purely philosophical basis—and consequently he was forced to continue the use of epicycles to account for the variation in apparent velocity and brightness of the planets from the earth.

The next step, and a very significant one, was taken by Johannes Kepler. During the years 1576–1596 a Swedish astronomer, Tycho Brahe, had collected masses of observational data on the motion of the planets. Kepler inherited Brahe's records and undertook to modify Copernican theory to fit these observations. He was particularly bothered by the orbit of Mars, whose large eccentricity made it very difficult to fit into circular orbit-epicycle theory. He was eventually led to make a very creative step, a complete break with the circular orbit hypothesis. He posed as a model for the motions of the planets the following three "laws":

1. The planets revolve around the sun in elliptical orbits with the sun at one focus (1609).
2. The radius vector from the sun to the planet sweeps out equal areas in equal times (1609).
3. The squares of the periods of revolution of any two planets are in the same ratio as the cubes of their mean distances to the sun (1619).

These laws are simply statements of observed facts. Nevertheless, they are perceptive and useful formulations of these observations. In addition to discovering these laws, Kepler also attempted to identify a physical mechanism for the motion of the planets. He hypothesized a sort of force emanating from the sun which influenced the planets. This model described very well the accumulated observations and set the stage for the next refinement, due to Isaac Newton.

All models developed up to the middle of the seventeenth century involved geometrical representations with minimal physical interpretation. The fundamental universal law of gravitation provides at once a physical interpretation and a concise and elegant mathematical model for the motion of the planets. Indeed, this law, when combined with the laws of motion, provides a description of the motion of all material particles. The law asserts that every material particle attracts every other material particle with a force which is directly proportional to the product of the masses and inversely proportional to the square of the distance between them. In this framework the motion of a planet could be determined by first considering the system consisting only of the planet and the sun. The latter problem involves only two bodies and is easy to solve. The resulting predictions, the three laws of Kepler, are good first approximations since the sun is the dominant mass in the solar system and the planets are widely separated. However, the law of gravitation asserts that each planet is, in fact, subject to forces due to each of the other planets, and these forces result in perturbations in the predicted elliptical orbits. The mathematical laws proposed by Newton provide such an accurate mathematical model for planetary motion that they led to the discovery of new planets.

One could examine the orbit of a specific planet and take into account the influence of all the other planets on this orbit. If discrepancies were observed between the predictions and observations, then one could infer that these discrepancies were due to another planet, and estimates could be obtained on its size and location. The planets Uranus, Neptune, and Pluto were actually discovered in this manner. However, even this remarkable model does not account for all the observations made of the planets. Early in this century small perturbations in the orbit of Mercury, unexplainable in Newtonian terms, provided some motivation for the development of the theory of relativity. The relativistic modification of Newtonian mechanics apparently accounts for these observations. Nevertheless, one should not view this model as ultimate, but rather as the best available at the present time.

The laws of Newton, viewed as a mathematical model, have provided an extremely effective tool to the physical sciences. The concepts of force, mass, velocity, etc., can be made quite precise and the model can be studied from a very abstract point of view. Although the social and life sciences do not yet have their equivalents of Newton's laws, the utility of mathematical models in the physical sciences gives hope that their use may contribute to the development of other sciences as well.

Daniel P. Maki and Maynard Thompson, *Mathematical Models and Applications*, Prentice-Hall, Englewood Cliffs, N.J., 1973, pp. 7–9. Reprinted by permission of Prentice-Hall, Inc.

MODELS IN SCIENCE

We have made virtue of necessity. The features of the axiomatic method forced on us by logic turn out to be exactly the features desired for a scientific model.

Because some terms are left undefined, an abstract axiomatic structure is adaptable to the needs of scientists with widely varying interests. Points and lines may be variously interpreted as representatives and committees, cities and routes, dots and segments, stakes and chains, stars and light paths, etc.

A great step forward was taken in science when it became clear that the principles guiding work in a given area of science are not laws of the universe, but assumptions we are making. They are not self-evident truths; they are our present perceptions. This attitude toward basic principles is obviously in accord with the idea of writing down some axioms to begin with. It is instructive to note that this is precisely the way in which Newton began his epoch-making work, *Principia*. He refused, even when urged, to offer explanations as to why certain assumptions (for example, his dictum, "To every action there is an equal and opposite reaction") should be so. He patterned his work after Euclid, merely stating his principles as axioms and then showing their consequences.

We have emphasized that the axiomatic method is utterly dependent on human reason in deriving the implications of the axioms. Some philosophers of science have seen this as a limitation, arguing that the universe may be a chaotic place and that our method of studying it limits us to understanding only those things that happen to fall into patterns agreeable to our thinking processes. But others see faith in an orderly universe as the very backbone of science. They point out that one of the compelling reasons for abandoning the Ptolemaic model of astronomy was that, with its increasingly complex system of 40 to 60 epicycles for each planet, there developed a feeling that things just couldn't be that complicated. Poincaré adds to the argument for simplicity the example of someone who has taken a series of readings and plotted points (as in the figure). "Who," he asks, "would fail to draw through those points as smooth a curve as possible?" And what does this smooth curve represent except the experimenter's fundamental faith in an orderly universe?

The possibility of deducing implications merely by thinking is of course one of the principal attractions of the axiomatic method to a scientist. The discoveries of the planets Uranus, Neptune, and Pluto by analysis of Newton's model stand as remarkable examples of the power of the method. Not to be overlooked, however, are the hundreds of less celebrated accomplishments achieved in the same way. The marvels of electronics in communications owe much to the analysis of models; so do bridges, electric coffee pots, internal-combustion engines, vaccines, atomic power, wristwatches, etc. We seldom appreciate the power of a method that enables us to predict how things will turn out without always having to try them.

Let us comment on the tendency in axiomatic thinking to want to minimize the number of axioms we use. While no logical difficulty is encountered if we state as axioms things that can be proved as theorems, this is not esthetically pleasing; and much effort has been directed toward reducing the number of axioms needed in various mathematical structures. Such efforts have practical benefits of course; when a structure is being considered as a possible model in science, the fewer axioms to be checked as to whether they seem to be true, the better. But beyond this, the effort to state the minimal number of axioms goes to the heart of what many see as the ultimate aim of science. To those who believe that science is possible because the universe is orderly, no goal less than discovery of the basic (and presumably simple) underlying causes of what we observe is worthy of the scientific effort.

Without the belief that it is possible to grasp the reality with our theoretical constructions, without the belief in the inner harmony of our world, there could be no science.
L. Infeld and A. Einstein

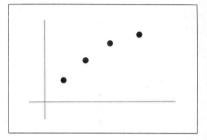

More laws are vain where less will serve.
Robert Hooke

The grand aim of all science is to cover the greatest possible number of empirical facts by logical deduction from the smallest possible number of hypotheses or axioms.
A. Einstein

AND FINALLY

For illustrative purposes, we have confined our discussion of modeling to either artificially simple situations or to the classic examples. Most scientists are concerned with something in between. An engineer for a power utility wants to know before a multimillion dollar atomic generating plant is built on a certain river whether or not the plant can be operated without heating the river above levels specified by environmental control regulations. This task requires more than well-intentioned rhetoric about the value of rivers; it calls for a model that will predict the river temperature under a variety of conditions. The variables that may be involved are numerous (air temperature, river temperature upstream from the plant, humidity and wind conditions, cloud cover, accumulated ground heat, river flow in cubic units per second, electrical demands at various times of the day, etc.), and the way in which they enter into determining the river temperature is open to speculation. (How much attention, if any, must be paid to surface evaporation on a dry, windy day?) Anyone facing such a problem develops a quick appreciation for the art of making simplifying assumptions, a willingness to accept approximate solutions, and an admiration for those who have solved similar problems with enough generality so that their methods can be easily adapted to a new situation.

Modeling plays an increasingly important role in the social sciences. Pollsters and political analysts try to develop models that will enable them to predict public preferences, whether for toothpaste or for political candidates. Economists seek models for an economy that will enable them to predict the effect of certain monetary policies. Psychologists want models for human learning patterns (so for some reason they study rats); and on it goes.

The mere collection of facts is not science. An investigation takes on the characteristics of a scientific endeavor when enough is known so that certain hypotheses can be formulated, that is, when a model can be constructed and subjected to testing.

SUMMARY

A mathematical model is obtained from an axiomatic system by assigning meanings to the undefined terms in such a way as to make the axioms appear to be true statements. Axiomatic systems do not grow in a vacuum; they commonly develop as attempts to model something observed in the world around us.

SELF–TEST

Choose the completion that seems best to you. If none of them seem quite correct to you, be prepared to tell why and to supply an ending of your own in class discussion.

1. Which of the following situations does not illustrate that people respond to a "belief in the inner harmony of our world"?
   (a) The reasons for abandoning the Ptolemaic model of the motions of the planets and stars.
   (b) The idea that we perceive order in the world because our thinking processes are orderly.
   (c) The tendency to draw as smooth a curve as possible through plotted points obtained as readings in an experiment (and perhaps even to apply a "fudge factor" to the points that don't "line up").
2. Scientific investigation
   (a) is ultimately based on assumptions.
   (b) is the one area of human activity in which we can have a high degree of certainty.
   (c) is so highly specialized that work in one area rarely relates to work in another area.
3. The distinction between the pure mathematician who works on axiomatic systems and the applied mathematician who uses the axiomatic systems to model natural phenomena
   (a) was easier to see in historic times when day-to-day affairs were not so intertwined with technology.
   (b) is more evident since our understanding of the axiomatic method has isolated it as a topic of study.
   (c) has never been accurate, and serves only as a convenience for discussing the nature of scientific inquiry.

PROBLEM SET 11.4

1. Using a simplified model, calculate the
   (a) Volume of your arm from your wrist to your elbow.
   (b) Volume of your head.
2. Using a simplified model, calculate the
   (a) Area of your arm (wrist to elbow).
   (b) Area of your head.
| c | 3. Estimate the number of apples in a bushel.
| c | 4. Estimate the number of marbles with a $\frac{1}{2}$ inch diameter that could be placed into a 1 gallon jug.
| c | 5. Jack Strop finds that it takes 45 seconds to fill a $2\frac{1}{2}$-gallon bucket with his garden hose. How long will it take him, using the same hose, to fill his new circular 30-foot-diameter plastic pool to a depth of 4 feet?

SOME COMMON FORMULAS
A cube of side a has surface area S and volume V given by

$$S = 6a^2$$
$$V = a^3$$

A sphere of radius r has surface area S and volume V given by

$$S = 4\pi r^2$$
$$V = \frac{4}{3}\pi r^3$$

A cylinder of radius r and height h has surface area S (not counting the circular bases) and volume V given by

$$S = 2\pi r h$$
$$V = \pi r^2 h$$

---

1 bushel = 2150 in.^{3}
1 gallon = 277 in.^{3}

c 6. Estimate the cost of fuel needed to drive a medium size automobile from Chicago to Los Angeles.

7. A globe is of course a model of the earth. Use a globe to estimate the shortest distance (if you are flying) between the following cities.
   (a) San Francisco and Cape Town.
   (b) Calcutta and Los Angeles.
   (c) New York and Sydney.
   (d) New Orleans and Tokyo.

8. Again using a globe, estimate the distance, traveling by boat, between the cities listed in Problem 7.

9. A theorem of geometry tells us that, if the sides of a triangle have lengths a, b, and c related by $c^2 = a^2 + b^2$, the angle opposite the side of length c is a right angle. This fact is used in many models. How can "point" and "line" be interpreted in the following situations?
   (a) A carpenter is setting forms in which to pour concrete for a patio and wishes to verify that the planks are at right angles.
   (b) Several of the neighbors have gotten together to lay out a Little League baseball field. They seek a method for making sure the foul lines are at right angles.

10. We introduced our discussion of finite geometries by talking about rules for governing a dormitory, then by discussing bus company regulations, and finally by talking about points and lines. By assigning still other meanings to "point" and "line," other potential models can be obtained. Do so. (For example, would the axioms stated at the beginning of Section 10.2 appear to be sensible statements if "points" represented letters and "lines" represented envelopes?)

Chapter 12
Geometries as Models

*What are we to think of the question: Is Euclidean
geometry true? It has no meaning. . . . One geometry cannot
be more true than another; it can only be more convenient.*
H. POINCARÉ

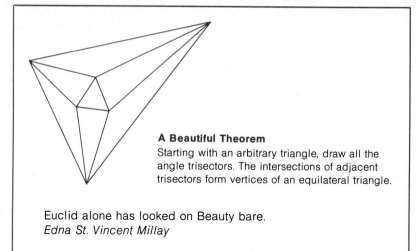

A Beautiful Theorem
Starting with an arbitrary triangle, draw all the angle trisectors. The intersections of adjacent trisectors form vertices of an equilateral triangle.

Euclid alone has looked on Beauty bare.
Edna St. Vincent Millay

The thirteen books of Euclid must have been a tremendous advance, probably even greater than that contained in the Principia of Newton.
Augustus DeMorgan

12.1 / Euclid's Work

Except for the Bible, probably no book has had a greater influence on Western culture than Euclid's *Elements*. And, like the Bible, it is a book that modern readers are more likely to praise than to open. Our purpose here therefore is to open Euclid's book, at least to the extent of summarizing part of what it says.

The first surprise to most readers, if they were to actually look at Euclid's *Elements,* would be to find that geometry is not the only topic treated. Euclid was a collector rather than an originator. Surely he was a man of mathematical talent, but his great contribution was to organize previously known mathematics so that everything followed from a few assumptions. Euclid's *Elements* is really a collection of 13 books, only 5 of which are devoted to plane geometry. Other topics include proportions, number theory, and solid geometry.

Even the idea of beginning with assumptions did not originate with Euclid. Aristotle had discussed the nature of axioms somewhat earlier, and his influence can be seen in Euclid's work. For instance, Aristotle pointed out that one need not worry about the truth of the assumptions, but both he and Euclid surely felt that the things about which they wrote were the stuff of reality. If one did not need to worry about whether or not the assumptions were true, it was because the derived consequences could be checked to see if they agreed with reality.

Whatever he thought about their truth, Euclid did have in mind the goal of making as few assumptions as possible. This is clear because he went to the trouble of proving many theorems that seem no less self-evident than the statements he took as assumptions.

Euclid failed, however, to appreciate the necessity of leaving some terms undefined. Book I begins with 23 definitions, and his troubles are apparent from reading the first few:

> A point is that which has no part.
> A line is a breadthless length. The extremities of a line are points.
> A straight line is a line that lies evenly with the points of itself.

Do these definitions really define? What is a "breadthless length"? What does it mean to "lie evenly"? And if these questions are answered, will the answers not use other terms needing definition, eventually leading us right back where we started?

A definition of a straight line that leads us in circles is indeed an unfortunate beginning for a book destined to serve for centuries as a model of logical thinking. This makes it easier for us, however, to agree with the modern viewpoint; some basic terms must be left undefined.

EUCLID'S POSTULATES (AXIOMS)

After listing a barrage of definitions, Euclid wrote down 10 grand assumptions. He called the first 5 "postulates" and the second 5 "common notions." We use the modern word "axiom" to describe them all. It is Euclid's choice of axioms that has stamped his name in history. Here they are, stated in their simple elegance.

Axiom 1. A straight line can be drawn from any point to any point.

Axiom 2. It is possible to extend a finite straight line indefinitely in a straight line.

Axiom 3. A circle can be drawn with any point as center and any radius.

Axiom 4. All right angles are equal.

Axiom 5 (Euclid's famous fifth, the Parallel Postulate). If a straight line intersects two straight lines so that the interior angles on the same side of it sum to less than two right angles, then the two lines, if extended indefinitely, will meet on that side on which the two angles are less than two right angles.

Axiom 6.   Things that are equal to the same thing are also equal to one another.

Axiom 7.   If equals be added to equals, the wholes are equals.

Axiom 8.   If equals be subtracted from equals, the remainders are equal.

Axiom 9.   Things that coincide with one another are equal to one another.

Axiom 10.   The whole is greater than the part.

For 300 B.C., they are remarkably clear. But to help our readers, we suggest that Euclid meant Axiom 1 to say that a *unique* line (i.e., just one line) can be drawn between any two different points. Similarly Axiom 2 should say that any finite straight line has a *unique* extension. Finally Axiom 5 (the parallel postulate) became the subject of a controversy which continued for 2000 years, a controversy we discuss in Section 12.2. It was restated by John Playfair in a form most of us find more understandable.

Axiom 5 (after John Playfair).   Given a line *L* and a point *P* not on *L*, there exists one and only one line *M* containing *P* and no points of *L*.

It is interesting to note that Euclid clearly distinguished between the definition of a figure and the proof from his axioms that such a figure can be constructed. This seemingly fine distinction led to some of the most famous problems in the history of mathematics, for it caused people to ask whether such and such a thing can be drawn using only a straightedge (to join any two points or to extend a segment indefinitely) and a compass (to draw a circle centered at any point, using any radius). Three of the most famous problems of this type are:

a. Trisect a given angle.
b. Construct a square with an area equal to the area of a given circle.
c. Construct a cube with a volume that is twice the volume of a given cube.

These problems attracted the efforts of many famous mathematicians, and many amateurs as well. Attempts to solve them have

To say, after you have tried and failed to do something, that no one will ever do it seems a bit arrogant. In mathematics, however, we can prove that some things cannot be done (assume they can be done, derive a contradiction). That is why no knowledgeable mathematician any longer tries to trisect angles. And it is the failure of nonmathematicians to understand this concept of impossibility which accounts for their willingness to keep working on the problem.

led to some very deep and useful mathematical ideas. They have also led to many wrong "solutions" and much frustration, for they are very hard. They are in fact worse than hard. We now know that they are all impossible. No one will ever solve any of them.

EUCLID'S THEOREMS

We list below several of Euclid's theorems. In stating these theorems, we assume the reader is familiar with definitions from elementary geometry (perpendicular and parallel lines, isosceles and equilateral triangles, etc.), and we have phrased them in ways that should sound familiar to the modern ear. Otherwise, the theorems and the order in which they are listed follow Euclid's *Elements*.

We have included proofs of several of the theorems for purposes of illustration. Most proofs are left as exercises for the reader.

Theorem 1. On a given finite straight line, an equilateral triangle can be constructed.
PROOF. Let the given line segment be *AB*. With *A* as center and *AB* as radius, construct a circle; construct a similar circle with *B* as center (Axiom 3). Let the circles intersect at *C*. Then *ABC* is the required triangle.

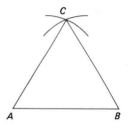

Theorem 4 (Side Angle Side). If two sides and the included angle of one triangle are equal to the corresponding parts of a second triangle, the triangles are congruent.
PROOF. Let the two triangles be designated ΔABC and $\Delta A'B'C'$, the notation chosen so that $\angle A = \angle A'$, $AB = A'B'$ and $AC = A'C'$. Imagine ΔABC moved so that the two equal sides and the equal included angles coincide. In particular, then *B* will coincide with *B'* and *C* with *C'*. Through these two points only one line can be drawn (this is where Euclid needs Axiom 1 in a form stating the uniqueness of the line). $BC = B'C'$ and the triangles coincide, as was to be proved.

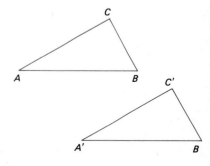

Theorem 5. The base angles of an isosceles triangle are equal.
PROOF. Let the given triangle be designated ΔABC, with $AB = AC$. The triangle can be "flipped" so that angle *B* lies where *C* was, and conversely. Then side *AB* corresponds to *AC*; and side *AC* corresponds to *AB*. Moreover, these corresponding sides and their included angles, being angle *A* in each case, are equal. Thus, by Theorem 4, the triangles ΔABC and ΔACB are congruent. It follows that the corresponding angles $\angle ABC$ and $\angle ACB$ are equal.

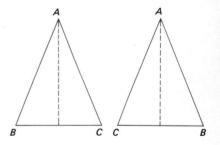

Theorem 6.   If two angles of a triangle are equal, then the sides opposite these angles are also equal.

Theorem 8 (Side Side Side).   If three sides of a triangle are equal to three corresponding sides of a second triangle, the triangles are congruent.

Theorem 9.   Any angle can be bisected.

Theorem 10.   Any given line segment can be bisected.

Theorem 11.   If *p* is a point on a line *l*, then at least one line can be drawn through *p* perpendicular to *l*.

Theorem 12.   If *p* is a point not on a line *l*, then at least one line can be drawn through *p* perpendicular to *l*.

Theorem 13.   If two lines intersect, any two adjacent angles have a sum of 180° (two right angles).

Theorem 15.   If two lines intersect, then the opposite angles (sometimes called vertical angles) are equal.

Theorem 16.   If one side of a triangle is extended, then the exterior angle is greater than either of the opposite interior angles.

Theorem 17.   The sum of any two angles of a triangle is less than 180° (two right angles).

Theorem 26 (Angle Angle Side).   If two angles and a side of one triangle are equal to the corresponding parts of a second triangle, then the triangles are congruent.

Theorem 27.   If a straight line falling on two other straight lines makes the alternate interior angles equal, then the two straight lines are parallel.

Theorem 29.   If a straight line falls on two parallel lines, it makes the alternate interior angles equal to each other.

Theorem 30. Lines parallel to the same line are parallel to each other.

Theorem 32. The sum of the interior angles of a triangle is 180° (two right angles).

Theorem 47 (Pythagoras). In a right triangle, the square of the length of the hypotenuse is equal to the sum of the squares of the other two sides.

This last theorem is certainly one of the best known and most useful theorems in plane geometry. There are several independent proofs attributable to ancient writers, mathematicians ranging from the most prominent to the obscure, and even an American president.

TWO CONSEQUENCES

Students in a plane geometry class are often required to use a format which emphasizes that each step must be justified by an axiom, a definition that has been given, or a theorem already proved. The method is illustrated by our proof that if it is given in the figure to the right that $AC = BC$ and $AD = BE$, then it follows that $CD = CE$.

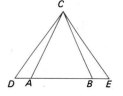

Statement	Reason
1. $AC = BC$ and $AD = BE$	1. Given
2. $\triangle ABC$ is isosceles	2. Definition of isosceles
3. $\angle CAB + \angle CAD = 180°$ $\angle CBA + \angle CBE = 180°$	3. Theorem 13
4. $\angle CAB + \angle CAD = \angle CBA + \angle CBE$	4. Axiom 6
5. $\angle CAB \qquad = \angle CBA$	5. Theorem 5
6. $\qquad \angle CAD = \qquad \angle CBE$	6. Axiom 8
7. $\triangle CAD \cong \triangle CBE$	7. Theorem 4
8. $CD = CE$	8. Definition of congruent triangles

We conclude this section with an application that deserves to be better known. Eratosthenes (275–194 B.C.) made, so far as we know, the first reasonably accurate calculation of the size of the earth. His argument not only applies geometry to a practical problem, but it nicely illistrates the way in which use is made of his model of the universe.

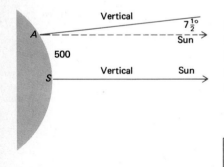

The model Eratosthenes used began with a spherical earth. Light was assumed to travel in rays (straight lines), and the sun was considered to be so far away that its rays could be taken to be parallel.

He then reasoned as follows. The city of Syene was 500 miles south of Alexandria. When the sun was directly overhead in Syene, the sun appeared at Alexandria to be $7\frac{1}{2}°$ off the vertical. Using the parallel lines provided by the sun's rays, the equality of vertical angles (Theorem 15), and the equality of alternate interior angles (Theorem 29), he arrived at the figure indicated.

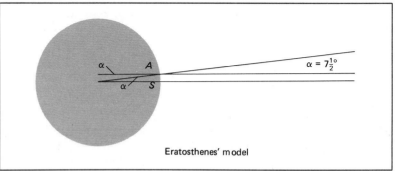

Eratosthenes' model

The angle at the center of the earth intercepts an arc of 500 miles. Working with proportions then, one writes

$$\frac{7\frac{1}{2}}{360} = \frac{500}{\text{circumference}}$$

or

$$\text{Circumference} = \frac{500(360)}{7\frac{1}{2}} = 24{,}000 \text{ miles}$$

(Modern geographers believe the earth to be pumpkin-shaped but, taking it as a sphere for easy calculation, the accepted value for the circumference today is 24,890 miles.)

SUMMARY

Most people in our society have heard of Euclid's *Elements,* but their misconceptions about it are many. You should now be able to give better answers to the questions below than might be expected from folks having only an ordinary education (i.e., not having read this book, or one like it).

> What is the subject of Euclid's *Elements*?
> Were most theorems of plane geometry known before the time of Euclid?
> What is Euclid's chief contribution to the development of mathematics?
> Does Euclid's work, as is sometimes alleged, continue to set the standard for logical exposition?
> In Euclid's eyes, what was the point?

In addition to providing some insights into Euclid's work, this chapter also serves as a reminder of certain concepts such as congruent triangles, isosceles triangles, and right triangles, and facts associated with them.

SELF–TEST

Choose the completion that seems best to you. If none of them seem quite correct to you, be prepared to tell why and to supply an ending of your own in class discussion.

1. Euclid proved some statements that are no less obvious than others that he took for axioms.
   (a) From this we deduce that Euclid understood that nothing is obvious.
   (b) It is therefore clear that not everything which seems obvious to the modern mind seemed obvious in Euclid's time.
   (c) Euclid does seem to have had in mind the goal of making as few assumptions as possible, and then proving everything else.
2. Euclid's first theorem asserted that on a given finite straight line, an equilateral triangle could be constructed.
   (a) We now know that there is really nothing to prove here; this is simply a restatement of the definition of an equilateral triangle.
   (b) This exemplifies the subtle but important distinction that Euclid made between defining a figure and proving that such a figure does exist.
   (c) This theorem is technically wrong since two distinct equilateral triangles can be constructed, one on each side of the line.
3. Attempts by amateurs to trisect an arbitrary angle using only a compass and straight-edge are routinely ignored by mathematicians
   (a) because it is unlikely that an amateur will solve a problem that has defied mathematicians for centuries.
   (b) because they know without reading them that they are wrong.
   (c) because we now have highly accurate tools to do the job, rendering the restriction to compass and straight-edge obsolete.

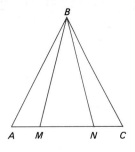

PROBLEM SET 12.1

In the following problems, use any of Axioms 1 through 10 and any of the stated theorems to prove the result.

1. Two isosceles triangles are congruent if a leg and the angle opposite the base of one are equal to the corresponding parts of the other.
2. Two right triangles are congruent if a side and the hypotenuse of one equal the side and hypotenuse of the other.
3. If *AM* is the perpendicular bisector of *BC*, then $\triangle ABC$ is isosceles.
4. The medians to the equal sides of an isosceles triangle are equal.
5. If $\triangle ABC$ of the top figure in the margin is isosceles and $\angle ABM = \angle CBN$, then $\triangle BMN$ is isosceles.

6–17. *Prove Theorems 6 through 17 in this section. In proving any theorem on the list, you may assume previously listed theorems but not subsequent ones.*

18. In the bottom figure in the margin, $\angle TRS = \angle TSR$, and *SJ* and *RK* bisect angles *S* and *R*, respectively. Prove that $\angle TJS = \angle TKR$.
19. In the bottom figure in the margin, *RK* and *SJ* bisect the base angles of the isosceles triangle *RST*. Prove that $RJ = SK$.

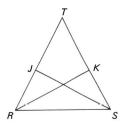

THE PARALLEL ROAD TO DESTRUCTION

Wolfgang Bolyai, father of John Bolyai who did pioneering work in non-Euclidean geometry, did nothing to encourage his son's efforts. Quite to the contrary! He once wrote to his son as follows.

> You must not attempt this approach to parallels. I know this way to its very end. I have traversed this bottomless night, which extinguished all light and joy of my life. I entreat you, leave the science of parallels alone.... I thought I would sacrifice myself for the sake of the truth. I was ready to become a martyr who would remove the flaw from geometry and return it purified to mankind. I accomplished monstrous, enormous labors; my creations are far better than those of others and yet I have not achieved complete satisfaction.... I turned back when I saw that no man can reach the bottom of the night. I turned back unconsoled, pitying myself and all mankind.
>
> I admit that I expect little from the deviation of your lines. It seems to me that I have been in these regions; that I have traveled past all reefs of this infernal Dead Sea and have always come back with broken mast and torn sail. The ruin of my disposition and my fall date back to this time. I thoughtlessly risked my life and happiness—aut Caesar aut nihil.

12.2 / Non-Euclidean Geometry

In many ways, the development of non-Euclidean geometry is one of the most significant of human achievements. Not only did it transform our understanding of geometry, but it affected our whole concept of science, and ultimately of reality.

DIFFICULTIES WITH THE PARALLEL POSTULATE

From the very beginning, Euclid's fifth axiom was troublesome. The most casual reader of his *Elements* is certain to notice it just because it is much longer than the others; and if he goes the extra step of actually reading the axioms, he will see that it is much less intuitive than the other nine. Euclid himself leaves evidence of being dissatisfied with the parallel postulate, since he proceeds as

The mathematician Jean le Rond d'Alembert wrote in 1759 that the problem of the parallel postulate was "the scandel of the elements of geometry." In 1763 a mathematician named Klügel wrote a doctoral thesis exposing errors in 28 different "proofs" of the parallel postulate.

far as he can without using it. Those who came after Euclid were also unhappy with this postulate, and their efforts to remedy the situation serve in some ways to underscore Euclid's genius in including it in the form that he did.

There were two basic approaches to the parallel postulate from 300 B.C. to 1800 A.D. One idea was to deduce it as a theorem from the other nine axioms. The other was to replace it by something that would more directly appeal to intuition. It was the latter effort that led Playfair to propose in 1795 the form mentioned in Section 12.1, and repeated here for convenience.

Playfair's Axiom.   Given a line *L* and a point *P* not on *L*, there exists one and only one line *M* containing *P* and no points of *L*.

One of the most systematic, and in a sense most fruitful, approaches to the problem was made by a Jesuit professor of mathematics named Girolamo Saccheri (1667–1733). His idea was to begin with all of Euclid's assumptions, except the troublesome Axiom 5, and to prove as many theorems as possible. This leads to what is called **absolute geometry.** The first 28 theorems of Euclid in which he avoided using Axiom 5 are therefore theorems of absolute geometry. Proceeding in this way, Saccheri eventually defined what is now called a **Saccheri Quadrilateral.** This is a quadrilateral *ABCD* in which angle *A* and angle *B* are right angles and the lengths of sides *AD* and *BC* are equal; that is, *AD* = *BC*. The reader's immediate reaction is probably to call such a figure a rectangle. That, however, is a response that comes either from what seems apparent from a picture, or else from a theorem of Euclidean geometry. Neither reason is allowed here. The picture is, we recall, just one of many possible models of the axiom system; and to prove that the figure is a rectangle, one needs the very axiom that has been omitted in absolute geometry. Working with only the nine axioms of absolute geometry (which, as we have observed, allow us to use the first 28 theorems of Euclid), we prove the following.

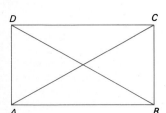

Theorem (Saccheri).   If *ABCD* is a Saccheri Quadrilateral, $\angle C = \angle D$. PROOF.   We can draw lines *AC* and *BD* (Axiom 1). Then $\triangle ABC$ is congruent to $\triangle ABD$ (Theorem 4), and so we have $\angle CAB = \angle DBA$, $\angle ACB = \angle BDA$, and *AC* = *BD*. Using the information that $\angle CAB = \angle DBA$ and the given fact that $\angle DAB = \angle CBA$, together with Axiom 8, we have $\angle DAC = \angle CBD$. Another appeal to Theorem 4 then gives $\triangle AACD$ congruent to $\triangle BDC$, hence $\angle ACD = \angle BDC$. Finally, since $\angle ACB = \angle BDA$ and $\angle ACD = \angle BDC$, we have from Axiom 7 that $\angle BCD = \angle ADC$; i.e., $\angle C = \angle D$.

With this theorem proved, Saccheri moved on to "prove" the

troublesome Axiom 5 by considering three possibilities: (1) the equal angles C and D are right angles; (2) they are obtuse angles; or (3) they are acute angles. He then showed that case 1 can be true if and only if Euclid's Parallel Postulate holds. It remained to show that both case 2 and case 3 led to contradictions, and he was able to do this for case 2. Angles C and D could not be obtuse, for a contradiction would result. He was one step from the long-sought proof. He only had to show that case 3 also led to a contradiction. But, try as he would, no contradiction could be obtained. He did get a series of theorems which seemed ridiculous. For instance, he proved that, if case 3 could be established, then

1. In the Saccheri Quadrilateral, $CD > AB$.
2. In any right triangle, the sum of the interior angles is less than 180°.

One must ask, however, why these theorems seem ridiculous. The answer is disconcerting. They are ridiculous because they violate what seems apparent from a picture, or because they contradict theorems from Euclidean geometry. These are the very arguments we rejected when the Saccheri Quadrilateral first appeared to us to be a rectangle. If we would not allow them before, we should not allow them now. Saccheri was apparently tiring, however (he died the same year his book was published), because after several theorems of this type he concluded that case 3 was impossible, since it led to conclusions that were "repugnant to the nature of straight lines." He published his work in a book modestly titled, *Euclid Freed of Every Defect.*

Saccheri's conclusion was the only unfortunate part of his work. The fact is that Saccheri's theorem is a theorem of absolute geometry, but that the axioms of absolute geometry do not allow one to decide in favor of either

> The best of thinkers are on occasion guilty of those lapses in logic that allow them to prove what they believe.

1. The equal angles C and D are right angles, or
3. The equal angles C and D are acute angles.

One of these can be chosen arbitrarily and added to the list of axioms. If case 1 is chosen, we are back to Euclidean geometry; if case 3 is chosen, we get a different geometry, but nevertheless a geometry that is self-consistent.

It had, as a matter of fact, occurred to several people in the late eighteenth century that it might be possible to obtain a logically consistent geometry even if one replaced Euclid's Parallel Postulate by something contradictory to it. Three names associated with this revolutionary idea are J. H. Lambert (1728–1777), C. F. Schweikart (1780–1859), and F. A. Taurinus (1794–1874). But these three apparently never thought that such axiom tampering could be related to the physical world, which they still believed to be just the way Euclidean geometry described it. It was the great mathematician Gauss who first grasped the full significance of what was blowing in the wind. He traced his ideas to a period when he was about 15 and wrote to a friend, "I am becoming more and more convinced that the [physical] necessity of our [Euclidean] geometry cannot be proved, at least not by nor for human reason." Gauss did not publish his thinking on this subject, but his ideas probably had some influence on the work of two younger men who did publish their ideas.

LOBATCHEVSKY (1793–1856) and BOLYAI (1802–1860)

Nikolai Lobatchevsky was a Russian who had for one of his teachers a good friend of Gauss. He wrote several papers about geometry before writing the full-scale treatment in 1840 that came to Gauss's attention and subsequently made Lobatchevsky famous. John Bolyai was the son of the Hungarian mathematician Wolfgang Bolyai. John began to fill in as lecturer to his father's university classes when he was 13, was an accomplished violinist, and once accepted the challenge to duel 13 fellow soldiers—the understanding being that he could have a short period of repose with his violin between each duel. R. Bonola records that he left all 13 lying in the square. Mightier than his sword was his pen, however, for in a 20-page appendix to one of his father's books, he immortalized himself with his "Essays on Elements of Mathematics for Studious Youths" (1832). It is significant to note that his father was also a friend of Gauss.

Since both Lobatchevsky and Bolyai were indirectly influenced by Gauss, it is perhaps not surprising that they had parallel ideas. Indeed, Gauss wrote (without much generosity, it is sometimes said) of John Bolyai's work that he was unable to praise it, for to do so would be to praise his own work. Bolyai and Lobatchevsky did develop the details of this new geometry independently, however. In fact, John Bolyai thought when he first read Lobatchevsky's work that it had been copied from him. The almost simultaneous appearance of their work is often cited as an example of the kind of thing that happens after the groundwork has been laid for a new idea.

The geometry of Lobatchevsky and Bolyai is obtained by adding to the axioms of absolute geometry one more axiom.

Lobatchevskian Parallel Axiom.   Given a line *l* and a point *p* not on *l*, there exist at least two lines containing *p* and no points of *l*.

The "repugnant" theorems of Saccheri then appear (along with the theorems of absolute geometry) as theorems in this new geom-

> Like many a son, John Bolyai chose to ignore his father's advice, and his progress on "this infernal Dead Sea" soon had his father writing in an altogether different vein.
> "If you have really succeeded in the question, it is right that no time be lost in making it public for two reasons: first because ideas pass easily from one to another, who can anticipate its publication; and secondly, there is some truth in this, that many things have an epoch, in which they are found at the same time in several places, just as the violets appear on every side in the spring."

etry. For instance, in Lobatchevskian geometry we are able to prove the following.

Theorem. The sum of the interior angles of a triangle is less than 180°.

We can almost sense that by now the reader is ready, logic or no logic, to side with Saccheri; these results violate good sense. Once again therefore we pause to remind the reader that "line" and "point" are undefined concepts, and that he or she is not to place on them the usual associations. Let us emphasize this by suggesting meanings that can be assigned to these words that will enable us to draw pictures in which the Lobatchevskian Parallel Axiom seems to be true. We restrict our attention to what, in familiar Euclidean terms, represents the inside of a circle (and not the boundary or rim). By "point" we mean a dot inside the circle. A "line" is a chord of the circle (minus its end points of course), and a "line segment" is the usual Euclidean "streak" joining any two (interior) points of the circle. It is easily seen that the nine axioms of absolute geometry seem to be true and, if we define parallel lines as lines having no point in common, it will be clear that through a point p not on a line l at least two lines (such as m_1 and m_2 in the figure) can be drawn parallel to l.

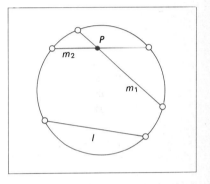

It now appears that there are an infinite number of lines through p that are parallel to l. This is so. In Lobatchevskian geometry one has the following.

Theorem. Given a line l and a point p not on l, there exist an infinite number of lines through p parallel to l.

Surely this is a new geometry. We stress that it is as logical as Euclid's geometry.

RIEMANN (1826–1866)

Once it was understood that one could, without jumping the tracks of straight thinking, abandon Euclid's postulate on parallel lines, others were quick to branch off. Bernhard Riemann was a student of (guess) Gauss; he was to make major contributions to many areas of mathematics, but it was in geometry that he gave the lecture that was to qualify him to teach at Göttingen in 1854. His idea was to take a direction opposite that of Lobatchevsky in replacing the troublesome parallel postulate.

Riemannian Parallel Axiom. There are no parallel lines.

Since the existence of parallel lines can be proved in absolute geometry, it is clear that Riemann could not make this assumption without abandoning more than Euclid's Parallel Postulate. This can be done in any of several ways. We outline only one. The idea is to restate Axiom 1 so that the line determined by two points need not be unique.

Axiom 1 (Riemann). Given any two points, there is at least one line containing them.

Riemann's statement has the virtue of saying exactly what he meant; he means to allow the possibility of more than one such line. With these changes, we again obtain a consistent geometry, and we again get theorems that differ in obvious ways from those of Euclidean geometry. In particular, having followed the work of Saccheri and his concern with the way in which interior angles add up, we note that we now have the "impossible" situation in which the two upper angles in the Saccheri Quadrilateral are obtuse angles.

Theorem. The sum of the interior angles of any triangle exceeds 180°.

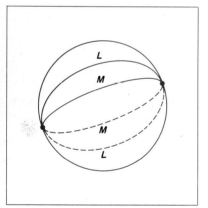

Once again we can assign meanings to the words "point" and "line" in such a way as to give us a picture in which the Riemannian axioms appear to be true. We think of the surface of a ball. Points are dots on this surface, and lines are great circles (the intersection formed by the ball and a plane cutting through the center). All the axioms appear to be true. In particular, there are no parallel lines (lines having no points in common); in fact, all lines intersect twice. Moreover, while any two points lie on at least one line (great circle), some pairs of points — those lying at opposite ends of a diameter — lie on an infinite number of distinct lines.

SUMMARY

Two basic approaches were used in attacking the "scandal" of Euclid's Parallel Postulate:

Deduce it from the other nine.
Replace it by something more intuitive.

Following the first line of attack, Saccheri defined a quadrilateral later associated with his name. With only the axioms of absolute geometry, he showed that either the parallel postulate could be derived or "repugnant" results were derived. Two mathematicians, Lobatchevsky and Bolyai, both having connections with Gauss, showed that an axiom contradicting the parallel postulate led to a consistent geometry and enabled one to prove the "repugnant" theorems of Saccheri. Riemann later started with still another substitute for Euclid's Parallel Postulate and obtained another self-consistent geometry. We have stated the substitute axioms and described for each geometry a model that satisfies the axioms.

SELF-TEST

Choose the completion that seems best to you. If none of them seem quite correct to you, be prepared to tell why and to supply an ending of your own in class discussion.

1. Euclid's treatment of the fifth axiom
   (a) is now known to be wrong.
   (b) shows that he had difficulty in treating an idea if he could not visualize it.
   (c) exhibits his genius, first because he recognized that this needed to be an axiom, and second because he stated it without making ambiguous or foolish statements about infinity.
2. The development of non-Euclidean geometries is often ranked as one of the great intellectual achievements of mankind. One reason for this is
   (a) it explained why Euclidean geometry could not be used for navigation on a spherical earth.
   (b) it dramatically demonstrated that the results we obtain depend upon the assumptions we make, and that replacing an assumption with its negation can result in a system no less logical than the original.
   (c) it demonstrated that even so revered a work as Euclid's *Elements* might have logical errors.
3. Efforts to prove Euclid's fifth axiom from the other nine
   (a) were doomed to failure because no such proof will ever be found.
   (b) may be resumed if geometry ever becomes a "fashionable" topic again.
   (c) resulted in numerous "proofs" that turned out to have errors in them.

PROBLEM SET 12.2

1. What evidence is there that Euclid was dissatisfied with his fifth postulate, the Parallel Postulate?
2. What reasons can you give for thinking that Gauss was the person principally responsible for the development of non-Euclidean geometry?
3. Can you think of instances in other branches of science where essentially the same idea is "found at the same time in several places, just as the violets appear on every side in the spring"?

4. The finite geometries discussed in Section 11.2 are non-Euclidean geometries. Contrast the axioms of finite geometries with those of Euclidean geometry. How many are the same? Where do they differ? Do you suppose finite geometries appeared before or after the work of Bolyai and Lobatchevsky?

Problems 5 and 6 relate to Lobatchevskian geometry in which we assume all the results of absolute geometry plus the Lobatchevskian Parallel Axiom, which, it will be recalled, leads to the following.

Theorem.   The sum of the interior angles of a triangle is less than 180°.

5. Use the theorem quoted above to show that in Saccheri's Quadrilateral the equal angles C and D must be acute.
6. Are the following statements true or false in Lobatchevskian geometry? (Refer to the model provided in the text.)
   (a) If two distinct lines m and n are both parallel to a third line, lines m and n are parallel to each other.
   (b) Rectangles do not exist.
   (c) Parallel lines are everywhere equidistant.
7. Are the following statements true or false in Riemannian geometry? (Refer to the model provided in the text.)
   (a) Two distinct triangles can have the same three vertices.
   (b) Two points determine either one line or infinitely many lines.
   (c) A triangle can have two angles of 90°.

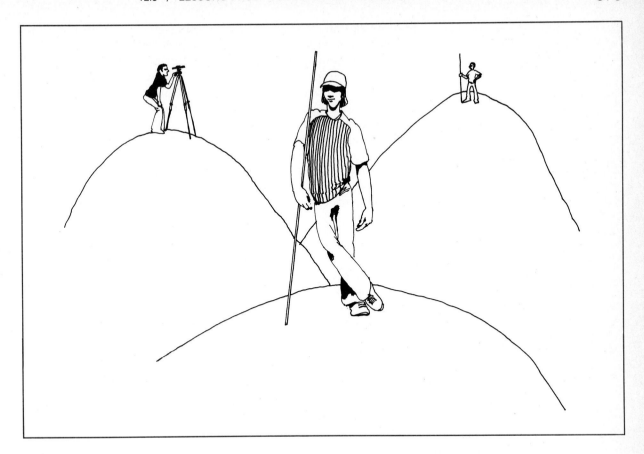

12.3 / Lessons from Non-Euclidean Geometry

We can almost hear the reader saying, "Well, your point is made. One does not have to assume that, through a point outside a line *l*, one and only one parallel to *l* can be drawn. One can assume with Lobatchevsky that millions of such lines are possible, or with Riemann that no such lines are possible. Since none of these assumptions lead to contradictions, any of them can be used to provide drill work for people who like logical drills. Very cute! Now let's get back to the plane facts of Euclid which correspond to the world the way it really is."

Such a response would be understandable; we may indeed appear to have belabored our point. At the same time, we as authors can be excused if we take some satisfaction in such a response, for it suggests that our readers have acquired an understanding of geometry that goes quite beyond that of most people. Yet we aim for more. After all, Lambert, Schweikart, and Taurinus apparently realized about 1800 that games could be played with the axioms in such a way that they would still yield logically consistent results. Our goal is to understand Gauss's observation that "the physical necessity of Euclidean geometry cannot be proved."

Any attempt to verify that a certain geometry is the so-called true geometry must identify the terms of geometry (points, lines, etc.) with objects in our world of experience. Then the following difficulties are encountered:

1. The truth of even the simplest axiom is difficult to ascertain.
2. The crucial axiom about parallel lines, depending as it does on extending lines indefinitely, reaches beyond the limits of our finite experience.
3. Attempts to check theoretical results (theorems) against facts are inevitably limited by our inability to measure things without error.
4. Ascertaining facts in the real world often involves assumptions about the real world which, if wrong, may explain why we have a lack of agreement between theoretical predictions and observations.

We illustrate each of these difficulties. With respect to the first, consider the common attempt to identify "points" with dots, and "lines" with streaks on a piece of paper. Look at two dots with a magnifying glass. Is it clear that only one line can be drawn through them? Is there any difference between passing through a point and touching a point?

Two tracks, laid for a railroad car, never touch of course. But they follow the curvature of the earth, so they aren't straight lines (not even in the sense of Riemann's geometry as interpreted on the surface of a ball). Now imagine a plane that extends into space.

Lay two tracks on it for a railroad car; that is, lay them so they never touch. Are these lines straight? How do you know? Can you—can anyone—follow such lines indefinitely far?

We have the following contrasting theorems in three geometries.

Euclidean: The interior angles of a triangle sum to 180°.

Lobatchevskian: The interior angles of a triangle sum to less than 180°.

Riemannian: The interior angles of a triangle sum to more than 180°.

Neither of the last two theorems tells us what deviation from 180° is to be expected. Having already discussed and experimented with summing the interior angles of a triangle (Section 10.1), we are well aware of the difficulty. Suppose we measure and get 179°59′58″. Is Lobatchevskian geometry thereby established as the true geometry? An experiment of this kind was once attributed to Gauss who, the story went, stationed observers on three mountain peaks and had them measure the angles of the triangle so formed. The figure 179°59′58″ given was the one attributed to them. Though it now seems certain that the story was not true, it illustrates the difficulty. Is an observed deviation due to observational error, or is the geometry of Lobatchevsky the real geometry? When telescope making became more refined, a similar experiment was performed using stars as three points of a triangle. This experiment, following the theoretical work of Einstein, suggested that a form of Riemannian geometry may be the true geometry.

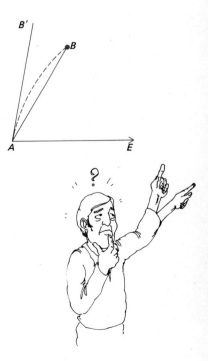

Finally, we note that any of the measurements that try to use large triangles, whether in the clouds or in the stars, rely on the assumption that light travels in a straight line. What if it doesn't? What if it bends? That is, what if light coming from *B* follows the dotted line to *A*? In measuring the angle from due east to *B*, the observer then looks in the direction *AB′* to see *B*. A grievous error has been made if the observer thinks that *EAB* has been measured.

It should now appear as something of an understatement to say that it is difficult to determine what is the "true" geometry. The better view seems to be that we should think of any geometry as a model, and that we should at any given time use the model that gives results most in accord with observations.

LEARNING FROM NON-EUCLIDEAN GEOMETRY

The introduction of non-Euclidean geometry laid to rest once and for all the notion that Euclid's postulate about parallel lines could be proved. And it did more than this. It forced on us the idea that axioms are not self-evident truths; and consequently it resulted

> If Euclid could be wrong, so
> might anyone else.

in a reevaluation of the relation between scientific theories and reality. We have come to see any theory as a model which we use as a tentative explanation of things that can be observed.

Besides altering the way in which we understand the world around us, non-Euclidean geometries have made us think about the very nature of human reason. We have more clearly distinguished between validity and truth. It has become more obvious that our conclusions depend on the initial assumptions, and that these are more arbitrary than was once thought, even if the goal is to obtain results useful in understanding what we observe around us.

> The conclusions of right thinking
> are valid, but they may be
> wrong.

Finally, non-Euclidean geometry provided the foundation for a concept of space and time that was to become very important for Einstein's work. This is the one result least likely to be understood (we, for example, have not even hinted at just how Riemannian geometry is related to the theory of relativity.) Yet, it is probably this "practical result" that would most persuasively justify non-Euclidean geometry in the eyes of most people (if they knew there were such a thing as non-Euclidean geometry). That which produces visible power unfortunately obscures the power of ideas that affect our whole understanding of the world around us.

> I attach special importance to
> the view of geometry I have just
> set forth, because without it I
> should have been unable to
> formulate the theory of relativity.
> A. Einstein

SUMMARY

We have listed four difficulties that stand in the way of deciding which geometry is the "true" geometry. These same problems occur whenever we attempt to verify a scientific theory and, when this lesson was driven home by the study of geometry, it changed our entire understanding of science. Gone is the expectation of finding a law that describes once and for all the way the world really is. With new force, it has become clear that "now we see through a glass darkly," and that any theory should be regarded as a tentative model.

SELF–TEST

Choose the completion that seems best to you. If none of them seem quite correct to you, be prepared to tell why and to supply an ending of your own in class discussion.

1. Experiments to accurately measure the sum of the interior angles of a triangle are of interest because
   (a) it is known that the sum is 180°, so the experiments afford an opportunity to check the accuracy of the measuring instruments.
   (b) a reliable result might settle the argument as to whether "real" geometry is Euclidean, Lobatchevskian, or Riemannian.
   (c) the foundations of Euclidean geometry would be destroyed if the sum could be shown to be different from 180°.

2. Non-Euclidean geometry
   (a) has no practical applications.
   (b) has stymied progress in geometry because it cast doubts on the foundations of the subject.
   (c) has helped us to understand that our assumptions are not so self-evident as we once thought.

PROBLEM SET 12.3

1. Discuss the nature of scientific laws such as those below. How are they verified? What possible questions exist about their applicability? Within what limits can they be used?
   (a) The force exerted by a spring is proportional to the distance it has been displaced (Hooke's Law).
   (b) To every action there is an equal and opposite reaction (Newton's Third Law).
   (c) An attack of appendicitis is accompanied by a rise in the white corpuscle count.

2. Think of some simple statements of fact about which you have recently had difficulty in coming to agreement with someone. For example:

   a. This board will be strong enough.
   b. There's plenty of gas to get to the next town.
   c. The radiator won't freeze tonight.

   Short of subjecting oneself to the inconvenience if not the danger of being wrong, how are such disputes settled?

3. Suppose that, as we move further from the center of the earth, the size of all material objects decreases proportionally. Could we detect this? How? Would this cause a finite universe to look infinite?

4. In a novel called *Flatland,* (Dover Publications, New York, 1952) Edwin A. Abbott describes a world in which everything happens in a plane and the characters have no perception of a third dimension. The two-dimensional characters can be perceived from our vantage point outside the plane as circles (the educated class—well rounded), triangles, etc., but they fail to see any of this. Can you think of any way they could learn to distinguish between characters of different shapes?

5. Try to recall and state in your own words the four difficulties that we have said confront anyone who wishes to decide on the "true" geometry. Can you restate these without referring to geometry so that they are applicable to attempts to verify any theory? Do these difficulties apply, for example, to one who wishes to prove that there is no God?

> **AN EXAGGERATION**
>
> It has been customary when Euclid, considered as a textbook, is attacked for his verbosity or his obscurity or his pedantry, to defend him on the ground that his logical excellence is transcendent, and affords invaluable training to the youthful powers of reasoning. This claim, however, vanishes on a close inspection. His definitions do not always define, his axioms are not always indemonstrable, his demonstrations require many axioms of which he is unconscious. A valid proof retains its demonstrative force when no figure is drawn, but very many of Euclid's earlier proofs fail before this test . . . . The value of his work as a masterpiece of logic has been very grossly exaggerated.
> *Bertrand Russell (1902)*

12.4 / Lessons from Euclidean Geometry

Many claims have been made about the value of studying Euclidean geometry. In this section we highlight what we believe are the main contributions, apart from the actual facts of plane geometry, that it makes to a general education.

EUCLID'S WEAKNESSES

Judged by present standards and understanding, there are defects in Euclid's *Elements*. It is our purpose to identify some of these, not to criticize Euclid (as some have) but to set in bold relief things we think have been learned.

We have already pointed out that Euclid never realized that some terms must be left undefined. His attempts stand in fact as a classic example of the futility of trying to define everything. This realization came rather late in intellectual history, however, and it is not surprising that Euclid stumbled over the idea. Now we do realize it, however, so there is no reason for an educated person, whatever his or her field, to think that every term should be defined. The political scientist can choose, without apology, to use a few basic terms (freedom, democracy, etc.) without defining them, as if everyone understood them. The theologian need not blush over his or her inability to define God. For some terms, there can be no definition except associations that develop through the

> Let no one ignorant of geometry enter this door.
> *Plato*

> We should not try to define every term.

ways in which the term is used (the assumptions that are made about it).

It is also likely that Euclid believed his assumptions to be true statements about the world as it really is. One appreciates the nature of his starting assumptions only after seeing that it is possible to start with different assumptions, that the different assumptions lead to a different but completely consistent system, and that the new system may also explain in a useful way things we observe in the world about us. Once this is seen, the lesson ought not to be lost on an educated person. The Declaration of Independence of the United States reasons from principles the writers took to be self-evident. The economic theory one accepts ultimately rests on a view of humans and how they can or will behave. The Christian reads in the Bible that "he that cometh to God must believe that He is," and similar assumptions undergird any system of religious thought. Any effort to develop a theory or explanation of anything ultimately rests on assumptions that cannot be proved.

> Any logical argument ultimately rests on assumptions.

Euclid's biggest problem, however, involved not his attitude toward the axioms and postulates he stated, but the many assumptions he made without realizing he had done so. This remark will be better understood after we examine the following attempt to prove what most people agree is a ridiculous assertion.

Theorem.   Every triangle is isosceles.
PROOF.   Beginning with an arbitrary triangle $\triangle ABC$, we draw the bisector of the angle at A (Theorem 9) and the perpendicular bisector of BC (Theorems 10 and 11), designating their intersection by K. Next draw BK and CK (Axiom 6), and from K drop perpendiculars KN and KR to sides AB and AC, respectively (Theorem 12). Then $\triangle BMK$ and $\triangle CMK$ are congruent (SAS), so $BK = CK$. Also $\triangle AKN$ and $\triangle AKR$ are congruent (AAS), so $NK = RK$. Thus $\triangle BKN$ is congruent to $\triangle CKR$ (Problem 2, Section 11.1), giving us $BN = CR$. Using once again the fact that $\triangle AKN$ is congruent to $\triangle AKR$, we have $NA = RA$. Thus $BA = CA$ (Axiom 7).

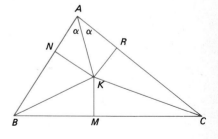

You probably think something is wrong with the proof. Try to find it.

Sooner or later, you will come to the conclusion that the picture is drawn incorrectly. And it is true that, if you start with the segments joining the dots A, B, and C shown on this page, and if you construct with a straightedge and compass the bisector of the angle at A and the perpendicular bisector of segment BC, you will find that they intersect outside the triangle.

You may, indeed you should, object that proofs are not supposed to depend on pictures. Euclid thought his proofs did not

David Hilbert

One must take great care if all the assumptions being made are to be identified.

Euclid understood that the goal is to make the minimal possible number of assumptions.

depend on the pictures. Yet the only thing we seem to find wrong with the proof is that the picture appears to be incorrect.

Actually, the argument above is not easily fixed up by drawing a better picture; indeed, with nothing more than the axioms stated by Euclid to guide you, it is not possible to decide whether a particular picture is right or wrong. To resolve the difficulty, nothing less will do than to add another axiom to Euclid's list.

Euclid did not of course try to prove this theorem. He just naturally (and unconsciously) made such further assumptions as he needed to exclude ridiculous theorems.

Thus is illustrated the biggest flaw in Euclid's work. Time and again he unconsciously used assumptions he had not stated. Our purpose is not to identify all his hidden assumptions. That is no easy job. The celebrated mathematician David Hilbert undertook this project, publishing his first effort in 1899; in 1930 he was still at it, that being the year in which he brought out his seventh revision of the project. When he had finished, the undefined terms had increased to 8 in number. He had grouped axioms according to the topic they dealt with, listing a total of 20 in five different groupings. It suffices to say that Euclidean geometry has turned out to be more complicated than Euclid expected.

Again there is a warning to the person who would be well-educated. It takes a great deal of effort to identify all the assumptions we are making.

EUCLID'S STRENGTHS

The endurance of his work is more elegant testimony to Euclid's strength than anything we can say. Nevertheless, we wish to comment on a few matters in particular.

Euclid's first theorem states that, on a finite straight line, an equilateral triangle can be constructed. If one thinks in terms of the "dots" and "streaks" usually used to represent points and lines, this assertion is hardly less obvious than Euclid's Axiom 3 which assures us that a circle can be drawn with any point as center and with any radius. Yet Euclid stated one assertion as a postulate and the other as a theorem. Clearly, Euclid was not trying to prove only things not obvious. He did have in mind the idea of assuming as little as possible and then deriving the rest.

The fifth postulate, Euclid's Parallel Postulate, has the mark of genius on it. In the first place, Euclid saw that such an assumption was necessary. As we have seen, generations of mathematicians have doubted this, and some have even offered (incorrect) proofs that this assumption is unnecessary. Euclid was right! Second, while aware that he needed such a postulate, he also recognized that its character was somehow different, and he avoided using it as long as possible. Finally, in his statement of the axiom, he carefully avoided mentioning infinity or otherwise introducing much of the vagueness (and sometimes outright nonsense) which even to this day appears in the writing of mystics who wish to relate the fate of parallel lines to immortality, etc.

Finally, let us recognize that, while modern critics can identify flaws in his work, Euclid wrote a book singularly influential in the area of science. Centuries later, Newton modeled his *Principia* after the style of Euclid's *Elements,* and it is still the basis of geometry courses taught in modern high schools. Moreover, the work of correcting such flaws as Euclid left has itself led to a deeper, more useful understanding of the nature of science.

> Euclid understood that a postulate about parallel lines was necessary, potentially troublesome, and possible to state unambiguously.

> Euclid understood what a valid proof is, and he wrote a book as enduring as human reason itself. The work he left has inspired some of the greatest of humankind's intellectual achievements.

SUMMARY

The criticisms made of Euclid's work merely illustrate what we have been saying throughout this part of the book. He did not seem to realize that some terms must be left undefined, and he probably felt that his axioms described the world the way it really was. His most serious error, however, was in failing to identify all his assumptions. We illustrated this by showing that, if we use only the assumptions he listed (instead of relying, as he did, on others besides), there is no easy way to exclude ridiculous theorems. However, we have called attention in marginal notes to some profound ideas that Euclid handled with a skill that fully justifies his reputation.

HATS OFF TO THIS MAN

Morris Kline (*Mathematics for Liberal Arts,* Addison-Wesley, Reading, Mass. 1967, p. 51) attributes the following story to Charles W. Eliot, former President of Harvard University.

President Eliot entered a crowded restaurant and handed his hat to the attendant in the check room. He got no hat check, and so when he returned sometime later to reclaim his hat, he was quite amazed to see the attendant at once pick his hat out of scores of hats on the racks. "How did you know that was my hat?" asked President Eliot. "I didn't," replied the attendant. "Why, then, did you hand it to me?" was Eliot's second question. "Because," said the attendant, "you handed it to me, sir."

SELF–TEST

Choose the completion that seems best to you. If none of them seem quite correct to you, be prepared to tell why and to supply an ending of your own in class dicsussion.

1. This section contains a proof, based on Euclid's axioms, that every triangle is isosceles. This proof is included
   (a) to show that even Euclid was capable of some pretty silly mistakes.
   (b) to show that disclaimers notwithstanding, Euclid's work did depend on the pictures.
   (c) to emphasize that Euclid failed to identify all the assumptions that he made.
2. Which of the following statements is not true of Euclid?
   (a) He made a futile effort to define all of his terms.
   (b) He thought his axioms were true statements about the world which all reasonable people would take to be self-evident.
   (c) He saw no reason to prove a statement that appeared to be self-evident.
   (d) He unconsciously assumed many things that were not stated in his listed axioms.
3. Which of the following is the greatest liability of Euclid's *Elements?*
   (a) The awe inspiring authority of his work inhibited explorations of alternatives, thus unwittingly holding back progress in the study of geometry.
   (b) It might have had far greater effect if the results could have been presented in a more readable style that played down the role of rigorous deduction.
   (c) Based on what he wrote, there is no way to exclude from the list of possible theorems some that are perfectly ridiculous.

PROBLEM SET 12.4

In Problems 1 through 4, list the implicit assumptions being made.

1. My son got a poor grade in mathematics, so I called his teacher. The teacher feels that the kid works hard and that he is probably doing about as well as possible in view of his low aptitude scores in mathematics.
2. My daughter got a poor grade in band. I guess that's to be expected. No one in our family has much ability in music.
3. I would have been embarrassed about not writing home for so long were it not for the fact that my folks are terrible correspondents themselves.
4. He should be ashamed of himself. He took advantage of Jim's ignorance when he sold him that lemon of a car.

Problems 5–8 are continuations of the correspondingly numbered problems in Problem Set 11.3. In each case we have given an invalid conclusion which may appear to some to be true. What minimal addition or change in the hypotheses will make the conclusion valid?

5. This course can be useful in preparing for certain jobs.
6. Some television serials are useful because they help us escape from the problems of real life.
7. Homer wants to learn.
8. Mr. Jones has a kind nature.
9. It can be proved in Euclidean geometry that, if a triangle is inscribed in a semicircle (understanding that two vertices lie at the end points of the diameter), the triangle is a right triangle. Using this theorem, we prove the following.

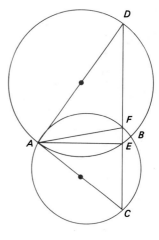

Theorem.   There exists a triangle having two right angles.
PROOF.   Draw two circles meeting at points *A* and *B*. Draw diameters *AC* and *AD* of the respective circles and designate by *E* and *F* the points where the line *CD* intersects the two circles. Now △*AFC* is a right triangle, as is △*AED* (because both are inscribed in semicircles). Then △*AEF* is the required triangle having two right angles.

What is wrong with this proof?

The interplay between generality and individuality, deduction and construction, logic and imagination—this is the profound essence of live mathematics. Any one or another of these aspects of mathematics can be at the center of a given achievement. In a far reaching development all of them will be involved. Generally speaking, such a development will start from the "concrete" ground, then discard ballast by abstraction and rise to the lofty layers of thin air where navigation and observation are easy; after this flight comes the crucial test of landing and reaching specific goals in the newly surveyed low plains of individual "reality." In brief, the flight into abstract generality must start from and return again to the concrete and specific.   *Richard Courant*

Part IV
Abstracting from the Familiar

ABSTRACTIONS

We point to three bulbs in a ceiling fixture and tell the toddler, "Three." We say the same thing about the cat's kittens, some envelopes on the table, and the eggs boiling on the stove. From what is common about the bulbs, the kittens, the envelopes, and the eggs, the child is expected to perceive the meaning of "three." The concept of number is an abstraction.

Back in Chapter 11, we introduced finite geometries by referring first to the governance of a dormitory and then to the regulation of a bus company. From what was common to the rules and regulations, we abstracted a set of statements about pokes and lokes, later changed to points and lines.

Emphasizing that the logic of our arguments was not to depend on preconceived notions about points and lines, we derived consequences that can apply to any number of specific situations, depending on what real objects are identified with the terms "point" and "line." Our reasoning was abstract, a matter of the logical relationships between the terms, and not related to the meaning of the terms.

Abstraction is a central feature of mathematics. We ask what essential properties are shared by different mathematical systems. At the same time we look for basic properties that make two systems distinct. Finally, when the essential properties that characterize a system have been identified, one looks for the inescapable implications of these properties.

In this, the fourth part of the book, we examine the basic properties of the number systems with which we are familiar, and then those of some less familiar systems. Against the background of these examples, we state the axioms for two important algebraic structures called rings and fields (no connection with circus rings or farmers' fields). In proving several theorems that must hold in these abstract settings, we obtain insights into why the familiar rules of arithmetic must be as they are. We also see that algebra, no less than geometry, is an axiomatic system in which all the results rest on a sparse set of assumptions we call axioms.

Emmy Noether (1882–1935)
Courtesy Bryn Mawr College Archives

Fraulein Noether was the most significant creative mathematical genius thus far produced since the higher education of women began. In the realm of algebra, in which the most gifted mathematicians have been busy for centuries, she discovered methods which have proved of enormous importance in the development of the present-day younger generation of mathematicians. Pure mathematics is, in its way, the poetry of logical ideas. One seeks the most general ideas of operation which will bring together in simple, logical and unified form the largest possible circle of formal relationships. In this effort toward logical beauty spiritual formulae are discovered necessary for the deeper penetration into the laws of nature.

Born in a Jewish family distinguished for the love of learning, Emmy Noether, who, in spite of the efforts of the great Göttingen mathematician, Hilbert, never reached the academic standing due her in her own country, none the less surrounded herself with a group of students and investigators at Göttingen, who have already become distinguished as teachers and investigators. Her unselfish, significant work over a period of many years was rewarded by new rulers of Germany with a dismissal, which cost her the means of maintaining her simple life and the opportunity to carry on her mathematical studies. Farsighted friends of science in this country were fortunately able to make such arrangements at Bryn Mawr College and at Princeton that she found in America up to the day of her death not only colleagues who esteemed her friendship but grateful pupils whose enthusiasm made her last years the happiest and perhaps the most fruitful of her entire career.

A. Einstein

Chapter 13
Number Systems

Numbers are an indispensable tool of civilization, serving to whip its activities into some sort of order. . . . The complexity of a civilization is mirrored in the complexity of its numbers.

PHILIP J. DAVIS

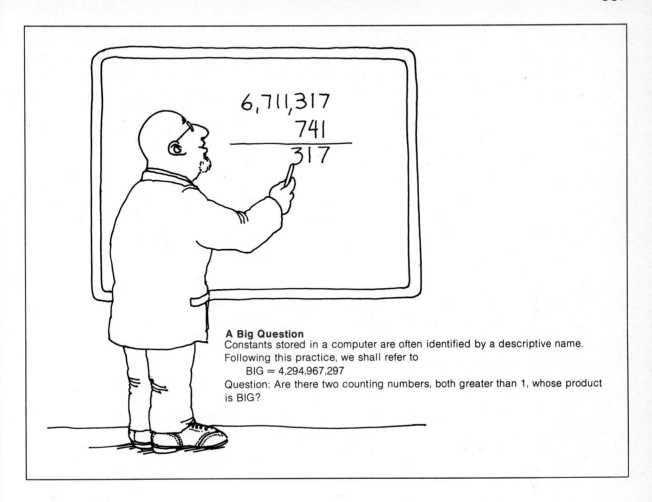

A Big Question
Constants stored in a computer are often identified by a descriptive name. Following this practice, we shall refer to
BIG = 4,294,967,297
Question: Are there two counting numbers, both greater than 1, whose product is BIG?

13.1 / The Counting Numbers

We have pointed out in our opening comments that a number like three is an abstraction. The collection of all such numbers (one, two, . . ., nine, ten, . . .) called the **counting numbers,** is the first example of an abstract mathematical system that we encounter in life. And though we meet it early, it seems to hold a fascination for people of all ages. Some, to be sure, are more fascinated than others. The most zealous come to be known as number theorists, or students of number theory.

The questions asked in number theory often appear to have no possible application and to be deceptively easy, and the outsider may legitimately wonder why they are asked at all. In the case of

our opening question about BIG, for example, it can be anticipated that some will respond, "Who cares?" It so happens, as we
shall see, that an impressive list of people have cared about this and
related questions. With the hope that you too may find some intriguing questions, we begin our discussion of abstract mathematical systems with the system of counting numbers.

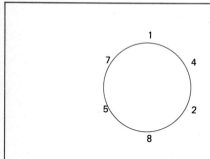

Around and Around We Go
Beginning at 1, and reading clockwise, we have 142857.

$$\begin{array}{r} 142857 \\ \times 2 \\ \hline 285714 \end{array}$$

Beginning at 2, and reading clockwise, we have 285714

Similarly, each of the following products can be read around the circle.

$$\begin{array}{r} 142857 \\ \times 3 \\ \hline 428571 \end{array} \qquad \begin{array}{r} 142857 \\ \times 4 \\ \hline 571428 \end{array} \qquad \begin{array}{r} 142857 \\ \times 5 \\ \hline 714285 \end{array} \qquad \begin{array}{r} 142857 \\ \times 6 \\ \hline 857142 \end{array}$$

DIVISORS AND MULTIPLES

We start with a review of some elementary notions. If three counting numbers are related by

$$ab = c$$

c is a **multiple** of a, and a is a **divisor** or **factor** of c. Similarly of
course c is a multiple of b, and b is a divisor of c.

Given a set of two or more counting numbers, they may or may
not have a common divisor greater than 1. The largest counting
number that is a divisor of all of them (even if that largest number
is 1) is called the **greatest common divisor** (g.c.d.) of the set. For
example, the g.c.d. of

$$A = \{8, 12, 20\}$$

is 4, while the g.c.d. of

$$B = \{12, 20, 35\}$$

is 1.

Every finite set of numbers has a common multiple. Thus a
common multiple of set A is $8 \cdot 12 \cdot 20 = 1920$, and a common multiple of set B is $12 \cdot 20 \cdot 35 = 8400$. Once a multiple has been
found, there is no trouble finding a bigger one; just double or
triple or whatever. The challenge lies in the other direction. What
is the **least common multiple** (l.c.m.)? That is, what is the smallest
number that is a multiple of each number in the set? For set A, the
answer is 120; try to find the answer for B.

The problem of finding the g.c.d. and the l.c.m. for a set be-

> Take any three-digit number,
> say 241 for example. Write it
> down again to obtain a six-
> digit number, such as 241,241.
> This six-digit number is always
> divisible by 7, 11, and 13. Why?

> The l.c.m. is first encountered
> when one is learning to add
> fractions. In that setting, one
> needs the l.c.m. of the
> denominators, more commonly
> called the least common
> denominator.

comes increasingly difficult as the numbers get larger, or the set contains more numbers, or both. Try, for example, to find the g.c.d. and the l.c.m. of the set

$$C = \{42, 63, 273, 364\}$$

To proceed most efficiently with such problems, it is best to express each member of the given set as a product of the primary building blocks of multiplication, the prime numbers. A **prime number** is a counting number greater than 1 that is divisible only by itself and by 1. The first few primes are

$$2, 3, 5, 7, 11, 13, 17, 19, 23, 29, \ldots$$

A counting number greater than 1 is either prime or it isn't. If it isn't, it has divisors which in turn are either prime or not prime. Building on this fact, we see that any number can be written as a product of primes. For example, the members of set C can be written

$$42 = 6 \cdot 7 = 2 \cdot 3 \cdot 7$$
$$63 = 3 \cdot 21 = 3 \cdot 3 \cdot 7$$
$$273 = 3 \cdot 91 = 3 \cdot 7 \cdot 13$$
$$364 = 2 \cdot 182 = 2 \cdot 2 \cdot 91 = 2 \cdot 2 \cdot 7 \cdot 13$$

The last paragraph is not likely to startle anyone and, to be truthful, neither is this one. This however, is in a sense a great tribute to the way in which arithmetic has been developed and taught, for we are dealing here with nothing less than the fundamental idea of arithmetic—and, behold, it seems obvious! After pointing out that a nonprime can be expressed as a product of primes, it remains only to say that, except for the order in which these primes are written down, this expression is unique; that is, there is only one such expression for the number.

The Fundamental Theorem of Arithmetic. Every nonprime counting number (greater than 1) can be written as a product of primes and, except for the order in which the factors are written, this expression is unique.

The meaning of the statement about uniqueness can be grasped by contrasting this theorem with a statement about sums. Every nonprime counting number can be written as a sum of primes, but in this case the expression is not unique. For example,

$$27 = 23 + 2 + 2$$
$$= 19 + 5 + 3$$
$$= 7 + 7 + 7 + 3 + 3$$

Once the members of a set of counting numbers have each been expressed as a product of primes (or simply left alone in the case of numbers that are prime to start with), the g.c.d. is easily determined. If no prime appears as a factor of each number in the set (as in set B below), the g.c.d. is 1. If there is a prime p that occurs in every number, p is a factor of the g.c.d. If there is only one such p (as in set C below), p is the g.c.d. If there is another prime q that occurs in every number (and we allow the possibility that $q = p$, as in set A below where every number has 2 as a factor twice), the g.c.d. has factors of p and q. The g.c.d. is the product of all primes that occur in every number, with a prime repeated k times in the g.c.d. if it occurs k times as a factor of each number in the given set.

For the sets A, B, and C, we have

$A = \{8, 12, 20\} = \{2 \cdot 2 \cdot 2, 2 \cdot 2 \cdot 3, 2 \cdot 2 \cdot 5\}$
$\text{g.c.d.} = 2 \cdot 2 = 4$
$B = \{12, 20, 35\} = \{2 \cdot 2 \cdot 3, 2 \cdot 2 \cdot 5, 5 \cdot 7\}$
$\text{g.c.d.} = 1$
$C = \{42, 63, 273, 364\} = \{2 \cdot 3 \cdot 7, 3 \cdot 3 \cdot 7, 3 \cdot 7 \cdot 13, 2 \cdot 2 \cdot 7 \cdot 13\}$
$\text{g.c.d.} = 7$

The prime factorization makes it easy to find the l.c.m. of a given set too. Begin by writing down the product of all distinct primes that appear in any of the given numbers. For set C, this means that we begin with the product of 2, 3, 7, and 13. Now consider the first prime p written in the product. Does it occur more often than once as a factor of some number in the given set? If it occurs twice (or three times or k times) as a factor of some number in the given set, it must occur twice (or three times or k times) in the l.c.m. Similarly consider each of the distinct primes.

Having written down 2, 3, 7, and 13 for the set C, we next observe that 2 occurs twice as a factor in one of the given numbers; so does 3. Hence, for the set C, we have an l.c.m. of $2 \cdot 2 \cdot 3 \cdot 3 \cdot 7 \cdot 13$. We use the same method to write down the l.c.m. for the sets A and B:

$\text{l.c.m. for } A = 2 \cdot 2 \cdot 2 \cdot 3 \cdot 5 = 120$
$\text{l.c.m. for } B = 2 \cdot 2 \cdot 3 \cdot 5 \cdot 7 = 420$
$\text{l.c.m. for } C = 2 \cdot 2 \cdot 3 \cdot 3 \cdot 7 \cdot 13 = 3276$

QUESTIONS ABOUT PRIMES

Can you find five consecutive nonprime numbers? How about seven?

No even number greater than 2 is prime; no multiple of 3 greater than 3 is prime. Each time we find a prime, all larger multiples of it are nonprime. It seems that it is very hard for a large number to be prime. Could it be that there is only a finite number of primes? No! Almost 2300 years ago, Euclid found an ingenious way of showing that there are infinitely many primes. His argument has come to serve as a classic example of proof by contradiction; it is outlined in Problem 20.

Consecutive odd primes like 11 and 13, 29 and 31, etc. are called twin primes. Are there infinitely many twin primes? Euclid didn't know. Neither does anyone else.

The numbers 1, 4, 7, 10, . . . are generated by the formula $t_n = 3n - 2$, $n = 1, 2, 3, \ldots$. Is there a similar formula that generates all the primes? Attempts have been made and can be summarized concisely; they have simply failed.

Well, if we can't find a formula that generates all the primes, can we find one that generates nothing but primes? Some attempts to answer this question have not simply failed; they have failed in notable ways.

To illustrate what we are after, consider

$$p_n = n^2 - n + 41$$

The first few values are shown at the right. It looks promising. In fact, the formula works for $n = 1, 2, \ldots, 40$. But alas, $p(41) = 1684 = (41)(41)$. See "How Many Are Enough" in Section 10.1.

There is a more notable failure. A lawyer named Pierre de Fermat (1601–1665), probably the most famous amateur mathematician who ever lived, once suggested that the formula

$$f_n = 2^{2^n} + 1$$

always produces primes. He had some evidence, for

$$f_1 = 5 \qquad f_2 = 17 \qquad f_3 = 257 \qquad f_4 = 65{,}537$$

are all primes. Besides this he had enormous prestige. But he was wrong. In 1732, Leonhard Euler, the gifted Swiss mathematician, discovered that

$$f_5 = 4{,}294{,}967{,}297 = (641)(6{,}700{,}417)$$

This number is BIG (see beginning of section).

For different reasons, mathematicians have studied the formula

$$g_n = 2^n - 1$$

It is conjectured that, while it doesn't always generate primes, at least it generates infinitely many. Modern electronic computers have been used to test the primeness of g_n for unbelievably large values of n. In 1963, the University of Illinois announced with great pride that $2^{11213} - 1$ is prime. But in 1971, Bryant

TABLE 13-1	
n	p_n
1	41
2	43
3	47
4	53
5	61
6	71
7	83
.	.
.	.
.	.
40	1601
41	1681

Pierre de Fermat (1601–1665) spent his entire working career in the service of the state. He never achieved great heights in his profession, but in his avocation of mathematics, it was quite another story. The simple elegance of his work in number theory at first obscures the depth of his thinking. In analysis he anticipated some of the ideas that Newton was to use so successfully, and he is generally acknowledged to have been one of the greatest mathematicians of the seventeenth century.

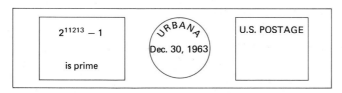

$2^{11213} - 1$

is prime

URBANA
Dec. 30, 1963

U.S. POSTAGE

Tuckerman did even better, showing that $2^{19937} - 1$ is prime. The latter number has 6002 digits, making it the largest prime known at the time.

SUMMARY

We have reviewed the definitions of several terms and perhaps introduced a few new ones. You should now be able to write down a short, clear sentence that defines each of the following:

Counting numbers
Multiple
Divisor (or factor)
Greatest common divisor (g.c.d.)
Least common multiple (l.c.m.)
Prime

The most important theorem mentioned in this section, indeed one of the most important in elementary mathematics, is the Fundamental Theorem of Arithmetic. You should be able to state it and explain what it means.

We have also learned some things about prime numbers: There is an infinite number of primes; no simple formula is known that will generate all the primes, nor is a simple formula known that will generate only primes, though there have been some notable conjectures along these lines. The number we called BIG is not prime, but some known primes are very big.

PROBLEM SET 13.1

1. Find the g.c.d. and l.c.m. for each of the following.
   (a) {10, 12, 16}   (b) {9, 15, 18}   (c) {8, 9, 25}
2. Find the g.c.d. and l.c.m. for the following sets.
   (a) {4, 5, 6}   (b) {6, 7, 8}   (c) {8, 10, 16}
3. Write the prime factorization for each of the following numbers.
   (a) 40   (b) 52   (c) 78   (d) 60   (e) 70
   (f) 126   (g) 252
4. Write the prime factorization for each of the following.
   (a) 36   (b) 63   (c) 84   (d) 40
   (e) 140   (f) 175
5. For each of the following sets, find the g.c.d. and the l.c.m. (Use the information in Problem 3).
   (a) {40, 52, 78}   (b) {60, 70, 126, 252}
6. For each of the following sets, find the g.c.d. and the l.c.m. (Use the information in Problem 4).
   (a) {36, 63, 84}   (b) {40, 140, 175}
7. The notion of least common denominator is useful in adding fractions. Perform the following operations and reduce your answer to the simplest form.

(a) $\frac{1}{4} + \frac{1}{6} + \frac{3}{8}$ (b) $\frac{3}{4} + \frac{9}{16} - \frac{5}{24}$

(c) $\frac{17}{30} + \frac{9}{14}$ (d) $\frac{7}{24} + \frac{11}{60} - \frac{7}{90}$

8. Follow the directions in Problem 7.

   (a) $\frac{1}{5} + \frac{3}{10} + \frac{8}{15}$ (b) $\frac{1}{6} + \frac{3}{8} + \frac{11}{12}$

   (c) $\frac{5}{90} + \frac{38}{600}$ (d) $\frac{8}{9} - \frac{3}{65} + \frac{13}{79}$

9. Here are some helpful divisibility rules.

   1. A number is divisible by 2 if and only if its last digit is even.

   2. A number is divisible by 3 if and only if the sum of its digits is divisible by 3; for example, 14,562 is divisible by 3 because $1 + 4 + 5 + 6 + 2 = 18$ is divisible by 3.

   3. A number is divisible by 5 if and only if its last digit is 0 or 5.

   Use these facts to help you find the prime factorization of

   (a) 10,800   (b) 52,650   (c) 38,775

10. Follow directions in Problem 9 for the following numbers.

   (a) 3572   (b) 11,775   (c) 29,920

11. Find the smallest number with eight different divisors.

12. Find all prime number years between 1950 and 2000.

13. What is the first value of $n > 1$ for which $g_n = 2^n - 1$ is not prime?

14. What is the first value of n for which $k_n = n^2 + n + 17$ fails to be prime?

15. Suppose we want to tile a floor measuring 12 by 15 feet. Only square tiles are available in sizes 4, 5, 8, or 9 inches on a side. If we wish to use only whole tiles, all of the same size, what size should we order?

16.* The prime factorization of a number N can be written

$$N = p_1^{a_1} p_2^{a_2} \cdots p_n^{a_n}$$

Determine the p's and a's for $N = 30$, 48, and 180. Let d_N be the number of divisors of N. Find d_N for $N = 30$, 48, and 180. Note that, in these three cases,

$$d_N = (a_1 + 1)(a_2 + 1) \cdots (a_n + 1)$$

Try to prove that this is true in general.

17.* The Sieve of Eratosthenes is a method for finding all the primes less than a given number n. The procedure is as follows. Write down in order all the counting numbers from 2 through n. Circle 2 and cross out all other multiples of 2 in the list. Then circle the next prime, 3, and cross out all other multiples of 3 in the list. Continue this process, moving each time to the next prime (the next number in order not already crossed out). The process is stopped when a number greater than $\sqrt{n}$ has been circled. The primes are those numbers not crossed out (whether circled or not). Use this method to determine the prime numbers less than 200.

18.* We discussed attempts to find an elementary formula generating all the primes (simple failure) and attempts to find a formula generating nothing but primes (some notable failures). There is a formula due to P. G. L. Dirichlet (1805–1859) known at least to generate an infinite number of primes. It says that, if a and b are counting numbers with no common divisor greater than 1, then

$D_n = an + b$ generates infinitely many primes. Show that, under the same conditions, D_n also generates an infinite number of nonprime numbers. Note that $F_n = 2n - 1$ also generates an infinite number of primes. Why do you suppose, then, that anyone should think Dirichlet's Theorem is worth attention?

19* In discussing the Fundamental Theorem of Arithmetic, we pointed out that it would be easy to express any number as a sum of primes. Consider this question: Can any even integer greater than 2 be expressed as a sum of two primes? After you have thought about it, go to a library, find a book on number theory, and look up Goldbach's conjecture.

20* Here, in outline form, is Euclid's argument showing that the number of primes is infinite. Suppose it is not. Then all the primes can be listed: 2, 3, 5, 7, 11, . . . , n, where n is the largest prime. Form the number

$$k = (2 \cdot 3 \cdot 5 \cdot 7 \cdot 11 \cdots n) + 1$$

Observe that 2 is not a factor of k, since if it were, that is, if $k = 2r$ for some integer r, we would have

$$2r - 2 \cdot 3 \cdot 5 \cdot 7 \cdot 11 \cdots n = 1$$
$$2(r - 3 \cdot 5 \cdot 7 \cdot 11 \cdots n) = 1$$

This is impossible. (Why?) Similarly, 3 is not a factor of k. Neither is any other of the listed primes. Thus there is some prime greater than n, perhaps k itself.

21* Choose some set $a_1, a_2, . . . , a_{101}$ of 101 numbers from the set $k = \{1, 2, . . . , 199{,}200\}$. Then for some i and j, a_i divides a_j. Hint: write each of the 101 selected numbers as a power of 2 (perhaps the 0 power) times an odd integer. That is, write $a_i = 2^{P_i}q_i$. Each of the 101 odd integers q_i is chosen from the set $\{1, 3, 5, . . . , 199\}$ of 100 odd integers. Use the pigeonhole principle.

GETTING TO THE ROOT OF THE THING

The algorithm on the right can be used to calculate square roots. Can you see how it works? Try to calculate the next two digits. Then square your answer to see how close it is to 2.

If carried out far enough, will you obtain a repeating decimal?

```
 1                    1. 4  1  4
(×2                  √2.00 00 00
  2 4                 1
   ×2                 1 00
 2 8 1                  96
   ×2                    400
 2 8 2 4                 281
                       11900
                       11296
                         604
```

13.2 / Extending the Number System

One way to introduce the solving of equations to youngsters is to ask what number, when added to 3, gives 7. Then repeat the question while writing

$$3 + ? = 7$$

Later still, since mathematicians believe that x is superior to other letters that might be employed, write

$$3 + x = 7$$

It is not unreasonable to think that from among the **counting numbers**

$$1, 2, 3, 4, \ldots,$$

youngsters in the early grades, perhaps aided by their fingers, will be able to select the correct answer.

Unless they have been introduced to the concept of negative numbers, however, it would be unreasonable to expect an answer to the question, what number added to 3 gives 2?

$$3 + x = 2$$

The youngsters we have in mind can be forgiven if they say that it can't be done. The solution of this problem requires that students have more at their fingertips, namely, the **integers**

$$\ldots, -4, -3, -2, -1, 0, 1, 2, 3, 4, \ldots$$

The integers, required to expand the class of equations that can be solved, are said to be an extension of the system of counting numbers. The integers, in the same way, must be extended to a

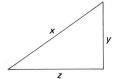

larger system if we are to solve

$$3x = 2$$

For this purpose, we need the **rational numbers** (quotients of integers) including such numbers as

$$\ldots, -\frac{3}{2}, -\frac{4}{3}, -1, -\frac{2}{3}, -\frac{1}{2}, -\frac{1}{3}, 0, \frac{1}{3}, \frac{1}{2}, \frac{2}{3}, 1, \frac{4}{3}, \frac{3}{2}, \ldots$$

A rational number does not have a closest neighbor since between any two of them, there is always another one; for example, between $\frac{3}{80}$ and $\frac{4}{80}$ we have

$$\frac{\frac{3}{80} + \frac{4}{80}}{2} = \frac{3 + 4}{2(80)} = \frac{7}{160}$$

The system is said to be dense; if every rational number were represented on the number line by a red dot, the result would appear to us as a solid red line.

  As plentiful as the rational numbers seem to be, their inadequacies were a matter of concern even to ancient mathematicians. They knew, from the Theorem of Pythagoras, that the lengths y and z of the legs of a right triangle are related to the length x of the hypotenuse by $x^2 = y^2 + z^2$. This meant that if $y = z = 1$, then the length of the hypotenuse had to satisfy

$$x^2 = 2$$

The problem was, however, that no rational number, when squared, equaled 2. Our proof of this fact, a classic example of proof by contradiction, is a variation of a proof given by Euclid.

  We begin by assuming that there is a rational number r which, when squared, gives 2. This number r is by definition the quotient of two positive integers m and n. That is, we suppose

$$r = \frac{m}{n} \quad \text{and} \quad \frac{m^2}{n^2} = 2$$

Multiplying both sides of the latter by n^2 gives

$$m^2 = 2n^2$$

Now by the Fundamental Theorem of Arithmetic (Section 13.1), m and n can each be written as a product of a unique set of prime numbers. Thus $m = p_1 \cdot p_2 \cdots p_j$, and $n = q_1 \cdot q_2 \cdots q_k$, where the p's and q's are prime numbers (not necessarily all different). When we substitute these expressions in $m^2 = 2n^2$, we get

$$p_1 \cdot p_1 \cdot p_2 \cdot p_2 \cdots p_j \cdot p_j = 2 \cdot q_1 \cdot q_1 \cdot q_2 \cdot q_2 \cdots q_k \cdot q_k$$

Next we count the number of times the prime number 2 appears on each side of the equation. On the left where each prime factor of m appears twice, the prime number 2 must occur an even number of times (possibly zero times); on the right, there is an odd number of 2's. This is impossible. Our original assumption must have been wrong. The number r cannot be rational.

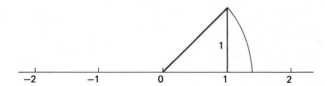

Here then, is the problem that troubled the ancient mathematicians. Return to the line on which all the rational numbers have been colored red. Using 0 as one end point, lay off the hypotenuse of the right triangle with equal sides of length 1. The end of the hypotenuse will not coincide with a red dot. If one only knows about rational numbers (which is what the ancients had to work with), what number measures the length of the hypotenuse? Such line segments came to be known as incommensurables.

Clearly we again need a larger system of numbers if we are to be able (conceptually) to label all of the points on the number line. The collection of numbers that is large enough to assign a unique number to each point on the line is called the set of **real numbers,** and the labeled line is called the real line. It must include all the rationals, since they certainly label points. But it must also include $\sqrt{2}$ (the number which when squared gives 2), π (pi, the circumference of a circle of diameter 1), and many more numbers.

The real numbers are conveniently represented using decimal notation. Recall that the familiar division algorithm enables us to represent any rational number by a decimal:

$$
\begin{array}{r}
0.77272 \\
22\overline{)17.000000} \\
15\,4 \\ \hline
1\,60 \\
1\,54 \\ \hline
60 \\
44 \\ \hline
160 \\
154 \\ \hline
60 \\
44 \\ \hline
16
\end{array}
\qquad
\frac{17}{22} = 0.772\overline{72}
$$

The bar indicates that the pattern of 72 repeats indefinitely. We say that $\frac{17}{22}$ is represented by a **repeating decimal.**

As a matter of fact, any rational number can be represented by a repeating decimal. The repeating pattern may be very simple, as in

$$\tfrac{1}{4} = 0.25000 \ldots$$
$$\tfrac{1}{3} = 0.3333 \ldots$$

or it may be more complicated, as in the case of

$$\tfrac{2}{7} = 0.2857 \ldots$$

where the repeating sequence is not yet evident. We are certain that it will repeat, however, because, if we examine the remainders in the division algorithm (shown in boldface below),

$$
\begin{array}{r}
0.2857 \\
7\,\overline{)\,2.0000} \\
\underline{1\,4} \\
\mathbf{60} \\
\underline{56} \\
\mathbf{40} \\
\underline{35} \\
\mathbf{50} \\
\underline{49} \\
\mathbf{1}
\end{array}
$$

it is clear that there are only seven possibilities: 0, 1, 2, . . ., 6. Thus, after no more than seven steps, we will find a remainder already obtained—at which point the process will begin to repeat. Actually, as you will discover (Problem 3), the repetition occurs in this example after six steps. A similar argument for any rational number a/b shows that its decimal representation has to repeat.

As indicated in Problems 7 through 10, the converse is also true. Any decimal that ultimately falls into a repeating sequence of digits represents a rational number.

The nonrational real numbers (called the **irrational numbers**) are, therefore, the numbers that are represented by nonrepeating decimals. Our opening problem, "Getting to the Root of the Thing," indicates in a rough (some would say lumpy) manner how to find the decimal representation for $\sqrt{2}$. We claim, though we do not intend to prove, that this procedure will produce a decimal—a nonrepeating decimal—whose square is 2 (Why must the decimal representation be nonrepeating?).

The real numbers, defined above as the collection of numbers required to (conceptually) label every point on a line, may also be described as the collection of numbers that can be represented using decimal notation, either as repeating or nonrepeating decimals. We ask our readers not to worry too much about several logical points. How exactly is an unending decimal to be understood? How are such things to be added or multiplied? These are hard questions which we must dismiss with a simple statement of assurance ("proof" by authority) that they do have perfectly logical answers.

We return to the recurring question of whether we now have enough numbers to solve any equation. One is not long in discovering that the real numbers do not suffice to solve $x^2 = -1$. For this, the so-called **imaginary number** i has been invented. It has the property that $i^2 = -1$, and it is used in combination with real numbers a and b to form the system of **complex numbers,** numbers of the form $a + bi$. Such numbers are added and multiplied according to the rules

$$(a + bi) + (c + di) = (a + c) + (b + d)i$$
$$(a + bi)(c + di) \quad = (ac - bd) + (ad + bc)i$$

The second product is remembered by writing

$$
\begin{aligned}
(a + bi)(c + di) &= a(c + di) + bi(c + di) \\
&= ac + adi + bci + bdi^2 \\
&= ac + (ad + bc)i + bd(-1) \\
&= (ac - bd) + (ad + bc)i
\end{aligned}
$$

You may assume by now that this is an unending process; invent more numbers, find an equation that still can't be solved, invent more numbers, etc. The remarkable fact about algebra is, however, that the process stops once the complex numbers have been invented.

Fundamental Theorem of Algebra. Any equation of the form

$$a_n x^n + \cdots + a_1 x + a_0 = 0$$

has a solution in the system of complex numbers.

SUMMARY

We have progressively extended the collection of numbers available to solve equations. Each new system is said to be an extension of the previous system because the new system always includes the previous one as a subset.

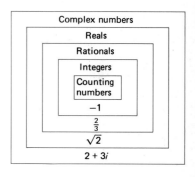

Equation	System Needed for Solution
$3 + x = 7$	Counting numbers: 1, 2, 3, 4, . . .
$3 + x = 2$	Integers: . . . , −4, −3, −2, −1, 0, 1, 2, 3, 4, . . .
$3x = 2$	Rationals: any number expressible as a quotient a/b of two integers, $b \neq 0$
$x^2 = 2$	Reals: any number expressible using decimal notation
$x^2 = -1$	Complex numbers: Numbers of the form $a + bi$ where a and b are real, $i^2 = -1$

The rational numbers are the real numbers that have a repeating decimal representation. The real numbers represented by non-repeating decimals are called irrational. Complex numbers of the form $a + bi$ where $b \neq 0$ are called imaginary numbers. According to the fundamental theorem of algebra, the table above is complete. No equation of the form $a_n x^n + \cdots + a_1 x + a_0 = 0$ will require for its solution a larger system than the complex numbers.

PROBLEM SET 13.2

1. Perform the indicated operations.
   (a) $0.375 + 12.24$ (b) $12.24 - 0.375$
   (c) $(0.375)(0.15)$ (d) $0.375/0.15$
2. Perform the indicated operations.
   (a) $56.13 + 1.245$ (b) $1.245 - 56.13$
   (c) $(0.037)(6.42)$ (d) $0.3144/0.24$
3. Continue the division process for $\frac{2}{7}$, which was started in the text, until the repeating pattern is evident. Then write $\frac{2}{7}$ as a decimal using the bar notation.
4. Find the decimal representation of $\frac{2}{13}$.
5. Write each of the following as a repeating decimal using the bar notation.
   (a) $\frac{3}{8}$ (b) $\frac{19}{8}$ (c) $\frac{5}{11}$ (d) $\frac{47}{12}$
6. Write as repeating decimals using the bar notation.
   (a) $\frac{3}{16}$ (b) $\frac{17}{16}$ (c) $\frac{5}{9}$ (d) $\frac{31}{9}$
7. Every repeating decimal represents a rational number. To show that this is true for $0.\overline{36}$, let $x = 0.\overline{36} = 0.363636\ldots$. Then $100x = 36.363636\ldots$. Subtract x from $100x$ and simplify as follows.

$$\begin{aligned} 100x &= 36.363636\ldots \\ x &= 0.363636\ldots \\ \hline 99x &= 36 \end{aligned}$$

$$x = \frac{36}{99} = \frac{9 \cdot 4}{9 \cdot 11} = \frac{4}{11}$$

We multiplied x by 100 because x was a decimal that repeated in a two-digit group. If the decimal had repeated in a three-digit group, we would have multiplied by 1000. Try this technique to obtain the rational number corresponding to $0.\overline{147}$.

8. Follow the pattern indicated in Problem 7 to find the rational number represented by $0.153\overline{153}$.
9. Find the rational numbers represented by the given repeating decimals.
   (a) $0.575757\ldots$ Hint: Multiply by 100.
   (b) $0.2575757\ldots$ Hint: Multiply by 10 and then use what you learned in part (a).
10. Use the hints in Problem 9 to find rational numbers represented by the following.
   (a) $0.696969\ldots$ (b) $0.5696969\ldots$

11. Consider the infinite decimal 0.121121112 . . ..
    (a) Following the pattern that you see, what are the next 10 digits?
    (b) Does this infinite decimal represent a rational or irrational number?
12. Write down the first few digits (enough to establish a pattern) of an infinite decimal that you are sure represents an irrational number.
13. The sum of two rational numbers is always rational. Is the sum of two irrational numbers always irrational?
14. The product of two rational numbers is always rational. Is the product of two irrational numbers necessarily irrational?
15. Show that $\sqrt{2} + \frac{3}{4}$ is irrational. Hint: Let $\sqrt{2} + \frac{3}{4} = r$ and suppose r is rational. Look for a contradiction.
16. Show that the sum of an irrational number and a rational number is necessarily irrational. See the hint in Problem 15.
17. Show that the product $\frac{3}{4} \cdot \sqrt{2}$ is irrational.
18. Show that the product of a nonzero rational number and an irrational number is always irrational.
19. Which of the following are rational and which are irrational? (See Problems 13 through 18.)
    (a) $\sqrt{2} + 1$ (b) $3\sqrt{2}$
    (c) $\sqrt{2}(\sqrt{2} + 1)$ (d) 0.12
    (e) $0.\overline{12}$ (f) $0.123456789101112 \ldots$
    (g) $\sqrt{2}(0.\overline{25})$ (h) $(0.\overline{25})(0.\overline{34})$
20. What is the smallest positive integer?
21. Write a positive rational number smaller than 0.0000001. What is the smallest positive rational number?
22. $(0.0000001)\sqrt{2}$ is irrational. Write a smaller positive irrational number. What is the smallest positive irrational number?
23. Show that between any two different rational numbers there is another rational number. (Actually, there are infinitely many.)
24. Show that between any two rational numbers there is an irrational number.
25. It is known that π is irrational. What does this mean about its decimal expansion?
26. Here is the beginning of the decimal expansion of π:

$$\pi = 3.14159 \ldots$$

Is $\pi - \frac{22}{7}$ positive or negative?
27.* Use the Pythagorean Theorem to show that $\sqrt{5}$ measures a length.
28.* Use the algorithm displayed at the beginning of this section to find the first four digits in the decimal expansion of $\sqrt{5}$.
29.* Verify by substitution that $1 + i$ is a solution to $x^2 - 2x + 2 = 0$.
30.* Verify by substitution that $1 - i$ is a solution to $x^2 - 2x + 2 = 0$.

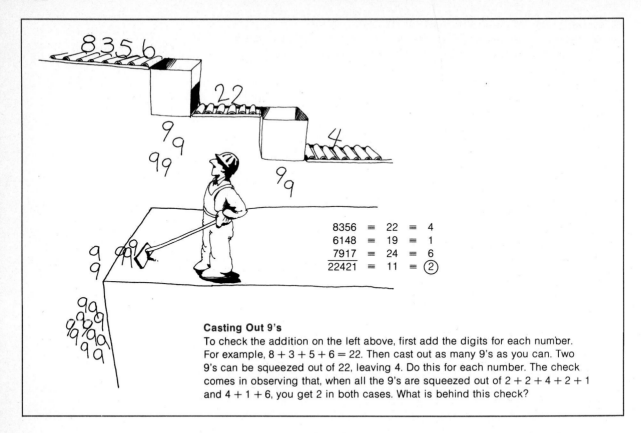

8356	≡	22	≡ 4
6148	≡	19	≡ 1
7917	≡	24	≡ 6
22421	≡	11	≡ ②

Casting Out 9's
To check the addition on the left above, first add the digits for each number. For example, 8 + 3 + 5 + 6 = 22. Then cast out as many 9's as you can. Two 9's can be squeezed out of 22, leaving 4. Do this for each number. The check comes in observing that, when all the 9's are squeezed out of 2 + 2 + 4 + 2 + 1 and 4 + 1 + 6, you get 2 in both cases. What is behind this check?

13.3 / Modular Number Systems

Suppose that a doctor gives you a pill at 11:00 A.M. and tells you to take one every 4 hours. What time should you take the next pill?

Suppose you are canning pickles, following a recipe that calls for soaking cucumbers in brine for 30 hours. If you put them into brine at 10:00 A.M., at what time the next day should you take them out?

These are common problems we all face, and somehow we solve them without writing anything down. If we did, we might be a bit surprised.

$$11 + 4 = 3 \quad \text{(for pills)}$$
$$10 + 30 = 4 \quad \text{(for pickles)}$$

What is going on? Actually the procedure is quite simple. One adds just as usual but then casts away as many 12's as possible. Here in a familiar setting are the ingredients of a strange new mathematical system; we'll call it **clock arithmetic.** It merits investigation.

Right away we notice the special role of 12. It acts like 0; that is, you can add 12 to any number on the clock and you are right back where you started. For this reason, we'll rub out 12 on our clock and replace it with 0. But having done this, we observe that 12 is not the only number that behaves like 0; so do -12, 24, 36, 48, ..., and a host of others. Similarly, -9, 3, 15, 27, etc., all act like 3 in clock arithmetic.

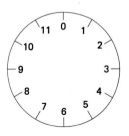

The situation reminds us of one that we faced with fractions. Recall that $\frac{3}{4}$, $\frac{6}{8}$, $\frac{9}{12}$, ... all represent the same number. In any calculation, we can use any one of these representatives as well as

TABLE 13-2											
+	**0**	**1**	**2**	**3**	**4**	**5**	**6**	**7**	**8**	**9**	**10**	**11**
0	0	1	2	3	4	5	6	7	8	9	10	11
1	1	2	3	4	5	6	7	8	9	10	11	0
2	2	3	4	5	6	7	8	9	10	11	0	1
3	3	4	5	6	7	8	9	10	11	0	1	2
4	4	5	6	7	8	9	10	11	0	1	2	3
5	5	6	7	8	9	10	11	0	1	2	3	4
6	6	7	8	9	10	11	0	1	2	3	4	5
7	7	8	9	10	11	0	1	2	3	4	5	6
8	8	9	10	11	0	1	2	3	4	5	6	7
9	9	10	11	0	1	2	3	4	5	6	7	8
10	10	11	0	1	2	3	4	5	6	7	8	9
11	11	0	1	2	3	4	5	6	7	8	9	10

TABLE 13-3											
×	**0**	**1**	**2**	**3**	**4**	**5**	**6**	**7**	**8**	**9**	**10**	**11**
0	0	0	0	0	0	0	0	0	0	0	0	0
1	0	1	2	3	4	5	6	7	8	9	10	11
2	0	2	4	6	8	10	0	2	4	6	8	10
3	0	3	6	9	0	3	6	9	0	3	6	9
4	0	4	8	0	4	8	0	4	8	0	4	8
5	0	5	10	3	8	1	6	11	4	9	2	7
6	0	6	0	6	0	6	0	6	0	6	0	6
7	0	7	2	9	4	11	6	1	8	3	10	5
8	0	8	4	0	8	4	0	8	4	0	8	4
9	0	9	6	3	0	9	6	3	0	9	6	3
10	0	10	8	6	4	2	0	10	8	6	4	2
11	0	11	10	9	8	7	6	5	4	3	2	1

another. Actually we tend to favor $\frac{3}{4}$; it's the reduced form of the fraction. Similarly, in clock arithmetic, we may think of −9, 3, 15, 27, . . . as representing the same number, with 3 being the reduced form or, as we prefer to say, the **principal representative.**

To put it slightly differently, we are using the clock to group the integers into 12 classes, each with its principal representative (shown in boldface below).

$$\ldots, -36, -24, -12, \quad \mathbf{0}, 12, 24, 36, \ldots$$
$$\ldots, -35, -23, -11, \quad \mathbf{1}, 13, 25, 37, \ldots$$
$$\ldots, -34, -22, -10, \quad \mathbf{2}, 14, 26, 38, \ldots$$
$$\vdots \qquad\qquad \vdots$$
$$\ldots, -25, -13, -\ 1, \quad \mathbf{11}, 23, 35, 47, \ldots$$

The members of any one class differ by multiples of 12.

Using the principal representatives, we can construct addition and multiplication tables for clock arithmetic. We simply add and multiply as usual and then cast away enough 12's to get back into the principal set.

Now we can amuse ourselves with all kinds of questions. For example, can we solve equations in this new mathematical system? The answer is, sometimes yes; but the reader should be prepared for some surprises. At the moment, it is best to search for solutions by trial and error. Try substituting each of the numbers from 0 to 11 in the equations

a. $x + 11 = 3$
b. $5x = 8$
c. $9x + 4 = 10$
d. $4x + 1 = \ 6$

If you were careful, here is what you found out. The number 4 is the only solution to equations a and b; the numbers 2, 6, and 10 all satisfy equation c; and equation d doesn't have any solution at all.

ARITHMETIC MODULO *m*

Arithmetic based on the clock is amusing and has practical uses (recall the pills and the pickles). However, its real importance lies more in what it suggests than in what we have seen so far. First, there is nothing sacred about the number 12. Dividing the clock into 12 units was an arbitrary decision, extending back into ancient time. There is nothing wrong with 8-, 10-, or 50-hour clocks. And thinking about other possibilities will provide us with a tool for solving a class of problems called Diophantine equations (Section

13.4). Second, the various systems provide very simple illustrations of the abstract mathematical systems we intend to study in Chapter 15.

Before we plunge in all the way, consider one more familiar example. Suppose that, for a year that starts on Tuesday, we make a mammoth calendar numbering the days from 1 to 365.

TABLE 13-4						
S	M	T	W	T	F	S
		1	2	3	4	5
6	7	8	9	10	11	12
13	14	15	16	17	18	19
20	21	22	23	24	25	26
27	28	29	30	31	32	33
34	35	36	37	38	39	40
$\vdots$	$\vdots$	$\vdots$	$\vdots$	$\vdots$	$\vdots$	$\vdots$
363	364	365				

Which numbers represent the same day of the week, say Sunday? Clearly it is the numbers 6, 13, 20, 27, . . ., numbers that differ by a multiple of 7. How do we know that the day numbered 365 is a Tuesday? Because 365 differs from 1 by a multiple of 7, that is,

$$365 - 1 = 7 \cdot 52$$

The dictionary says that to modulate is to regulate or tone down in accord with some rule. To modulate by 7 is to tone down by a multiple of 7. Here then is the precise definition toward which we have been leading. We say that **a is equal to b modulo 7** if $a - b$ is a multiple of 7; that is,

$$a \equiv b \bmod 7 \qquad \text{if } a - b = 7k$$

for some integer k. Note the triple bars ($\equiv$) which we use consistently to distinguish this new kind of equality from the ordinary one ($=$). Thus

$$27 \equiv 6 \bmod 7 \qquad [\text{since } 27 - 6 = 7(3)]$$
$$19 \equiv 5 \bmod 7 \qquad [\text{since } 19 - 5 = 7(2)]$$
$$-8 \equiv 13 \bmod 7 \qquad [\text{since } -8 - 13 = 7(-3)]$$

This new type of equality shares all the important properties of ordinary equality. For example,

> Hugo signs a 60-day note on Friday. On what day of the week will it come due?

1. $a \equiv a$.
2. If $a \equiv b$, then $b \equiv a$.
3. If $a \equiv b$ and $b \equiv c$, then $a \equiv c$.
4. If $a \equiv b$ and $c \equiv d$, then $a + c \equiv b + d$.
5. If $a \equiv b$ and $c \equiv d$, then $a \cdot c \equiv b \cdot d$.

Property 4 is the familiar statement, "Equals added to equals give equals." To illustrate, $27 \equiv 6 \bmod 7$ and $19 \equiv 5 \bmod 7$; sure enough $27 + 19 \equiv 6 + 5 \bmod 7$ (since $46 - 11 = 7 \cdot 5$).

Now what is done for 7 or 12 can be done for any modulus m. We write

$$a \equiv b \bmod m \qquad \text{if } a - b = m \cdot k$$

for some integer k, and properties 1 through 5 always hold (see Problems 14 through 17).

ARITHMETIC MODULO 9

$$
\begin{array}{r}
928 \\
9\,\overline{)\,8356} \\
81 \\
\hline
25 \\
18 \\
\hline
76 \\
72 \\
\hline
\textcircled{4}
\end{array}
$$

It is time to respond to the question in our opening cartoon. What is behind the arithmetic checking process called casting out 9's?

Let's begin by doing some reductions modulo 9. Take the number 8356, for example. Modulo 9, it is equal to one of the numbers from the principal set $\{0, 1, 2, 3, 4, 5, 6, 7, 8\}$. For such a large number one can remove multiples of 9 by long division. Just divide by 9 and find the remainder (see margin). Thus

$$8356 \equiv 4 \bmod 9$$

As the reader can check, this method works for any modulus whatever.

Long division is okay, but for arithmetic modulo 9 there is a better way. All our readers will recall that 8356 is just a compact way of writing

$$8(1000) + 3(100) + 5(10) + 6$$

But $10 \equiv 1 \bmod 9$, $100 = 10^2 \equiv 1^2 = 1 \bmod 9$, $1000 = 10^3 \equiv 1^3 = 1 \bmod 9$, etc. By the properties above, especially properties 4 and 5,

$$8(1000) + 3(100) + 5(10) + 6 \equiv 8 + 3 + 5 + 6 = 22$$

In fact, this reasoning shows that *any number is equal to the sum of its digits modulo 9*. This is particularly useful for large numbers, but of course it works for small numbers as well. Thus

$$22 \equiv 2 + 2 = 4$$

But then most of us can cast 9's out of 22 directly.

Now we see why the check in our cartoon worked. Here it is again:

$$8356 \equiv 22 \equiv 4$$
$$6148 \equiv 19 \equiv 1$$
$$\underline{7917 \equiv 24 \equiv 6}$$
$$22{,}421 \equiv 11 \equiv ②$$

It is really just that old friend, "Equals added to equals must give equals," interpreted for the new kind of equality. If we don't get equality in this process, we've made a mistake.

Property 5 means that casting out 9's can also be used to check multiplication. Here is an example:

$$287 \equiv 17 \equiv 8$$
$$\underline{37 \equiv 10 \equiv 1}$$
$$2009$$
$$\underline{861}$$
$$\overline{10619} \equiv 17 \equiv ⑧$$

> Remember that we are multiplying, so we multiply the 8 by the 1.

SUMMARY

Two integers a and b are said to be equal modulo the positive integer m if and only if there is an integer k such that

$$a - b = mk$$

Equality modulo m has all the properties normally associated with equality (equals can be added to or multiplied by equals, things equal to the same thing equal each other, etc.) The numbers $0, 1, 2, \ldots, m - 1$ are called principal representatives of numbers in the system modulo m and, in a given system, the reader should have no trouble representing a sum or a product by a principal representative; thus

$$17 + 12 \equiv 10 \bmod 19$$
$$17 \cdot 12 \equiv 14 \bmod 19$$

Finally, the reader should master the techniques of checking ordinary computations by making use of arithmetic modulo 9.

PROBLEM SET 13.3

1. Make addition and multiplication tables for a 7-hour clock. Now solve the following equations in this arithmetic.
   (a) $x + 4 = 3$
   (b) $2x = 3$
   (c) $2x + 3 = 4$
   (d) $3x + 2 = 1$

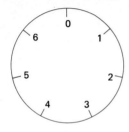

2. Make addition and multiplication tables for a 6-hour clock and solve the equations in Problem 1.

3. What day of the week will it be 93 days after Wednesday? 193 days after Wednesday? Hint: Use a 7-unit clock and think of Wednesday as the fourth day of the week.

4. What time will an ordinary clock show 93 hours after it shows 4? 193 hours after it shows 4?

5. Reduce each of the following, giving the answer from the principal set. Hint: To reduce 77 mod 8, divide 77 by 8, obtaining a remainder of 5. Then $77 \equiv 5 \bmod 8$.

    (a) 98 mod 5   (b) 981 mod 5
    (c) $(98 + 981) \bmod 5$ (d) $(98 \cdot 981) \bmod 5$
    (e) 492 mod 7   (f) 9811 mod 7
    (g) $(492 + 9811) \bmod 7$ (h) $(492 \cdot 9811) \bmod 7$

6. Follow the directions in Problem 5.

    (a) 47 mod 14   (b) 891 mod 14
    (c) $(47 + 491) \bmod 14$ (d) $(47 \cdot 491) \bmod 14$
    (e) 75 mod 8   (f) 750 mod 8
    (g) $(75 + 750) \bmod 8$ (h) $(75 \cdot 750) \bmod 8$

7. Find all solutions in the appropriate principal set to the following equations.

    (a) $x + 9 \equiv 2 \bmod 12$ (b) $3x \equiv 9 \bmod 12$
    (c) $5x + 1 \equiv 4 \bmod 12$ (d) $7x + 2 \equiv 6 \bmod 12$

8. Solve each of the equations in Problem 7 with modulo 12 replaced by modulo 11.

9. Use the easy method to reduce each of the following.

    (a) 3451 mod 9   (b) 623,852 mod 9
    (c) $(4562 + 7321 + 9876) \bmod 9$
    (d) $((62{,}381)(92{,}734)) \bmod 9$

10. Reduce each of the following.

    (a) 25,763 mod 9   (b) 742,316 mod 9
    (c) $(2576 + 4321) \bmod 9$ (d) $((52{,}345)(8743)) \bmod 9$

11. Some of the following calculations may be incorrect. Use the method of casting out 9's to identify them.

(a)	(b)	(c)	(d)
3417	9625	371	433
2985	7163	816	721
6321	3582	2226	433
4173	1473	371	866
16886	21843	2958	3031
		301736	312293

12. Follow the instructions in Problem 11.

(a)	(b)	(c)	(d)
8614	5312	4916	9162
2375	4871	27	38
1627	2638	34312	73296
3748	1549	9832	27386
16364	14270	132632	347156

13. The casting out 9's check does not detect certain calculation errors. Can you identify them?

14. Show that if $a \equiv b$ mod 12 and $b \equiv c$ mod 12, then $a \equiv c$ mod 12. Hint: By hypothesis, $a - b = 12k$ and $b - c = 12j$ for some integers k and j. Thus $a - c = a - b + b - c = 12k + 12j$. Now what?

15. Show that, if $a \equiv b$ mod m and $b \equiv c$ mod m, then $a \equiv c$ mod m. Hint: See Problem 14.

16. Show that, if $a \equiv b$ mod 12 and $c \equiv d$ mod 12, then $a + c \equiv b + d$ mod 12.

17. Show that, if $a \equiv b$ mod 7 and $c \equiv d$ mod 7, then $a \cdot c \equiv b \cdot d$ mod 7.

18. In ordinary arithmetic, $-a$ is the number that when added to a gives 0. For example, $-2 + 2 = 0$ and $-5 + 5 = 0$. We call $-a$ the **additive inverse** of a. Consider now a 7-hour clock with numbers 0, 1, 2, 3, 4, 5, 6. Note that 5 is the additive inverse of 2, since $5 + 2 = 0$. Find the additive inverse of
    (a) 1   (b) 3   (c) 4   (d) 5

19. In ordinary arithmetic, $1/a$ is the number that when multiplied by a gives 1. For example, $\frac{1}{4} \cdot 4 = 1$ and $\frac{1}{9} \cdot 9 = 1$. We call $1/a$ the **multiplicative inverse** of a. Consider a 7-hour clock with numbers 0, 1, 2, 3, 4, 5, 6. In this arithmetic, 3 is the multiplicative inverse of 5, since $3 \cdot 5 = 1$. Find the multiplicative inverse of
    (a) 2   (b) 3   (c) 4   (d) 6

20. On a 12-hour clock, try to find the multiplicative inverse of each of the numbers 0 through 11. You will discover that several of them do not have a multiplicative inverse.

21.* (Project.) Since casting out 9's doesn't detect errors caused by inverting digits (see Problem 13), a method called casting out 11's has been suggested. Can you devise this system? Hint: $10 \equiv -1$ mod 11.

22.* (Project.) For various choices of $a < 7$ compute a^7 mod 7. For various choices of $a < 5$ compute a^5 mod 5. Try other experiments that these computations suggest. Make a conjecture. Try to prove it. Look up Fermat's Theorem in a book on number theory.

23.* (Project.) We pose the following question. Can you discover a rule or rules that will enable us to tell, from looking at $a \neq 0$, b, and m, whether or not there are solutions to $ax \equiv b$ mod m? To begin with, look at some examples. You already have a multiplication table for arithmetic modulo 12. Make similar tables for modulo 5, 6, and 9. In many years of posing this problem for classes to work on, we have heard the following suggested rules. Try to decide which ones are true, and which are false.
    (a) If m is odd, then $ax \equiv b$ mod m will have a solution no matter how $a \neq 0$ and b are chosen.
    (b) There will always be a solution if a is odd.
    (c) There will always be a solution if a is odd and b is even.
    (d) There will always be a solution if m is prime.
    (e) We can be certain of a solution only if m is prime.
    Some of the above are false. Some are true. None give the entire story. Make a conjecture of your own. Discuss it with others.

A Regular Cut-Up

Hugo Hardback, the head librarian, reported to the board that in furnishing the new library, he had spent $3409 for tables and chairs. With the temerity that sometimes seizes people in positions of power, one board member asked how close they had come to the planned seating capacity. Hardback testily pointed out that, since the board had previously approved the purchase of tables costing $288 each and chairs costing $19 each, it should not be necessary to ask how many chairs had been purchased.

$$115 \text{ chairs @ } \$19 = \$2185$$
$$4\tfrac{1}{4} \text{ tables @ } \$288 = \underline{\$1224}$$
$$\$3409$$

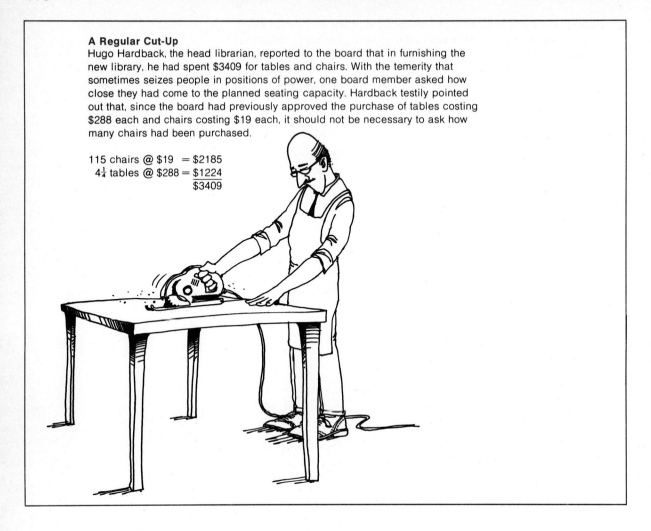

13.4 / Equations with Integer Answers

If we let s represent the number of chairs and t represent the number of tables bought by the librarian described in the problem above,

$$19s + 288t = 3409$$

There are plenty of solutions to this equation. For instance, if we choose $s = 115$, a little computation gives $288t = 1224$, or

$$t = \frac{1224}{288} = \frac{17}{4} = 4\frac{1}{4}$$

Since t represents the number of tables, however, this answer leaves something to be desired.

Sometimes a modest amount of thought about a problem makes it clear that only integer answers are acceptable. Such problems are called **Diophantine problems.** It will perhaps not surprise the reader to learn that modular arithmetic is often useful in solving such problems, and it is to a brief introduction of this whole idea that we now turn.

Suppose that in the equation

$$19s + 288t = 3409$$

we reduce everything modulo 19. Since $19 \equiv 0$, $288 \equiv 3$, and $3409 \equiv 8 \bmod 19$, this leads us to

$$3t \equiv 8 \bmod 19$$

Now we saw in Section 13.3 that not all such equations have a solution and that some have several. Happily, we can settle the matter in a reasonably small number of trials (in this case, no more than 19). In due time, after discovering that $0, 1, \ldots, 8$ do not work, we come to try $t = 9$. We get

$$3 \cdot 9 = 27 \equiv 8 \bmod 19$$

Thus $t \equiv 9 \bmod 19$; that is,

$$t = \ldots, -10, 9, 28, 47, \ldots$$

DIOPHANTUS

Diophantus was a Greek mathematician who lived about 250 A.D. While not all of his manuscripts have come down to us, we have enough of his work to know that he was interested in algebraic problems that have integral answers. Problems of this kind are for this reason called Diophantine problems. Of his personal life, we have only a description in the form of a problem. It seems that Diophantus spent $\frac{1}{6}$ of his life as a boy, grew a beard after $\frac{1}{2}$ more, and married after yet another $\frac{1}{7}$. A son was born 5 years later, but died in the prime of life. Four years after his son died, Diophantus died, having lived twice as long as his son had lived. Can you determine the number of years he lived?

---

should all work. It appears that there are many solutions to our original problem. Let us try $t = 9$. Then

$$19s + 2592 = 3409$$
$$19s \qquad = 817$$
$$s \qquad = 43$$

Certainly $t = 9$ and $s = 43$ is one solution.

Since negative answers make no more sense than fractional ones in our problem, we only need to look for values of s that correspond to positive choices of t. For $t = 28$, we get

$$19s + 8064 = 3409$$
$$s \qquad = -245$$

Since larger values of t will give us a number larger than 8064, it is clear that all other possible positive values of t will give negative values of s. The only possible conclusion is that the new library addition has 9 tables and 43 chairs.

Consider again the equation

$$19s + 288t = 3409$$

We proceeded by reducing everything in sight modulo 19. Given an equation, it would of course be correct to reduce both sides using any modulus we please. The incentive to use 19 is clear; in this way we made one of the unknowns drop out of the equation. The same thing could have been accomplished using 288. This would have given

$$19s \equiv 241 \bmod 288$$

The only drawback is that, since our only method of solution is

trial and error, we would be faced with 288 trials. Of course (since we know the answer — having worked it out above), if we proceed systematically, trying $s = 1, 2, \ldots$, we will be rewarded on the forty-third trial. This is scant comfort to anyone not being paid by the hour, and it explains why we used 19 as the modulus.

If you seek integer solutions to

$$rx + sy = t$$

where r, s, and t are given positive integers, select the smaller of r and s and then reduce the equation modulo this number. You will be left with an equation of the form

$$az \equiv b \bmod m$$

We know that such equations may have no solution, a unique solution, or several solutions in the set $\{0, 1, 2, \ldots, m-1\}$. If there are solutions, they can be found by trial and error.

EXAMPLE WITH TWO LARGE COEFFICIENTS

If we attempt to solve a problem such as

$$87x + 281y = 4983$$

the procedure outlined above results in

$$20y \equiv 24 \bmod 87$$

The number of possibilities to be tried is still too large to be taken seriously, so we appeal to the definition of equality modulo 87 and note that there must be an integer k such that

$$20y = 24 + 87k$$

If we apply the same techniques to this equation, reducing everything modulo 20 (the smallest coefficient), we get

$$0 \equiv 4 + 7k \bmod 20$$

Since $-4 \equiv 16 \bmod 20$, we have

$$7k \equiv -4 \equiv 16 \bmod 20$$

We can now find the solution (if there is one) by trial and error or, if 20 trials still seem more than we wish to try, the same procedure can be repeated again. Either way, we very shortly obtain

$$k \equiv 8 \bmod 20$$

Substituting the value $k = 8$ in $20y = 24 + 87k$ gives

$$20y = 24 + 87(8) = 720$$

or

$$y = 36$$

Thus the solution to $20y \equiv 24 \bmod 87$ is $y \equiv 36 \bmod 87$. Finally, substitution of $y = 36$ into the original problem gives

$$
\begin{aligned}
87x + 281(36) &= 4983 \\
87x &= -5133 \\
x &= -59
\end{aligned}
$$

From $y \equiv 36 \bmod 87$, we can obtain other solutions as well. A partial listing gives

y	$\ldots$	-51	36	123	$\ldots$
x	$\ldots$	222	-59	-340	$\ldots$

It is noted that, as was to be expected, the values obtained for x are equal modulo 281. If the problem giving rise to this equation were such that only positive integers would be acceptable answers, we would report that there are no answers, since in the pairing of answers indicated in the table, either x or y is always negative.

SUMMARY

Given an equation of the form

$$ax + by = c$$

in which a, b, and c are integers and in which we are interested only in finding integer values for x and y, we can use modular arithmetic to find solutions if they exist. We simply reduce the equation modulo a or modulo b, whichever is the most convenient (which usually means using the one closest to zero). The resulting modular equation in one unknown can then be investigated for possible solutions.

PROBLEM SET 13.4

1. Find all integer solutions to the following Diophantine equations.
   (a) $2x + 5y = 2$ (b) $15x + 16y = 17$
   (c) $37x + 25y = 8$ (d) $38x - 14y = 43$
   (e) $117x + 86y = 157$
2. Find all integer solutions.
   (a) $x - y = 7$ (b) $13x + 14y = 15$
   (c) $41x - 18y = 11$ (d) $26x + 54y = 119$
   (e) $87x - 137y = 908$
3. Find all integer solutions that satisfy both the equations

$$3x - y + t = 0$$
$$2x - 3y + 5t = 17$$

   Hint: Begin by eliminating t.
4. Find the integer solutions of the pair of equations

$$-2x + 5y - 3t = -3$$
$$3x - 2y + t = -2$$

   Hint: Begin by eliminating t.
5. Homer and Horatio put up a sign offering to wash and clean the interior of a car for $4, or to wash and wax it for $25. Homer takes all the $4 jobs, and Horatio the $25 jobs. They agree that each will take the money for his own customers, but they use a common box in which to keep their money. After a weekend of hard work they find that they have $143, but they can't remember who gets what. Can you help them?
6. The Mathematics Association of America sells books in the Studies in Mathematics Series to members for $5 per book. At the registration desk at a regional meeting, books in this series were on sale. At the same table, members paid for their registration and lunch ticket ($8). The person collecting the money reported that she had collected $866, only then to be told that the funds were to be kept separate, the $8 fees going to the local arrangements committee and the $5 payments going to the regional treasurer. She had seven checks for $5, 74 checks for $8, and four checks for $13. Can you

help her figure out how much money goes to the local committee? What if it can be ascertained from the box in which the books were brought to the meeting that no more than 20 books were sold?

7. The company that sells the tables and chairs that Hugo Hardback's library bought reports receipts of $20,603 for a given month. If the company expects orders of about six chairs to each table, what guess would you give as to the number of tables and chairs sold that month?

8. In another month (see Problem 7), receipts were $19,445. How many each of tables and chairs were sold that month?

9. In a short story entitled "Coconuts," author Ben Williams told of two businessmen bidding for the same contract. One, in order to distract the other, who had a penchant for mathematical puzzles, sent him the following puzzle just as the deadline for bids was approaching.

> Five men and a monkey were shipwrecked on a desert island, and they spent the first day gathering coconuts for food. They piled them all up together and then went to sleep for the night.
>
> But when they were all asleep one man woke up, and he thought there might be a row about dividing the coconuts in the morning, so he decided to take his share. So he divided the coconuts into five piles. He had one coconut left over, and he gave that to the monkey, and he hid his pile and put the rest all back together.
>
> By and by the next man woke up and did the same thing. And he had one left over, and he gave it to the monkey. And all five of the men did the same thing, one after the other; each one taking a fifth of the coconuts in the pile when he woke up, and each one having one left over for the monkey. And in the morning they divided what coconuts were left, and they came out in five equal shares. Of course each one must have known there were coconuts missing; but each one was as guilty as the other, so they didn't say anything. How many coconuts were there in the beginning?

The story, along with this now famous puzzle, appeared in the October 9, 1926, issue of *The Saturday Evening Post*. It is reported that some 2000 letters arrived in the week after the story appeared asking for the solution. It is not reported how many of these requests were thrown to a monkey, but editor George Lorimer did send the following wire to author Williams: "FOR THE LOVE OF MIKE, HOW MANY COCONUTS? HELL POPPING AROUND HERE:"

Chapter 14
The System of Matrices

It is true that Fourier has the opinion that the principal object of mathematics is the public utility and the explanation of natural phenomena; but a scientist like him ought to know that the unique object of science is the honor of the human spirit and on this basis a question of the theory of numbers is worth as much as a question about the planetary system.
C. G. J. JACOBI

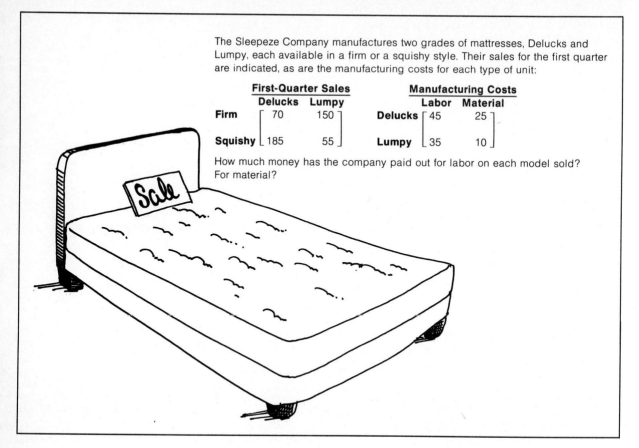

The Sleepeze Company manufactures two grades of mattresses, Delucks and Lumpy, each available in a firm or a squishy style. Their sales for the first quarter are indicated, as are the manufacturing costs for each type of unit:

First-Quarter Sales		**Manufacturing Costs**	
Delucks	Lumpy	Labor	Material
Firm 70	150	**Delucks** 45	25
Squishy 185	55	**Lumpy** 35	10

How much money has the company paid out for labor on each model sold? For material?

14.1 / Boxes of Numbers

The home handyman occasionally discovers among his collected "junk" something that can be fashioned into just the tool he needs. Scientists sometimes have the same luck. While working to develop a certain theory, they suddenly discover that the very mathematical tools they need have been available for a long time.

Matrices, the subject of this chapter, have been "discovered" several times. Much of what we know about the subject today was being used in the early 1800s by people using the tools of calculus and differential equations to study celestial mechanics. Then in 1858, Arthur Cayley wrote a paper often cited as the place where matrices were invented. Nevertheless, it apparently came as something of a surprise to Werner Heisenberg when he realized in 1925 that matrices were just what he needed to develop his idea of quantum mechanics. And it is safe to say that all of the early workers would be surprised if they could now see the uses to which matrices are put in modern physics, economics, business, statistics, and systems research.

Here, then, is an example that nicely portrays what we wish to say about the role of abstraction in mathematics. First arising in a specific application, later developed as an object of mathematical curiosity, a subject ultimately becomes an indispensable tool in applied mathematics. The abstraction lifts the subject from the context in which it first arises, identifies the crucial ideas, and develops their consequences, sometimes with and sometimes without reference to particular applications. Ultimate applications may not even be anticipated as the subject is developed.

It is the process of abstraction on which we now wish to focus attention. We encourage readers therefore, at least in the next two sections, not to bedevil themselves with the question, What's all this good for? Applications appear in Section 14.3, but for now we'll pursue the subject for its own intrinsic interest. This is a part of the spirit of mathematics.

SOME DEFINITIONS

A matrix is a rectangular array (box) of numbers arranged in rows and columns:

$$A = \begin{bmatrix} -1 & 3 & 0 \\ 4 & -2 & 5 \end{bmatrix} \qquad B = \begin{bmatrix} 2 & 0 \\ -3 & 1 \\ 0 & 4 \end{bmatrix}$$

Matrix A is said to be 2×3 (read "2 by 3"), meaning it has two rows and three columns; matrix B is 3×2. Two matrices are said to be equal if and only if they have the same dimensions (same number of rows and same number of columns) and if their corresponding entries are equal.

There are applications in which the matrices are very large, but we for the most part confine ourselves to 2×2 matrices like the ones used in the introduction to this section.

MULTIPLICATION OF MATRICES

The product of two 2×2 matrices is again a 2×2 matrix, but it is obtained according to a rule that seems peculiar when it is first encountered. The procedure is suggested by the problem posed at the beginning of this section. There, given the first-quarter sales and the manufacturing costs for various products of the Sleepeeze Company, we were asked to find the company's labor and material costs for each product sold. Let us write down all the information, together with the answers to the questions about costs.

	First-Quarter Sales			Manufacturing Costs	
	Delucks	**Lumpy**		**Labor**	**Material**
Firm	70	150	**Delucks**	45	25
Squishy	185	55	**Lumpy**	35	10

Suppose we wish to analyze the costs of producing the firm and the squishy mattresses sold in the first quarter. Since the manufacturing costs are the same, whether we produce a firm or a squishy mattress, we figure the labor costs for the firm mattresses sold during the first quarter as follows.

$$\begin{pmatrix}\text{Labor costs for}\\ \text{firm mattresses}\end{pmatrix} \text{ are } \begin{pmatrix}\text{number of}\\ \text{Delucks sold}\end{pmatrix}\begin{pmatrix}\text{labor cost}\\ \text{for Delucks}\end{pmatrix} \text{ plus}$$

$$= \qquad (70) \qquad (45) \qquad +$$

$$\begin{pmatrix}\text{number of}\\ \text{Lumpy sold}\end{pmatrix}\begin{pmatrix}\text{labor cost}\\ \text{for Lumpy}\end{pmatrix}$$

$$(150) \qquad (35)$$

Similar computations give us the other entries in what can be called the cost matrix:

	Total Costs	
	Labor	**Material**
Firm	$70(45) + 150(35)$	$70(25) + 150(10)$
Squishy	$185(45) + \ 55(35)$	$185(25) + \ 55(10)$

The pattern here is the one used to define the multiplication of any two 2×2 matrices:

$$\begin{bmatrix} a & b \\ c & d \end{bmatrix} \cdot \begin{bmatrix} A & B \\ C & D \end{bmatrix} = \begin{bmatrix} aA + bC & aB + bD \\ cA + dC & cB + dD \end{bmatrix}$$

The definition is difficult to remember without a few pointers. In this case, one's fingers will do. To obtain the entry in the first row and first column of the product, place the left index finger on a and the right one on A. Slide the left finger along the first (top) row and the right finger down the first (left) column. The fingers in turn point to aA and bC. Their sum is the desired entry.

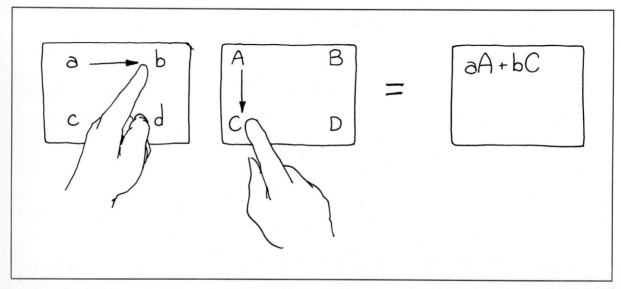

Similarly, to get the entry in the second row and first column of the product, place the left index finger on c, preparing to slide it along the second row. Place the right index finger on A, preparing to slide it down the first column simultaneously. This time the fingers pick out the products cA and dC.

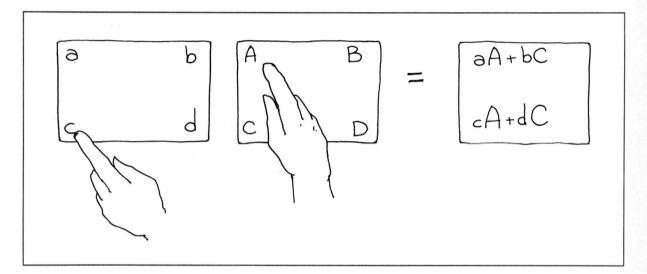

Use your fingers as well as your head to follow this numerical example:

$$\begin{bmatrix} 2 & 3 \\ 1 & 4 \end{bmatrix}\begin{bmatrix} -3 & -1 \\ 0 & 5 \end{bmatrix} = \begin{bmatrix} 2(-3)+3(0) & 2(-1)+3(5) \\ 1(-3)+4(0) & 1(-1)+4(5) \end{bmatrix} = \begin{bmatrix} -6 & 13 \\ -3 & 19 \end{bmatrix}$$

Though our definition covers only the multiplication of 2×2 matrices, the "finger rule" points the way to the multiplication of any two matrices in which the number of columns in the left matrix is the same as the number of rows in the right. Thus the matrices A and B used above to illustrate a 2×3 and a 3×2 matrix can be multiplied:

$$\begin{aligned} AB &= \begin{bmatrix} -1 & 3 & 0 \\ 4 & -2 & 5 \end{bmatrix}\begin{bmatrix} 2 & 0 \\ -3 & 1 \\ 0 & 4 \end{bmatrix} \\ &= \begin{bmatrix} (-1)(2)+3(-3)+0(0) & (-1)(0)+3(1)+0(4) \\ 4(2)+(-2)(-3)+5(0) & 4(0)+(-2)(1)+5(4) \end{bmatrix} \\ &= \begin{bmatrix} -11 & 3 \\ 14 & 18 \end{bmatrix} \end{aligned}$$

ADDITION OF MATRICES

Contrary to the rule for multiplication (actually, it is the rule for multiplication that is contrary), the definition for the addition of matrices is straightforward. Any two matrices with the same dimensions are added by simply adding corresponding terms. For the 2×2 matrices in which we are primarily interested,

$$\begin{bmatrix} a & b \\ c & d \end{bmatrix} + \begin{bmatrix} A & B \\ C & D \end{bmatrix} = \begin{bmatrix} a + A & b + B \\ c + C & d + D \end{bmatrix}$$

SUMMARY

In this section we have introduced the concept of a matrix, explained what we mean by the dimensions of a matrix, and stated the conditions under which we call two matrices equal. Definitions for addition and multiplication have been given, the first being straightforward and the second at first seeming peculiar. It should be pointed out that matrices, together with the definitions of addition and multiplication, acquaint us with another mathematical system.

This new mathematical system has some strange properties, especially with respect to multiplication. These have been left for you to discover, and it is therefore strongly recommended that you do at least all the odd-numbered problems in order to see what is to be discovered.

PROBLEM SET 14.1

$$A = \begin{bmatrix} 1 & 1 \\ 1 & 2 \end{bmatrix} \quad B = \begin{bmatrix} 5 & 7 \\ 2 & 3 \end{bmatrix} \quad C = \begin{bmatrix} 3 & -1 \\ -5 & 2 \end{bmatrix} \quad D = \begin{bmatrix} 5 & -3 \\ -3 & 2 \end{bmatrix} \quad I = \begin{bmatrix} 1 & 0 \\ 0 & 1 \end{bmatrix}$$

1. Find AB and BA. Find BC and CB. What have you learned?
2. Find AC and CA. Find CD and DC. What have you learned?
3. In Problem 1, you found AB and BC. Call the answers M and N, respectively. Find MC and AN.
4. In Problem 2 you found AC and CD. Call the answers R and S, respectively. Find RD and AS.
5. In Problem 3, you found $MC = (AB)C$ and $AN = A(BC)$. In the same way find $(BC)D$ and $B(CD)$.
6. In Problem 4, you found $RD = (AC)D$ and $AS = A(CD)$. In the same way find $(CD)B$ and $C(DB)$.
7. Find $B + C$, and then $A(B + C)$. Does this equal $AB + AC$?
8. Find $C + D$, and then $(C + D)B$. Does this equal $CB + DB$?
9. Find AI, IA, DI, and ID.
10. Find BI, IB, CI, and IC.
11. (a) Let $E = \begin{bmatrix} 2 & -1 \\ -1 & 1 \end{bmatrix}$. Find AE and EA.

    (b) Let $F = \begin{bmatrix} 2 & 1 \\ 5 & 3 \end{bmatrix}$. Find CF and FC.

12. (a) Let $G = \begin{bmatrix} 3 & -7 \\ -2 & 5 \end{bmatrix}$. Find BG and GB.

(b) Let $H = \begin{bmatrix} 2 & 3 \\ 3 & 5 \end{bmatrix}$. Find DH and HD.

13. Given a matrix K, the matrix L is called the inverse of K with respect to multiplication if $KL = I$. Try to find an inverse with respect to multiplication for each of the following.

(a) $\begin{bmatrix} 7 & 4 \\ 5 & 3 \end{bmatrix}$ (b) $\begin{bmatrix} 8 & 5 \\ 5 & 3 \end{bmatrix}$ (c) $\begin{bmatrix} 4 & 7 \\ 2 & 4 \end{bmatrix}$ (d) $\begin{bmatrix} 4 & 2 \\ 6 & 3 \end{bmatrix}$

14. As in Problem 11, try to find an inverse with respect to multiplication for each of the following.

(a) $\begin{bmatrix} 5 & 7 \\ 2 & 3 \end{bmatrix}$ (b) $\begin{bmatrix} 3 & 5 \\ 2 & 3 \end{bmatrix}$ (c) $\begin{bmatrix} 4 & 3 \\ 2 & 2 \end{bmatrix}$ (d) $\begin{bmatrix} 6 & 9 \\ 2 & 3 \end{bmatrix}$

15. Compute the following products.

(a) $\begin{bmatrix} 3 & 1 \\ 0 & 2 \\ 1 & 4 \end{bmatrix} \begin{bmatrix} 1 & 2 & 1 \\ 3 & 1 & 0 \end{bmatrix}$ (b) $\begin{bmatrix} 3 & 1 & 0 \\ 2 & 0 & 2 \end{bmatrix} \begin{bmatrix} 1 & -1 & 2 \\ 2 & 1 & 0 \\ 0 & 1 & -1 \end{bmatrix}$

16. Compute the following products.

(a) $\begin{bmatrix} 1 & 2 \\ 0 & 1 \\ 2 & 2 \\ 1 & 0 \end{bmatrix} \begin{bmatrix} 3 & 1 & 0 \\ 1 & 2 & 4 \end{bmatrix}$ (b) $\begin{bmatrix} 1 & 3 & -1 \\ 0 & 1 & 2 \end{bmatrix} \begin{bmatrix} 1 & 2 \\ 4 & 1 \\ -1 & 0 \end{bmatrix}$

17. What must be true of the dimensions k, l, m, and n of two matrices

$$k \begin{bmatrix} \\ \end{bmatrix}^{\displaystyle l} \qquad m \begin{bmatrix} \\ \end{bmatrix}^{\displaystyle n}$$

if they are to be multiplied? What will be the dimensions of the product?

18. Sales for the Sleepeeze Company for three quarters are indicated. Find the labor and material costs for each model in each quarter.

	Q_1		Q_2		Q_3	
	Delucks	Lumpy	Delucks	Lumpy	Delucks	Lumpy
Firm	70	150	80	135	75	165
Squishy	185	55	125	30	210	60

19. Find the total sales of the Sleepeeze Company for the three quarters indicated in Problem 18. Find the total labor and material costs for each model over this period. Compare with the answers obtained for Problem 18.

Something to Discover

$$M = \begin{bmatrix} a & b \\ c & d \end{bmatrix}$$

The determinant of M is defined
to be the number $ad - bc$.

$$A = \begin{bmatrix} 3 & 5 \\ 2 & 4 \end{bmatrix} \qquad B = \begin{bmatrix} 5 & 2 \\ 6 & 3 \end{bmatrix}$$

For instance, the determinant of A is 2, and the determinant of B is 3. Find the
product AB. What is its determinant?

14.2 / Properties of Matrix Multiplication

Perhaps it is appropriate to remind our readers again of the spirit
in which we are proceeding, for some may identify with the character in our cartoon above who wonders why anyone cares about
the properties of matrix multiplication. Research of the kind we

are trying to describe in this chapter often seeks the answer to a question, even when there is no apparent "payoff" for finding it. We are reminded of the time that we took a group of students on a tour of the Argonne National Laboratories. A resident scientist explained the elaborate apparatus he was constructing to determine the speed with which a particle was traveling. When he finished his lecture, a student asked the obvious question, "Why would anyone want to know?"

Slowly withdrawing the pipe from his mouth and using it to punctuate his remarks, he gave his answer. "I don't know why anyone would want to know, but someday, somebody, for some reason, might ask. And if they do" (here the pointer swung down with a flourish that almost left our scientist friend impaled on the stub end of his pipe) "I'm going to know the answer."

It is this spirit of curiosity, this willingness to pause and examine the unusual, that often carries science forward. And it is on this same curiosity that we now rely in urging the reader to further explore some of the peculiarities of matrix multiplication as they emerged from Problem Set 14.1.

One of the things quickly learned from a few examples is that there are matrices A and B for which $AB \neq BA$. That is, matrix multiplication is not **commutative.** We all know that it makes a difference as to whether we take off our slippers and step into the bath water, or vice versa, but it comes as a shock to most people to find that there are useful mathematical systems in which $AB \neq BA$.

> Matrix multiplication is not commutative.

Such unconventional behavior surely arouses suspicions about other laws long taken for granted. For example, is the following always true?

$$(AB)C = A(BC)$$

This is called the **associative** property, and it does hold for matrix multiplication. Though examples don't prove anything (a point we have stressed repeatedly), they do suggest things. In the case of Problems 3 through 6 in Problem Set 14.1, they suggest something that is true. Since the proof involves cumbersome notation, we ask the reader to take our word for it, even though such advice is contrary to our warnings about proof by authority.

> Matrix multiplication is associative.

The matrix

$$I = \begin{bmatrix} 1 & 0 \\ 0 & 1 \end{bmatrix}$$

plays an important role in matrix multiplication:

$$\begin{bmatrix} 1 & 0 \\ 0 & 1 \end{bmatrix}\begin{bmatrix} a & b \\ c & d \end{bmatrix} = \begin{bmatrix} a & b \\ c & d \end{bmatrix} = \begin{bmatrix} a & b \\ c & d \end{bmatrix}\begin{bmatrix} 1 & 0 \\ 0 & 1 \end{bmatrix}$$

$$I = \begin{bmatrix} 1 & 0 \\ 0 & 1 \end{bmatrix}$$

is the multiplicative identity.

It commutes with any matrix, and it acts the way the number 1 acts in ordinary multiplication. It is called the identity with respect to multiplication, or simply the **multiplicative identity.**

Given a matrix K, the matrix L is called the inverse of K with respect to multiplication (or the **multiplicative inverse**) if $LK = I$. Thus, since

$$\begin{bmatrix} 2 & -5 \\ -3 & 8 \end{bmatrix}\begin{bmatrix} 8 & 5 \\ 3 & 2 \end{bmatrix} = \begin{bmatrix} 1 & 0 \\ 0 & 1 \end{bmatrix}$$

we say that

$$\begin{bmatrix} 2 & -5 \\ -3 & 8 \end{bmatrix}$$

is the inverse of

$$\begin{bmatrix} 8 & 5 \\ 3 & 2 \end{bmatrix}$$

After observing a few matrices and their inverses, most students are ready to make a guess:

The inverse of $\begin{bmatrix} a & b \\ c & d \end{bmatrix}$ may be $\begin{bmatrix} d & -b \\ -c & a \end{bmatrix}$.

We have tried in the past to encourage guessing as a part of mathematics. We have also tried to encourage the checking of guesses against numerous examples. The guess above suggests that

The inverse of $\begin{bmatrix} 5 & 6 \\ 2 & 3 \end{bmatrix}$ may be $\begin{bmatrix} 3 & -6 \\ -2 & 5 \end{bmatrix}$.

It may be, but alas it isn't. The product gives

$$\begin{bmatrix} 3 & -6 \\ -2 & 5 \end{bmatrix}\begin{bmatrix} 5 & 6 \\ 2 & 3 \end{bmatrix} = \begin{bmatrix} 3 & 0 \\ 0 & 3 \end{bmatrix}$$

This has the merit of 0's in the right places, but it suffers the defect of having 3's where we wanted 1's. Some insight is gained by checking the guess in its general form, rather than for a particular matrix. That is, try multiplying

$$\begin{bmatrix} d & -b \\ -c & a \end{bmatrix}\begin{bmatrix} a & b \\ c & d \end{bmatrix} = \begin{bmatrix} ad - bc & 0 \\ 0 & ad - bc \end{bmatrix}$$

Evidently the guess works only so long as $ad - bc = 1$. This explains why it worked for some examples but not others in Problems 11 through 14 in Problem Set 14.1.

For the given matrix

$$K = \begin{bmatrix} a & b \\ c & d \end{bmatrix}$$

let us set $ad - bc = D$. The number D is called the **determinant** of matrix K. After some experimenting (yes, one can conduct experiments in mathematics), it usually occurs to the venturesome soul (the one who tries to find out instead of waiting to be told) to try what we shall call

$$K^{-1} = \begin{bmatrix} d/D & -b/D \\ -c/D & a/D \end{bmatrix}$$

as a possible inverse matrix. This in fact works as long as $D \neq 0$. When $D = 0$, there is no multiplicative inverse for K.

Since matrix multiplication is not commutative, it comes as something of a surprise to learn that K^{-1} works on either side; that is, $K^{-1}K = KK^{-1} = I$.

A word about notation is in order. Students who recall their algebra well enough may remember that 3^{-1} means $\frac{1}{3}$. It does not follow that K^{-1} means 1 over K. When K is a matrix, this would result in a very odd-looking creature. One should think of K^{-1} as the matrix that multiplies K to give I. It does not hurt to think of 3^{-1} as the number that multiplies 3 to give 1.

Let us state formally what we have learned about the multiplicative inverse of a matrix

$$K = \begin{bmatrix} a & b \\ c & d \end{bmatrix}$$

Theorem. Set $D = ad - bc$. If $D = 0$, no inverse exists. If $D \neq 0$, the inverse is

$$K^{-1} = \begin{bmatrix} d/D & -b/D \\ -c/D & a/D \end{bmatrix}$$

> We can find the multiplicative inverse if $D \neq 0$.

Once the right guess is made, the proof is easy. Simply multiply $K^{-1} \cdot K$ and $K \cdot K^{-1}$. The products will both be I, every time.

Finally we mention the distributive law which holds for matrices (as suggested by Problems 7 and 8 in Problem Set 14.1).

$$A(B + C) = AB + BC$$

The only caution to be exercised in using the distributive property is to be sure to pay attention to the order of multiplication. Thus

$$(B + C)A = BA + CA$$

> Matrices satisfy the distributive law.

but

$$(B + C)A \neq AB + AC$$

SUMMARY

We have stressed the following properties for 2×2 matrices:

1. Multiplication is associative; that is, for every A, B, and C, $A(BC) = (AB)C$.
2. Multiplication is not commutative; there exist matrices A and B for which $AB \neq BA$.
3. There is a multiplicative identity, which we denote by I, satisfying $AI = IA = A$.
4. If a matrix A has a nonzero determinant, there is a multiplicative inverse A^{-1} satisfying

$$A \cdot A^{-1} = A^{-1} \cdot A = I$$

5. The distributive laws hold:

$$A \cdot (B + C) = AB + AC$$
$$(B + C) \cdot A = BA + CA$$

The reader should, by choosing arbitrary 2×2 matrices, be able to illustrate any of these properties.

While we have not stressed the same properties for addition, it is easy to verify that matrix addition is associative and commutative, that there is an additive identity, and that every matrix has an additive inverse (see Problems 17 and 18).

PROBLEM SET 14.2

$$A = \begin{bmatrix} 1 & -2 \\ 2 & -3 \end{bmatrix} \quad B = \begin{bmatrix} -1 & 0 \\ 3 & 1 \end{bmatrix} \quad C = \begin{bmatrix} 2 & -1 \\ 1 & 4 \end{bmatrix} \quad D = \begin{bmatrix} 1 & 4 \\ 2 & 0 \end{bmatrix}$$

1. Use matrices A, B, and C to illustrate the associative law for multiplication.
2. Use matrices B, C, and D to illustrate the associative law for multiplication.
3. Use matrices A and B to show that multiplication is not commutative.
4. Use matrices C and D to show that multiplication is not commutative.
5. Use matrices A, B, and C to illustrate the distributive law $A(B + C) = AB + AC$.
6. Use matrices B, C, and D to illustrate the distributive law $B(C + D) = BC + BD$.
7. Find the multiplicative inverse of
   (a) A (b) C (c) AC
   Is the inverse of AC equal to $A^{-1}C^{-1}$?
8. Find the multiplicative inverse of
   (a) B (b) D (c) BD
   Is the inverse of BD equal to $B^{-1}D^{-1}$?

9. Find $C^{-1}A^{-1}$. Find the inverse of CA. Compare your answers with those for Problem 7.

10. Find $D^{-1}B^{-1}$. Find the inverse of DB. Compare your answers with those for Problem 8.

11. Let det A denote the determinant of A. Find det A, det B, and det AB.

12. Using the notation of Problem 11, find det C, det D, and det CD.

13. Use the distributive law to show that $(A + B)^2 = A^2 + AB + BA + B^2$. Illustrate this with matrices A, B, C, and D.

14. Use matrices C and D to show that $(C + D)^2 \neq C^2 + 2CD + D^2$.

15. Illustrate that $A(BC) = (AB)C$ using the 3×3 matrices

$$A = \begin{bmatrix} 1 & 0 & -1 \\ 2 & 1 & 1 \\ 3 & 0 & -2 \end{bmatrix} \quad B = \begin{bmatrix} 2 & 1 & 1 \\ 1 & 3 & 2 \\ -1 & 0 & 0 \end{bmatrix} \quad C = \begin{bmatrix} 0 & 2 & -2 \\ -1 & 1 & 0 \\ 2 & 0 & 3 \end{bmatrix}$$

16. Illustrate that $(BA)C = B(AC)$ using the 3×3 matrices in Problem 15.

17. What is the additive identity in the system of 2×2 matrices?

18. For a given 2×2 matrix

$$\begin{bmatrix} a & b \\ c & d \end{bmatrix}$$

what is the additive inverse?

19. Does det $(A + B) = $ det $A + $ det B?

Useless

I have never done anything "useful" . . . I have helped to train other mathematicians, but mathematicians of the same kind as myself, and their work has been, so far at any rate as I have helped them to it, as useless as my own. Judged by all practical standards, the value of my mathematical life is nil. . . . Time may change all this. No one foresaw the applications of matrices and groups and other purely mathematical theories to modern physics, and it may be that some "highbrow" applied mathematics will become useful in as unexpected a way; but the evidence so far points to the conclusion that in one subject as in another, it is what is commonplace and dull that counts for private life.

G. H. Hardy, A Mathematician's Apology, Cambridge University Press, 1967

G. H. Hardy, was generally acknowledged during his lifetime (1877–1947) to be one of the world's leading mathematicians. A student of number theory, he delighted in the esoteric quality of his work.

14.3 / Some Applications

Our presentation has been somewhat in the spirit of the historical development of matrices. Much was known about their properties before there were many applications, but the applications have been both numerous and important. This tendency for applications to follow theoretical developments is a feature that, for some people, justifies abstract research. To others, such justification is unnecessary, a kind of insult to the integrity of intellectual creativeness—like defending a symphony on the grounds that workers in a factory produce faster when it is played as background music.

Having lectured a bit in the last two sections on the subject of being willing to explore matrices in the almost total absence of any practical motivation, we now tip our hand by including at least a few applications in this closing section. Even at the risk of putting ourselves at odds with so eminent a mathematician as G. H. Hardy, we admit to feeling that the most interesting mathematics does have applications. It doesn't seem right, then, to leave the topic of matrices without pointing the way to at least some of the practical uses to which they can be put.

SOLVING SYSTEMS OF EQUATIONS

We once saw a professor sitting on his back porch (Section 3.3) deducing the number of boys and dogs in his yard by counting the number of heads and legs. This led to a set of two simultaneous equations:

$$x + y = 21$$
$$2x + 4y = 70$$

We are now able to rewrite these as a single matrix equation:

$$\begin{bmatrix} 1 & 1 \\ 2 & 4 \end{bmatrix} \begin{bmatrix} x \\ y \end{bmatrix} = \begin{bmatrix} 21 \\ 70 \end{bmatrix}$$

How do we solve an equation of the form $AZ = B$? The first answer to the question is usually, "Divide by A." The problem, however, if A, Z, and B are matrices, is that we don't have a definition for matrix division.

We do have a definition for multiplication. Can we solve $AZ = B$ by multiplication? Of course we can if we know the multiplicative inverse of A. And as it happens, we do know how to find A^{-1} for a 2×2 matrix if det $A \neq 0$. This suggests the pattern. Multiply both sides by A^{-1}:

$$A^{-1}(AZ) = A^{-1}B$$
$$(A^{-1}A)Z = A^{-1}B$$
$$IZ = A^{-1}B$$
$$Z = A^{-1}B$$

Note that we multiplied both sides of the equation on the left by A^{-1}. We did not write

$$AZA^{-1} = BA^{-1}$$

This would of course be correct, for it obeys the golden rule of algebra. It is correct, but it is not helpful, because we can't put A^{-1} next to A since A and Z don't commute.

Neither did we write

$$A^{-1}AZ = BA^{-1}$$

This wouldn't even be correct, for we have not obeyed the golden rule. We multiplied the left side by A^{-1} on the left, and the right side by A^{-1} on the right.

We are now ready to solve our matrix equation

$$\begin{bmatrix} 1 & 1 \\ 2 & 4 \end{bmatrix} \begin{bmatrix} x \\ y \end{bmatrix} = \begin{bmatrix} 21 \\ 70 \end{bmatrix}$$

The inverse of

$$\begin{bmatrix} 1 & 1 \\ 2 & 4 \end{bmatrix} \quad \text{is} \quad \begin{bmatrix} \frac{4}{2} & -\frac{1}{2} \\ -\frac{2}{2} & \frac{1}{2} \end{bmatrix}$$

Multiplying both sides on the left by this inverse, we have

$$\begin{bmatrix} \frac{4}{2} & -\frac{1}{2} \\ -\frac{2}{2} & \frac{1}{2} \end{bmatrix}\begin{bmatrix} 1 & 1 \\ 2 & 4 \end{bmatrix}\begin{bmatrix} x \\ y \end{bmatrix} = \begin{bmatrix} \frac{4}{2} & -\frac{1}{2} \\ -\frac{2}{2} & \frac{1}{2} \end{bmatrix}\begin{bmatrix} 21 \\ 70 \end{bmatrix}$$

$$\begin{bmatrix} 1 & 0 \\ 0 & 1 \end{bmatrix}\begin{bmatrix} x \\ y \end{bmatrix} = \begin{bmatrix} \dfrac{84 - 70}{2} \\ \dfrac{-42 + 70}{2} \end{bmatrix}$$

$$\begin{bmatrix} x \\ y \end{bmatrix} = \begin{bmatrix} 7 \\ 14 \end{bmatrix}$$

Again, as in Section 3.3, we get $x = 7$ and $y = 14$.

The method just described works with a system of three equations in three unknowns, and more generally for n equations in n unknowns. Our problem of course is that we only know how to find inverses of 2×2 matrices. Since our purpose here is just to show why people become interested in studying matrices, we pursue this no further.

BUSINESS

We turn now to another example in which matrices seem to be useful. Remember, our purpose is limited. We mean only to suggest that there is much under the surface which we scratch here.

In Problem 18 in Problem Set 14.1, we indicated the sales for the Sleepeeze Company for three quarters as:

	Q_1		Q_2		Q_3	
	Delucks	**Lumpy**	**Delucks**	**Lumpy**	**Delucks**	**Lumpy**
Firm	70	150	80	135	75	165
Squishy	185	55	125	30	210	60

Using the cost matrix

	C	
	Labor	**Material**
Delucks	45	25
Lumpy	35	10

we obtain the labor and material costs for each model in each quarter from the matrix products Q_1C, Q_2C, and Q_3C. Total sales for the three quarters are found by adding Q_1, Q_2, and Q_3, and the labor and material costs for each model over the three quarters is $(Q_1 + Q_2 + Q_3)C$. That these total costs equal the sum of the costs in each quarter is just common sense; it is also the distributive law extended to three terms:

$$(Q_1 + Q_2 + Q_3)C = Q_1C + Q_2C + Q_3C$$

SUMMARY

A system of n equations in n unknowns can be written as a matrix equation

$$AZ = B$$

If we know the multiplicative inverse A^{-1} of matrix A (being careful to multiply both sides on the left, since matrix multiplication is not commutative), we can solve the system:

$$A^{-1}AZ = A^{-1}B$$
$$Z = A^{-1}B$$

This is of practical use to us only in the case in which $n = 2$, since this is the only case for which we have learned how to find the multiplicative inverse A^{-1}. (You can find out about larger n's in books on matrix theory.)

There are numerous other applications in which the product of two matrices naturally displays desired information. These may involve multiplying nonsquare matrices, as some of the problems illustrate.

PROBLEM SET 14.3

1. Solve the following sets of equations by matrix methods.
   (a) $3x - 5y = 19$ (b) $3x + 4y = 4$ (c) $3x + 4y = 5$
        $4x - 7y = 26$ $x - 2y = -7$ $2x - 5y = -12$
2. Solve the following sets of equations by matrix methods.
   (a) $5x + 7y = -1$ (b) $4x - y = 8$ (c) $2x + 3y = 8$
        $2x + 3y = 0$ $6x + 3y = 3$ $5x - 2y = 1$
3. Use matrix methods to solve the systems
   (a) $x - 2y + 3z = 3$ (b) $x - 2y + 3z = 4$
        $x - y + 4z = -2$ $x - y + 4z = -3$
        $x \quad\ + 6z = 5$ $x \quad\ + 6z = 1$
   Hint: Multiply

$$\begin{bmatrix} -6 & 12 & -5 \\ -2 & 3 & -1 \\ 1 & -2 & 1 \end{bmatrix} \begin{bmatrix} 1 & -2 & 3 \\ 1 & -1 & 4 \\ 1 & 0 & 6 \end{bmatrix}$$

4. Use matrix methods to solve the systems
   (a) $2x + 9y + 13z = 4$ (b) $2x + 9y + 13z = -1$
        $3x + 5y + 7z = -2$ $3x + 5y + 7z = 7$
       $-x + 2y + 3z = 5$ $-x + 2y + 3z = 3$

Hint: Multiply

$$\begin{bmatrix} 1 & -1 & -2 \\ -16 & 19 & 25 \\ 11 & -13 & -17 \end{bmatrix} \begin{bmatrix} 2 & 9 & 13 \\ 3 & 5 & 7 \\ -1 & 2 & 3 \end{bmatrix}$$

5. The Kindle Company, which sold the tables and chairs to Librarian Hardback (Section 13.4), has like most companies raised prices periodically. The prices for various years are indicated in the table. Find the cost of Hardback's orders (9 tables and 43 chairs) for each of the years indicated.

	Table	Chair
1970	195	12
1975	239	16
1980	288	19

6. The Kindle Company (see Problem 5) has warehouses in Posthole (P), Center City (C), and Klondike (K). The number of tables and chairs commonly stocked by each warehouse is indicated in the table. Find the value of the inventory in each warehouse for the years indicated in Problem 5.

	P	C	K
Tables	10	65	15
Chairs	75	350	100

7. Suppose the Kindle Company (Problem 5) has warehouses in Atwater, Bayside, Crestline, and Deepwater. Suppose further that the four warehouses stocked respectively, 300, 150, 200, and 400 chairs and 40, 20, 25, and 60 tables. Find the money tied up in inventory in each of the warehouses in 1970, 1975, and 1980.

8. Using the figures in Problem 7, suppose the prices in 1972, 1977, and 1979 were, respectively, chairs $13, tables $205; chairs $18, tables $275; and chairs $20, tables $300. Find the amount of money tied up in inventory in each of the warehouses for these years.

9. A company manufactures three styles of kitchen cabinets, and each style comes in two grades. Style 1 requires 5 square units of plywood and 12 units of time to build; style 2 requires 3 units of plywood and 8 units of time; and style 3 requires 2 units of plywood and 6 units of time. For grade-A cabinets, wood costs $15 per square unit and labor costs $22 per unit of time. For grade B, with cheaper material and less experienced cabinetmakers, the cost drops to $12 per square unit and $17 per unit of time. Display these figures in two matrices in such a way that their product shows the cost of materials and time required for each grade of each style of cabinet.

10. Steady Eddy's Pizzeria offers four choices of pizza. The Cheese Special requires 1 unit of dough, 3 units of cheese, and 1 unit of tomato sauce. The Sausage requires 1 unit of dough, 1 unit of cheese, 1 unit of sausage, and 1 unit of tomato sauce. The Super Sausage calls for 1 unit of dough, 1 unit of cheese, $\frac{3}{2}$ units of sausage, 1 unit of mushrooms, and 1 unit of tomato sauce. The Large Special calls for 2 units of dough, $\frac{5}{2}$ units of cheese, 2 units of sausage, 2 units of mushrooms, and $\frac{5}{2}$ units of tomato sauce. Depending on whether Eddy buys supplies in small or large amounts he pays, per unit, for dough $0.35 or $0.28, for cheese $0.35 or $0.28, for sausage $0.79 or $0.63, for mushrooms $0.18 or $0.15, and for tomato sauce $0.12 or $0.09. Display these figures in two matrices in such a way that their product shows the cost of each choice of pizza for each of the two purchasing options.

Chapter 15
Algebraic Structures

*To carry out his role of abstractor, the mathematician must
continually pose such questions as "What is the common
aspect of diverse situations?" or "What is the heart of
the matter?" He must always ask himself, "What makes such
and such a thing tick?" Once he has discovered the answer
to these questions and has extracted the crucial simple parts,
he can examine these parts in isolation. He blinds himself,
temporarily, to the whole picture, which may be confusing.*
PHILIP DAVIS AND WILLIAM CHINN

SUBSTITUTION OF EQUALS

If x and y are different names for the same thing, can y be substituted for x in a discussion? This question was bothering Bertrand Russell when he contrived a puzzle similar to the following. We begin by supporting the truth of two statements:

1. Homer wonders if Amy is the woman he will marry.
2. Amy is in fact the woman he will marry.

Now we ask if, since statement 2 is true, we can substitute in statement 1, replacing "the woman he will marry" with "Amy." This leaves statement 1 reading

Homer wonders if Amy is Amy.

15.1 / Basic Concepts of Algebra

Matrix multiplication, we have seen, exhibits some properties that surprise us. We saw, for example, that for two matrices A and B it is generally the case that $AB \neq BA$. From examples such as

$$\begin{bmatrix} 2 & -1 \\ -4 & 2 \end{bmatrix}\begin{bmatrix} 1 & -1 \\ -1 & 3 \end{bmatrix} = \begin{bmatrix} 2 & -1 \\ -4 & 2 \end{bmatrix}\begin{bmatrix} 2 & -4 \\ 1 & -3 \end{bmatrix}$$

we learned that $AB = AC$ does not imply $B = C$. We have also seen situations such as

$$\begin{bmatrix} 2 & -1 \\ -4 & 2 \end{bmatrix}\begin{bmatrix} 3 & 1 \\ 6 & 2 \end{bmatrix} = \begin{bmatrix} 0 & 0 \\ 0 & 0 \end{bmatrix}$$

in which $AB = 0$ but $A \neq 0$ and $B \neq 0$.

These surprising properties are not unique to the system of matrices. In the system of integers modulo 12, we saw the nonsense that would result from canceling the 3's in the expression $3 \cdot 5 \equiv 3 \cdot 9$. And divisors of zero became commonplace as we got accustomed to such products as $3 \cdot 4 \equiv 6 \cdot 2 \equiv 9 \cdot 8 \equiv 0$.

We could introduce other algebraic systems exhibiting other surprising properties, but more examples are unlikely to change our general expectations. We continue to be surprised by anything that violates what our familiarity with arithmetic and high school algebra has conditioned us to expect. And this is wholesome. Our purpose here is not to confuse you about things you feel you understand, but to identify the basic principles that underlie both arithmetic and common algebra.

We have drawn our examples from the algebra of matrices and from modular arithmetic. These systems, as well as the number systems reviewed in Chapter 13 and common high school algebra, illustrate the concept of an algebraic structure. Like any system of deductive thought, algebraic structures begin with undefined terms and axioms. As was the case with the various geometric structures we studied (finite geometries and Euclidean and non-Euclidean geometries), most algebraic structures share a common terminology and some common axioms. And as was the case with geometries, algebraic structures differ from one another by virtue of certain variations in the axioms. Our purpose in this section is to introduce some of the terms and assumptions common to a variety of algebraic structures.

The **elements** of algebraic structures, that is, the objects about which we state axioms, are of course left undefined. They are designated by single letters, and one of the things you must keep in mind is that an element designated by b can be a matrix, a member of the set of integers modulo m, or something you've never heard of. Therefore you must be careful not to perform any operation on b (like assuming that $1/b$ makes sense) not specifically allowed by an axiom.

BINARY OPERATIONS

Not knowing what kind of elements we are talking about, it is hard to say how they are to be combined (added, multiplied, intersected, etc.), but it is generally the case that there is at least one way to combine two elements to get another. Without specifying just how this is to be done, i.e., leaving it undefined, we refer to this combining of two terms as a **binary operation.** Suppose, for example, that the elements with which we are working are the rational numbers. We can create a binary operation $*$ by defining

$$4 * 6 = 5$$
$$7 * 2 = \tfrac{9}{2}$$

$$a * b = \frac{a + b}{2}$$

You have not seen such a definition before and, to be truthful, once you have put aside this book, you'll probably not see it again. In the spirit of abstract mathematics, however, this should not prevent us from investigating the properties of $*$ (Problems 4 through 5). It does what a binary operation is supposed to do; given two elements of a set, it tells us how to combine them to get a third element.

It is possible that a binary operation between two elements of a set may produce an element not in the set. Thus, if we start with the set of integers and use the operation $*$ defined above, we may obtain an element no longer in the set ($7 * 2$ is not an integer). Again, beginning with the set of integers and using the operation ÷ of division, we are quickly taken outside our original system. Sometimes, however, we stay inside the system. The binary operation of ordinary addition, +, applied to any two integers always gives another integer. We say the integers are closed with respect to addition. A set S is **closed** with respect to a binary operation ⊛ if, for any two elements x and y in S, the element $x ⊛ y$ is again in S.

An analogy can be drawn between the notion of a closed set and the biblical notion in which two creatures coming together (a binary operation) produce a creature of their own kind. Mutations then correspond to nonclosed sets, in which a binary relationship produces an element not in the set. Since two odd integers, when they multiply, bring forth another odd integer, we say the odd integers are closed with respect to multiplication. This same set of odd integers is not closed, however, with respect to addition. Do you see why?

EQUALITY

Another idea common to many algebraic structures, indeed to most mathematical structures (geometry, logic, etc.) is the notion of equality. Again the specific meaning of the term depends on the context. A youngster in the sixth grade understands

$$\tfrac{2}{3} = \tfrac{10}{15}$$

but she would think it quite peculiar if she saw $15 = 3$ written on the blackboard. Yet in arithmetic modulo 12, this is correct (though we have preferred to write $15 \equiv 3$). It means one thing to say $A = B$ if A and B are matrices; it means something else if A and B represent lengths of sides of triangles.

Whatever the specific meaning attached to the relationship (which we here designate by $\mathscr{E}$), there are four properties (axioms) that mathematicians always expect of an **equivalence relation.**

Determinative: Given two elements r and s, we must have a clear rule which enables us to determine whether or not $r \mathscr{E} s$.

Reflexive: For all elements r, it must be that $r \mathscr{E} r$.

Symmetric: Whenever $r \mathscr{E} s$, it must be that $s \mathscr{E} r$.

Transitive: If $r \mathscr{E} s$ and $s \mathscr{E} t$, then $r \mathscr{E} t$.

When we begin with the set of all triangles, the idea of congruence is an equivalence relation. In this situation, we usually write $r \cong s$. If we begin with the same set of triangles, the notion of similarity will also pass the tests (satisfy the axioms) listed above. When this is the idea of equivalence that concerns us, we commonly write $r \sim s$.

WELL-DEFINED BINARY OPERATIONS

There is one more idea, common to many algebraic structures, that should be mentioned while we are discussing binary operations and notions of equality. We illustrate it by using a method for "adding" fractions which, while wrong by commonly accepted standards, is nevertheless popular. Consider the rule

$$\frac{a}{b} \text{``+''} \frac{c}{d} = \frac{a+c}{b+d}$$

Having said already that this rule is wrong by commonly accepted standards, let us point out that there are applications in which this is exactly the correct rule. Suppose, for example, that about midseason the mighty Casey has been at bat 261 times and has hit safely 88 times. His average is $88/261 = .337$. Suppose further that in the remainder of the season he goes to the bat 232 times and gets 65 hits. Then his average is found by writing

$$\frac{88}{261} \text{``+''} \frac{65}{232} = \frac{88+65}{261+232} = \frac{153}{493} = .310$$

This application of "addition" notwithstanding, it is still the case that most of the time we do not accept this definition as a reasonable way to add fractions, and there are several good reasons for this. One of them is that there are few situations in life in which this kind of "addition" gives results that correspond to our expectations. We wish, however, to stress another good reason for rejecting this definition. We noted above that

$$\frac{2}{3} = \frac{10}{15}$$

Now consider the consequences of "adding" $\frac{2}{5}$ to both sides. On the left we get

$$\frac{2}{3} \text{``+''} \frac{2}{5} = \frac{4}{8} = \frac{1}{2}$$

while on the right we get

$$\frac{10}{15} \text{``+''} \frac{2}{5} = \frac{12}{20} = \frac{3}{5}$$

Surely if we "add" the same element to equals, the result should be equal, and on these grounds (along with others, as we said before), this kind of addition is disappointing.

Thus is illustrated our last idea, namely, that a binary operation $\circledast$ between elements of a set in which there is an equivalence relation $\mathscr{E}$, should be **well-defined** with respect to $\mathscr{E}$. That is, if

$$r \mathscr{E} s$$

we definitely want it to be true that, for any element t,

$$(r \circledast t) \mathscr{E} (s \circledast t)$$

and

$$(t \circledast r) \; \mathscr{E} \; (t \circledast s)$$

Note the care with which we have stated our definition. We have *not* said that, if $r \; \mathscr{E} \; s$, then $(r \circledast t) \; \mathscr{E} \; (t \circledast s)$. What is the difference?

Our study of geometry conditions us to expect that theorems can be proved from axioms; that is, the axioms say more than is immediately evident. This is as true in algebra as elsewhere. Note that our definition of a well-defined operation requires only that we be able to operate (multiply, add, or whatever) on equals with the same element. However, this implies the commonly accepted proposition that we can operate on equals with equals (e.g., add equals to equals and multiply equals by equals). We conclude with a demonstration of this fact, our first proof of an abstract algebraic theorem.

Theorem. If $r \; \mathscr{E} \; s$, $t \; \mathscr{E} \; u$, and $\circledast$ is well-defined with respect to $\mathscr{E}$, then $(r \circledast t) \; \mathscr{E} \; (s \circledast u)$.

PROOF.　From the fact that $r \; \mathscr{E} \; s$,

$$(r \circledast t) \; \mathscr{E} \; (s \circledast t)$$

and from the fact that $t \; \mathscr{E} \; u$,

$$(s \circledast t) \; \mathscr{E} \; (s \circledast u)$$

Now from the transitive property of $\mathscr{E}$, we have

$$(r \circledast t) \; \mathscr{E} \; (s \circledast u)$$

SUMMARY

Algebraic structures consist of elements, one or more methods of combining these elements by a binary operation, and a notion of equivalence between elements. A set may or may not be closed with respect to a binary operation. Different sets of elements call for different concepts of equivalence, and there may be more than one concept of equivalence for the same set of elements. Any concept that is to be described as an equivalence relation, however, must satisfy four conditions: determinative, reflexive, symmetric, and transitive. Finally, we have noted that, given an equivalence relation $\mathscr{E}$, we always require that any binary operation that is to be used shall be well-defined, a requirement that underlies the common assertion that equals can be multiplied by (or added to) equals.

PROBLEM SET 15.1

1. Consider the binary operation of ÷ (division) defined on the set of rational numbers.
   (a) Is ÷ commutative?
   (b) Is ÷ associative?

2. Consider the binary operation of × (multiplication) defined on the integers modulo 12.
   (a) Is × commutative?
   (b) Is × associative?

3. Consider the operation ■ defined on the set of rationals by $a \blacksquare b = a + b + ab$ (so, for example, $2 \blacksquare 3 = 11$).
   (a) $(4 \blacksquare 7) \blacksquare -2 =$
   (b) $4 \blacksquare (7 \blacksquare -2) =$
   (c) Is ■ associative?
   (d) Is ■ commutative?

4. In this section we defined ∗ on the set of rationals by $a * b = \frac{1}{2}(a + b)$.
   (a) $(4 * 7) * -2 =$
   (b) $4 * (7 * -2) =$
   (c) Is ∗ associative?
   (d) Is ∗ commutative?

5. The operation "+" defined by (a/b) "+" $(c/d) = (a + c)/(b + d)$ was seen to be not well-defined since, while
$$\tfrac{2}{3} = \tfrac{10}{15}$$
$$\tfrac{2}{3} \text{ "+" } \tfrac{2}{5} \neq \tfrac{10}{15} \text{ "+" } \tfrac{2}{5}$$
   What about ∗ defined in Problem 4 above? Does $\frac{2}{3} * \frac{2}{5}$ equal $\frac{10}{15} * \frac{2}{5}$?

6. What about ■ as defined in Problem 3? Does $\frac{2}{3} \blacksquare \frac{2}{5}$ equal $\frac{10}{15} \blacksquare \frac{2}{5}$?

7. Is the set of positive integers closed under the following operations?
   (a) Ordinary addition, +.
   (b) Ordinary multiplication, ×.
   (c) The operation ∗ defined in Problem 4.
   (d) The operation ■ defined in Problem 3.

8. Is the set of integers that are multiples of 3 closed under the operations listed in Problem 7?

9. Consider the set of all cars parked in a certain lot. Which of the following relationships are determinative?
   (a) Has the same number of cylinders as.
   (b) Is the same color as.
   (c) Uses the same size tires as.
   (d) Is as pretty as.

10. Consider the set of all children in a certain classroom. Which of the following relationships are determinative?
    (a) Has the same color hair as.
    (b) Is as talented as.
    (c) Is as tall as.
    (d) Is the same sex as.

11. Consider the set of all U.S. citizens. Decide whether the listed relationships are reflexive; symmetric; transitive.
    (a) Enjoys the same hobby as.

(b) Is married to.

(c) Is a first cousin of.

(d) Is as tall as.

(e) Lives within a mile of.

(f) Has ridden in the same car pool as.

12. Consider the set of all counting numbers. Decide whether the listed relationships are reflexive; symmetric; transitive.

(a) Is a divisor of.

(b) Is a multiple of.

(c) Has as many distinct prime factors as.

(d) Is a factor of the same number as.

(e) Has the same number of distinct prime factors as.

13. Beginning with the set of all cities shown on a map of the United States, let us agree that city X is separated from city Y if the distance between them is at least 100 miles. Is the relation "separated from" reflexive? Symmetric? Transitive?

14. Again beginning with the set of all cities on a map, let us say that city X is a neighbor of city Y if cities X and Y are within 100 miles of each other. Is the relation, "is a neighbor of" reflexive? Symmetric? Transitive?

15. Can you think of a relationship between the cities on a map of the United States that is reflexive and transitive but not symmetric?

16. Can you think of a relationship between the children in an elementary school classroom that is symmetric and transitive but not reflexive?

17.* For a fixed integer k, let $\equiv$ designate the relationship of equality modulo k. Show that $\equiv$ is an equivalence relation.

18.* Consider the set of all points (a, b) in the plane. We will say that $(a, b) \approx (c, d)$ if and only if $a + d = b + c$. For example, $(3, 7) \approx (11, 15)$. Show that $\approx$ is an equivalence relation.

19.* Referring to Problem 15, show that multiplication is well-defined modulo k; that is, show that, if $a \equiv b$, then $ac \equiv bc$.

20.* Referring to Problem 16, define the multiplication of two points as follows.

$$(a, b) \cdot (e, f) = (ae + bf, af + be)$$

Show that multiplication is well-defined with respect to $\approx$.

> ARITHMETIC MODULO 30
> If 17 is added to 13, we get 0. If 9 is added to 21, we get 0.
>
> $$\begin{array}{r} 17 \\ \underline{9} \\ 153 \equiv 3 \end{array} \qquad \begin{array}{r} 13 \\ \underline{21} \\ 13 \\ \underline{26} \\ 273 \equiv 3 \end{array}$$
>
> Why do we get 3 in both cases?

15.2 / Mathematical Rings

The abstract theory of rings is entirely a product of the twentieth century. David Hilbert (1862–1943) coined the term "ring"; but it was Emmy Noether (1882–1935) who gave the theory an axiomatic basis and developed it to full flower. She is rightly called the founder of modern abstract algebra.

An algebraic structure consists of a set of elements with a notion of equality and at least one binary operation, all being subject to certain axioms. But what should we take as axioms? We are guided to an answer by examining the common properties of the number systems we studied in Chapters 13 and 14. This process of recognizing and isolating the structural features common to a variety of situations is the essence of what we call *abstraction*.

Recall five mathematical systems studied in those earlier chapters: the integers, the rational numbers, the real numbers, the modular systems, the system of 2×2 matrices. In each of them, we have a notion of equality and two binary operations called addition and multiplication. What basic properties do all these systems have in common?

First, we note that both operations are **associative,** that is,

$$a + (b + c) = (a + b) + c$$
$$a(bc) = (ab)c$$

The situation is quite different when we look at commutativity. For while the **commutative** law for addition

$$a + b = b + a$$

always holds, the corresponding law for multiplication may fail (recall that $ab \neq ba$ for matrices).

In each case there is an **additive identity,** that is, an element that acts like zero:

$$a + \mathbf{0} = \mathbf{0} + a = a$$

But we quickly point out that this zero may be different in different contexts. This is why we have used a boldfaced zero. It reminds us that we have to think of it abstractly. For in the integers $\mathbf{0}$ means the ordinary 0, whereas in the matrix system

$$\mathbf{0} = \begin{bmatrix} 0 & 0 \\ 0 & 0 \end{bmatrix}$$

Similarly, we always have a multiplicative identity $\mathbf{1}$ which acts like 1 and therefore satisfies

$$a\mathbf{1} = \mathbf{1}a = a$$

Here $\mathbf{1}$ is the ordinary number 1 when we are considering the integers. But when we are working with matrices, it is

$$\mathbf{1} = \begin{bmatrix} 1 & 0 \\ 0 & 1 \end{bmatrix}$$

When we come to inverses, we must be especially careful. The **additive inverse** of a is denoted by $\bar{a}$ (reminding us of negative a); it satisfies

$$a + \bar{a} = \bar{a} + a = \mathbf{0}$$

For the integers, $\bar{a}$ is simply $-a$, while if

$$a = \begin{bmatrix} -2 & -4 \\ 1 & 2 \end{bmatrix} \quad \text{then} \quad \bar{a} = \begin{bmatrix} 2 & 4 \\ -1 & -2 \end{bmatrix}$$

However, in the integers modulo 12, if $a \equiv 3$, then $\bar{a} \equiv 9$. We have to be careful, but in each of our five systems we always have $\bar{a}$.

When we consider the corresponding notion for multiplication, we have real trouble. Borrowing the notation we used for matrices, we want for each $a \neq \mathbf{0}$ an element a^{-1} called the **multiplicative inverse** satisfying

$$aa^{-1} = a^{-1}a = \mathbf{1}$$

We may not get it. In the integers modulo 12, 4 has no multiplicative inverse; neither does

$$\begin{bmatrix} -2 & -4 \\ 1 & 2 \end{bmatrix}$$

in the matrix system, since its determinant is 0. However, when m/n is a rational number, the reciprocal n/m is its multiplicative inverse.

Finally, we mention the two **distributive** laws which hold in all five systems:

$$a(b + c) = ab + ac$$
$$(a + b)c = ac + bc$$

The formulas enclosed in the boxes are shared by all our systems. These are the properties that we isolate and make the axioms for an abstract structure that we call a ring.

DEFINITION OF A RING

A mathematical **ring** is a set of elements with a notion of equality (=) that is closed with respect to two well-defined binary operations called addition, $\oplus$, and multiplication, $\otimes$. These operations are subject to the following axioms.

TABLE 15-1
AXIOMS FOR A RING

Name	Law for Addition	Law for Multiplication
Associative law	$a \oplus (b \oplus c) = (a \oplus b) \oplus c$	$a \otimes (b \otimes c) = (a \otimes b) \otimes c$
Commutative law	$a \oplus b = b \oplus a$	
Existence of identity	There is an element **0** satisfying $a \oplus 0 = 0 \oplus a = a$ for all a	There is an element **1** satisfying $a \otimes 1 = 1 \otimes a = a$ for all a
Existence of inverses	For each a, there is an element $\bar{a}$ satisfying $a \oplus \bar{a} = \bar{a} \oplus a = 0$	
Distributive laws	$a \otimes (b \oplus c) = (a \otimes b) \oplus (a \otimes c)$ $(a \oplus b) \otimes c = (a \otimes c) \oplus (b \otimes c)$	

There are two conspicuous gaps in the chart, corresponding to two properties which are missing in some of our examples, the commutative law of multiplication and the existence of multiplicative inverses. Filling in these two gaps is the main job in Section 15.3.

SOME THEOREMS THAT ARE TRUE IN ANY RING

As in geometry, so in algebra, axioms imply other rules—called theorems—which can be proved. Our first theorem deals with addition and justifies what is often called "canceling common terms."

Theorem 1.   If $r \oplus t = s \oplus t$, then $r = s$.

Since one of our goals is to emphasize how algebra, like geometry, rests on axioms, we stress this similarity by writing the proof in the style commonly used in elementary geometry (Section 12.1).

Statement	Reason
1. $r \oplus t = s \oplus t$	1. Given
2. $(r \oplus t) \oplus \bar{t} = (s \oplus t) \oplus \bar{t}$	2. Existence of additive inverse $\bar{t}$; add equals to equals
3. $r \oplus (t \oplus \bar{t}) = s \oplus (t \oplus \bar{t})$	3. Addition is associative
4. $r \oplus \mathbf{0} = s \oplus \mathbf{0}$	4. $\bar{t}$ is the additive inverse of t
5. $r = s$	5. $\mathbf{0}$ is the additive identity

We all know that in ordinary arithmetic zero times anything is zero. This fact holds in an arbitrary ring.

Theorem 2.   For any r, it is true that $r \otimes \mathbf{0} = \mathbf{0}$ and $\mathbf{0} \otimes r = \mathbf{0}$.

Statement	Reason
1. For any s, $\mathbf{0} \oplus s = s$	1. $\mathbf{0}$ is the additive identity
2. $\mathbf{0} \oplus (r \otimes \mathbf{0}) = r \otimes \mathbf{0}$	2. Since statement 1 is true for any s, it is true for $s = r \otimes \mathbf{0}$
3. $\mathbf{0} \oplus \mathbf{0} = \mathbf{0}$	3. Statement 1 is true for $s = \mathbf{0}$
4. $r \otimes (\mathbf{0} \oplus \mathbf{0}) = r \otimes \mathbf{0}$	4. Multiply equals by equals
5. $\mathbf{0} \oplus (r \otimes \mathbf{0}) = r \otimes (\mathbf{0} \oplus \mathbf{0})$	5. Transitive law for equality (things equal to the same thing, in this case $r \otimes \mathbf{0}$, are equal to each other)
6. $\mathbf{0} \oplus (r \otimes \mathbf{0}) = (r \otimes \mathbf{0}) \oplus (r \otimes \mathbf{0})$	6. Use the distributive law on the right side of statement 5
7. $\mathbf{0} = r \otimes \mathbf{0}$	7. Use Theorem 1 with $t = r \otimes \mathbf{0}$

This proves only half of the theorem, namely, $r \otimes \mathbf{0} = \mathbf{0}$. It is tempting to say that the other half, $\mathbf{0} \otimes r = \mathbf{0}$, follows from the commutative law, but alas we don't have the commutative law for multiplication in a ring. There is a better way (longer but correct). The demonstration above can be modified (a small change will do) so that it takes care of the second result. See if you can find the secret (Problem 11).

Now we are ready to give a logical demonstration of a fact we talked about back in Section 2.2 for the integers. There we claimed that $(-a)(-b) = ab$. That may seem quite obvious to you for the integers. But is the corresponding fact for matrices obvious? Suppose

$$a = \begin{bmatrix} 2 & 1 \\ -3 & 4 \end{bmatrix} \qquad b = \begin{bmatrix} 1 & -1 \\ -4 & 3 \end{bmatrix}$$

Then

$$ab = \begin{bmatrix} -2 & 1 \\ -19 & 15 \end{bmatrix}$$

The additive inverses of a and b are

$$\bar{a} = \begin{bmatrix} -2 & -1 \\ 3 & -4 \end{bmatrix} \qquad \bar{b} = \begin{bmatrix} -1 & 1 \\ 4 & -3 \end{bmatrix}$$

You should compute $\bar{a}\bar{b}$ to see that you really do get the same answer as ab. In any case, we are now going to give a proof, an incontestable demonstration, that this holds in any ring whatever. This time we write the proof in a format more common in algebra.

Theorem 3. $\bar{a} \otimes \bar{b} = a \otimes b$

PROOF.   First note that

$$\begin{aligned}
(\bar{a} \otimes \bar{b}) \oplus (\bar{a} \otimes b) &= \bar{a} \otimes (\bar{b} \oplus b) \qquad \text{(distributive law)} \\
&= \bar{a} \otimes \mathbf{0} \qquad \text{(additive inverse)} \\
&= \mathbf{0} \qquad \text{(Theorem 2)}
\end{aligned}$$

Also, and for exactly the same reasons,

$$\begin{aligned}
(a \otimes b) \oplus (\bar{a} \otimes b) &= (a \oplus \bar{a}) \otimes b \\
&= \mathbf{0} \otimes b \\
&= \mathbf{0}
\end{aligned}$$

Thus

$$(\bar{a} \otimes \bar{b}) \oplus (\bar{a} \otimes b) = (a \otimes b) \oplus (\bar{a} \otimes b)$$

since both are equal to the same thing, namely, $\mathbf{0}$. Finally, from Theorem 1,

$$\bar{a} \otimes \bar{b} = a \otimes b$$

Incidentally, the problem that opens this section is an illustration of Theorem 3. See if you can understand why.

Our list of theorems could go on indefinitely, but perhaps three are enough to indicate the flavor of modern abstract algebra. Readers who want to know more about rings can check any book on modern algebra or abstract algebra. Those who do, however, must be warned that while we have for simplicity followed a well known text [G. Birkhoff and S. MacLane, *A Survey of Modern Algebra,* 3rd ed., Macmillan, 1965, p. 346], most authors do not require that a ring must have a multiplicative identity.

SUMMARY

When we examine the five mathematical systems (integers, rationals, reals, modular numbers, matrices) for common basic properties, we are led to the axioms, which appear in the table on page 446. These axioms allow us to make most (but not all) of the manipulations that are familiar from arithmetic and elementary algebra. We must only make sure that we avoid anything that depends on the commutative law of multiplication and the existence of multiplicative inverses.

PROBLEM SET 15.2

1. The addition and multiplication tables for the integers modulo 5 are shown in the margin. If $x = 2$, $y = 3$, find
   (a) $\bar{x}$ (b) $\bar{y}$ (c) $\bar{x}\bar{y}$ (d) xy
2. Do the calculations in Problem 1 if $x = 3$ and $y = 4$. What theorem is illustrated by these calculations?
3. In the integers modulo 5, find, if possible,
   (a) 0^{-1} (b) 1^{-1} (c) 2^{-1} (d) 3^{-1} (e) 4^{-1}
4. In the integers modulo 3, find each of the following.
   (a) $\bar{1}$ (b) 1^{-1} (c) $\bar{2}$ (d) 2^{-1}
5. In the integers modulo 12, find, if possible,
   (a) $\bar{3}$ (b) 3^{-1} (c) $\bar{5}$ (d) 5^{-1} (e) $\overline{11}$ (f) 11^{-1}
6. Do Problem 5 in the integers modulo 15.
7. In the ring of 2×2 matrices, let

$$x = \begin{bmatrix} 1 & 2 \\ 1 & 3 \end{bmatrix} \qquad y = \begin{bmatrix} 5 & 5 \\ 1 & 1 \end{bmatrix}$$

   Find, if possible,
   (a) $\bar{x}$ (b) x^{-1} (c) $\bar{y}$ (d) y^{-1}
8. Do Problem 7 with

$$x = \begin{bmatrix} 1 & -1 \\ 1 & -3 \end{bmatrix} \qquad y = \begin{bmatrix} 2 & -2 \\ 1 & 4 \end{bmatrix}$$

9. The counting numbers are not a ring. Which ring axioms fail?
10. Which integers have multiplicative inverses that are integers?
11. In elementary algebra, we learn that

$$(a + b)(a - b) = a^2 - b^2$$

   Is this valid in any ring? Why or why not?
12. Why does

$$(a + b)^2 = a^2 + 2ab + b^2$$

   fail to be valid in an arbitrary ring?
13. We proved only half of Theorem 2. Complete the proof by showing that $\mathbf{0} \otimes r = \mathbf{0}$. Hint: Start with $s \oplus \mathbf{0} = s$ and then let $s = \mathbf{0} \otimes r$. Mimic the earlier proof.

TABLE 15-2

+	0	1	2	3	4
0	0	1	2	3	4
1	1	2	3	4	0
2	2	3	4	0	1
3	3	4	0	1	2
4	4	0	1	2	3

TABLE 15-3

×	0	1	2	3	4
0	0	0	0	0	0
1	0	1	2	3	4
2	0	2	4	1	3
3	0	3	1	4	2
4	0	4	3	2	1

14. Show that $\overline{a \oplus b} = \bar{a} \oplus \bar{b}$. Hint: Show first that
    $(\bar{a} \oplus \bar{b}) \oplus (a \oplus b) = \overline{(a \oplus b)} \oplus (a \oplus b)$. Then use Theorem 1.
15. Show that $a \otimes \bar{b} = \overline{a \otimes b}$ by justifying each of the following steps.
    $$\overline{(a \otimes b)} \oplus (a \otimes b) = \mathbf{0}$$
    $$(a \otimes \bar{b}) \oplus (a \otimes b) = a \otimes (\bar{b} \oplus b)$$
    $$= a \otimes \mathbf{0}$$
    $$= \mathbf{0}$$
    $$(a \otimes \bar{b}) \oplus (a \otimes b) = \overline{(a \otimes b)} \oplus (a \otimes b)$$
    $$a \otimes \bar{b} = \overline{a \otimes b}$$
16. Show that $\bar{a} \otimes b = \overline{a \otimes b}$. Hint: See Problem 15.
17. Show that $\bar{\bar{a}} = a$.
18.* Consider the collection of all subsets of the ordinary Euclidean plane. Let $\oplus$ be $\cup$ and $\otimes$ be $\cap$. Which of the ring axioms are satisfied?

THE UNIVERSAL LAW OF CANCELATION

If the same number appears several times in an expression, it can be canceled. Examples are

$$\frac{16}{64} = \frac{1}{4} \qquad \text{(cancel the 6's)}$$

$$\frac{9^3 + 4^3}{9^3 + 5^3} = \frac{9 + 4}{9 + 5} \qquad \text{(cancel the 3's)}$$

15.3 / Mathematical Fields

Sometimes we can cancel, and sometimes we can't. There is no universal law of cancelation. The examples above were carefully contrived. They are correct, but the slightest change in the numbers makes them false. However, there is one cancelation law that always holds; we learned it in Section 15.2. If $a + b = a + c$, then $b = c$. The corresponding cancelation law for multiplication may not hold. For we learned that, in the matrix system, $a \cdot b = a \cdot c$ does not imply $b = c$. However, if a, b, and c are ordinary numbers, even this multiplicative cancelation is okay. This leads us to a study of systems in which the cancelation of common factors is valid. It has something to do with the two gaps we left in the axiom table on page 446.

DEFINITION OF A FIELD

A mathematical **field** is a set of elements with a notion of equality
($=$), which is closed with respect to two well-defined binary operations called addition ($\oplus$) and multiplication ($\otimes$). These operations
obey the following axioms.

TABLE 15-4
AXIOMS FOR A FIELD

Name	Law for Addition	Law for Multiplication
Associative law	$a \oplus (b \oplus c) = (a \oplus b) \oplus c$	$a \otimes (b \otimes c) = (a \otimes b) \otimes c$
Commutative law	$a \oplus b = b \oplus a$	$a \otimes b = b \otimes a$
Existence of identity	There is an element **0** satisfying $a \oplus \mathbf{0} = \mathbf{0} \oplus a = a$ for all a	There is an element **1** satisfying $a \otimes \mathbf{1} = \mathbf{1} \otimes a = a$ for all a
Existence of inverses	For each a, there is an element $\bar{a}$ satisfying $a \oplus \bar{a} = \bar{a} \oplus a = \mathbf{0}$	For each $a \neq \mathbf{0}$, there is an element a^{-1} satisfying $a \otimes a^{-1} = a^{-1} \otimes a = \mathbf{1}$
Distributive laws	$a \otimes (b \oplus c) = (a \otimes b) \oplus (a \otimes c)$ $(a \oplus b) \otimes c = (a \otimes c) \oplus (b \otimes c)$	

Everything is as it was for a ring, except that we now require
the commutative law for multiplication and the existence of multiplicative inverses.

Have we met any mathematical systems that satisfy all these
axioms? We certainly have. The rational numbers, the real numbers, and the complex numbers are fields. So are the integers
modulo 7. However, the integers modulo 12 are not a field (some
numbers, such as 4 for instance, have no multiplicative inverse).
The system of matrices is not a field for two reasons; The commutative law fails, and multiplicative inverses don't always exist.

SOME THEOREMS THAT ARE TRUE IN ANY FIELD

Here's the theorem about cancelation of factors:

Theorem 1.   If $a \otimes b = a \otimes c$ and $a \neq \mathbf{0}$, then $b = c$.
PROOF.   Since $a \neq 0$, it has a multiplicative inverse a^{-1}. Multiply
by a^{-1} on the left of each side in the given statement:

$$a^{-1} \otimes (a \otimes b) = a^{-1} \otimes (a \otimes c)$$

Next use the associative law for multiplication:

$$(a^{-1} \otimes a) \otimes b = (a^{-1} \otimes a) \otimes c$$

Since $a^{-1} \otimes a = 1$, we get

$$1 \otimes b = 1 \otimes c$$

or

$$b = c$$

Our second result is sometimes referred to as "no divisors of 0."

Theorem 2.   If $a \otimes b = 0$, then either $a = 0$ or $b = 0$.

PROOF.   If both a and b are **0**, there is nothing to prove. If $a \neq 0$, it has an inverse a^{-1}. Multiply on the left in $a \otimes b = 0$ by a^{-1} to obtain

$$a^{-1} \otimes (a \otimes b) = a^{-1} \otimes 0$$
$$(a^{-1} \otimes a) \otimes b = a^{-1} \otimes 0$$
$$1 \otimes b = a^{-1} \otimes 0$$
$$b = 0$$

A completely similar argument works to show that, if $b \neq 0$, then $a = 0$.

Recall from your study of the rational numbers that $1/(1/b) = b$. Here is the corresponding fact, true in any field:

Theorem 3.   If $b \neq 0$, then $(b^{-1})^{-1} = b$.

PROOF.   To prove this, start with the fact that

$$b^{-1} \otimes (b^{-1})^{-1} = 1$$

Then multiply on the left by b and make the obvious simplifications.

AN EXAMPLE

We have claimed that the integers modulo 7 form a field. Perhaps a little more explanation should be given. Symbolize this system by $\{0, 1, 2, 3, 4, 5, 6\}$; we called these the principal representatives in Section 13.3. It is easy to make addition and multiplication tables for this system. They are shown below. Now we can verify all the axioms.

TABLE 15-5							
+	**0**	**1**	**2**	**3**	**4**	**5**	**6**
0	0	1	2	3	4	5	6
1	1	2	3	4	5	6	0
2	2	3	4	5	6	0	1
3	3	4	5	6	0	1	2
4	4	5	6	0	1	2	3
5	5	6	0	1	2	3	4
6	6	0	1	2	3	4	5

TABLE 15-6							
×	**0**	**1**	**2**	**3**	**4**	**5**	**6**
0	0	0	0	0	0	0	0
1	0	1	2	3	4	5	6
2	0	2	4	6	1	3	5
3	0	3	6	2	5	1	4
4	0	4	1	5	2	6	3
5	0	5	3	1	6	4	2
6	0	6	5	4	3	2	1

The question of inverses is always crucial. We suggest that you check on the following facts.

$$\overline{0} = 0 \qquad 1^{-1} = 1$$
$$\overline{1} = 6 \qquad 2^{-1} = 4$$
$$\overline{2} = 5 \qquad 3^{-1} = 5$$
$$\overline{3} = 4 \qquad 4^{-1} = 2$$
$$\overline{4} = 3 \qquad 5^{-1} = 3$$
$$\overline{5} = 2 \qquad 6^{-1} = 6$$
$$\overline{6} = 1$$

We have observed that the integers modulo 12 are not a field, since in this system some elements don't have multiplicative inverses. This suggests an interesting and hard question. For which m's are the integers modulo m a field? There is no reason to think the answer is simple. But it is. It is a question we explore in Problems 22 and 23.

SUMMARY

It has been said that, while memorization without understanding is useless, understanding without memorization is hopeless. A student who is to succeed in algebra must memorize the axioms. In the case of a field, this is easy because of the symmetry of the axioms with respect to the two operations. Just remember that we have associativity, commutativity, identities, and inverses for both operations, and that the distributive laws hold.

Finally, we suggest that terms such as "cross-multiply," "cancel," "transpose," etc., should not be used unless one is certain that, if pressed to do so, he or she can describe exactly what has been done in terms of the axioms.

PROBLEM SET 15.3

1. Find additive inverses for all elements and multiplicative inverses for all nonzero elements of $\{0, 1, 2, 3, 4\}$, the integers modulo 5. Then convince yourself that this system forms a field.
2. Follow the directions in Problem 1 for the integers modulo 11.
3. In the integers modulo 13, find
   (a) $\overline{4}$ (b) 4^{-1} (c) $\overline{5}$ (d) 5^{-1}
4. Convince yourself that the integers modulo 13 form a field.
5. Theorem 1 says that, if $a \otimes b = a \otimes c$ and $a \neq \mathbf{0}$, then $b = c$. This means that we can cancel on the left. Show that we can also cancel on the right; that is, if $b \otimes a = c \otimes a$ and $a \neq \mathbf{0}$, then $b = c$.
6. Show that, if $a \otimes b \otimes c = \mathbf{0}$ in a field, at least one of a, b, and c is $\mathbf{0}$.
7. Show that the familiar fact from elementary algebra

   $$(a + b)^2 = a^2 + 2ab + b^2$$

   is true in any field. Begin by writing it in the symbols we have been using in an abstract field.

8. Rewrite

$$(a + b)(a - b) = a^2 - b^2$$

in the symbols of an abstract field and then show that it is always correct.

In Problems 9 through 14, you are to think of sets as being the elements of an algebraic structure with the operations of union ∪ and intersection ∩. Think of ∪ as ⊕ and ∩ as ⊗. Let A, B, and C be the sets diagramed below.

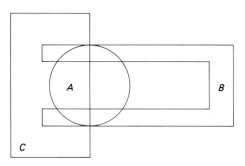

9. Make two copies of the set diagram above. On one shade (A ∪ B) ∪ C, and on the other A ∪ (B ∪ C). What law is illustrated?

10. Make two more copies of the set diagram. On one shade (A ∩ B) ∩ C, and on the other A ∩ (B ∩ C). What law is illustrated?

11. Again make two copies of the set diagram. On one shade A ∩ (B ∪ C), and on the other (A ∩ B) ∪ (A ∩ C). What law is illustrated?

12. Follow the directions in Problem 11, this time shading (B ∪ C) ∩ A and (B ∩ A) ∪ (C ∩ A).

13. If we consider the subsets of points on this page as the elements of an algebraic structure, what is the
    (a) Multiplicative identity?
    (b) Additive identity?

14. Refer to Problem 13. Does every subset have a multiplicative inverse? An additive inverse?

In Problems 15 through 20, we refer to the set C of all 2 × 2 matrices of the form

$$\begin{bmatrix} a & -b \\ b & a \end{bmatrix}$$

where a and b are real numbers. For example, C contains

$$\begin{bmatrix} 2 & -3 \\ 3 & 2 \end{bmatrix}, \quad \begin{bmatrix} 1 & 4 \\ -4 & 1 \end{bmatrix}, \quad \begin{bmatrix} -5 & \frac{1}{2} \\ -\frac{1}{2} & -5 \end{bmatrix}, \quad \begin{bmatrix} 1 & 0 \\ 0 & 1 \end{bmatrix}$$

15. Show that C is closed with respect to multiplication. Hint: Show that the product

$$\begin{bmatrix} a & -b \\ b & a \end{bmatrix} \begin{bmatrix} c & -d \\ d & c \end{bmatrix}$$

is again a matrix of a form belonging to C.

16. Show that C is closed with respect to addition. Notice the hint given in Problem 15.

17. In general, matrix multiplication is not commutative. However, show that, for matrices in C, it is.

18. Show that all nonzero matrices in C have multiplicative inverses.

19. Is C a field?

20. Show that $x^2 + 1 = 0$ has a solution in C.

21. By this time you should be able to make a conjecture about which m's make the integers modulo m a field. For example, we know that $m = 3, 5, 7,$ and 11 all work. What is your conjecture?

22.* We need a fact about the integers, the Euclidean Algorithm, which you can find in any book on number theory. It says that, if d is the g.c.d. of the positive integers a and b, there are integers r and s such that

$$d = ra + sb$$

(the algorithm also provides a method of finding r and s). In particular, if the g.c.d. is 1,

$$1 = ra + sb$$

From this fact, you should be able to prove that, if m is prime, the integers modulo m form a field. Hint: Let b be any nonzero element of the integers modulo m, where m is prime. Then there are integers r and s such that $rb + sm = 1$ or equivalently $rb - 1 = -sm$. Finish the argument.

23.* Show that, if m is not prime, the integers modulo m are not a field.

THERE ARE ONLY
$$7^4 = 2401 \text{ POSSIBILITIES } !$$

Matrices Modulo 7

Suppose the numbers entered in a matrix are in the system of integers modulo 7. Then, for example,

$$\begin{bmatrix} 2 & 6 \\ 4 & 3 \end{bmatrix}\begin{bmatrix} 1 & 5 \\ 3 & 2 \end{bmatrix} = \begin{bmatrix} 6 & 1 \\ 6 & 5 \end{bmatrix}$$

Find four positive integers (no negatives and no fractions), a, b, c, and d, so that multiplication modulo 7 gives

$$\begin{bmatrix} a & b \\ c & d \end{bmatrix}\begin{bmatrix} 1 & 5 \\ 3 & 2 \end{bmatrix} = \begin{bmatrix} 1 & 0 \\ 0 & 1 \end{bmatrix}$$

15.4 / Solving Equations

Consider the following questions.

1. Can you find a rational number x such that $3x = 7$?
2. Can you find a matrix

$$x - \begin{bmatrix} x_1 & x_2 \\ x_3 & x_4 \end{bmatrix}$$

   such that

$$\begin{bmatrix} 2 & -1 \\ 1 & 0 \end{bmatrix}\begin{bmatrix} x_1 & x_2 \\ x_3 & x_4 \end{bmatrix} = \begin{bmatrix} 3 & 1 \\ 2 & -2 \end{bmatrix}$$

3. Can you find a number x in the set $\{0, 1, \ldots, 11\}$ such that $4x \equiv 7 \bmod 12$?
4. Can you find a number x in the set $\{0, 1, \ldots, 10\}$ such that $4x \equiv 7 \bmod 11$?

Each problem really involves two related questions. The first is, Does a solution exist? (Since in some situations, like question 3, no solution exists.) The second is, If a solution exists, how do we find it?

In the setting of an abstract algebraic structure, each of the problems posed deals with an equation of the form $a \otimes x = b$. If the algebraic structure is a field, the two related questions can both be answered with dispatch. If $a \neq \mathbf{0}$, a solution exists, and we can give specific instructions for finding it. Simply multiply both sides by a^{-1}:

$$a^{-1} \otimes (a \otimes x) = a^{-1} \otimes b$$

Making obvious use of the axioms, we quickly see that

$$x = a^{-1} \otimes b$$

The key of course is the existence of a multiplicative inverse. In a field we are guaranteed that the inverse exists. We saw that the integers modulo 12 did not form a field, because not all elements had a multiplicative inverse, and this is precisely why we must answer no to question 3. There is no multiplicative inverse of 4 mod 12.

However, having observed that the integers modulo 11 form a field, we know that 4 does have a multiplicative inverse in this system. In fact, $3(4) \equiv 1 \bmod 11$. Therefore we answer question 4 by writing

$$3(4x) \equiv 3(7) \bmod 11$$
$$x \equiv 10 \bmod 11$$

The reader may feel that we have only shifted the problem to another, equally difficult question. Instead of asking for a solution to $a \otimes x = b$, we are asking for the multiplicative inverse of a. If, as in arithmetic modulo 11, we are reduced to trial and error to find a^{-1}, why not just solve the given problem by trial and error?

Several answers can be given. In the first place, we often have available methods for finding inverses when they exist. An algorithm is available to help us in modular arithmetic (Problems 11 and 12). We know how to find the inverse of a 2×2 matrix when it exists.

Second, when we work in a field, we can always answer at least the first question: Does an answer exist? Yes. The trial-and-error method is not quite so discouraging when we are assured that some trial will not be an error.

Third, the procedure described focuses our attention on the right question. Even when working in a system that is not a field, such as the set of all 2×2 matrices, we know how to proceed. In question 2, whether or not we can find the desired matrix depends on whether or not we can find an inverse of

$$A = \begin{bmatrix} 2 & -1 \\ 1 & 0 \end{bmatrix}$$

We can. In fact,

$$A^{-1} = \begin{bmatrix} 0 & 1 \\ -1 & 2 \end{bmatrix}$$

Therefore the equation can be solved, and we know how to do it.

$$\begin{bmatrix} 0 & 1 \\ -1 & 2 \end{bmatrix}\begin{bmatrix} 2 & -1 \\ 1 & 0 \end{bmatrix}\begin{bmatrix} x_1 & x_2 \\ x_3 & x_4 \end{bmatrix} = \begin{bmatrix} 0 & 1 \\ -1 & 2 \end{bmatrix}\begin{bmatrix} 3 & 1 \\ 2 & -2 \end{bmatrix}$$

$$\begin{bmatrix} x_1 & x_2 \\ x_3 & x_4 \end{bmatrix} = \begin{bmatrix} 2 & -2 \\ 1 & -5 \end{bmatrix}$$

It must be noted that, in the case of a matrix equation $AX = B$, it is essential to multiply by A^{-1} on the left: $A^{-1}AX = A^{-1}B$. Multiplying on the right does not work. Why?

Finally, we suggest that, even in the most familiar setting, the rational numbers, the best instruction for solving

$$\tfrac{15}{4}x = \tfrac{9}{14}$$

is, "Multiply by the inverse of $\tfrac{15}{4}$" (for, while most people remember that, to divide, one inverts and multiplies, nowhere near so fine a percentage remembers which fraction to invert). This leads directly to

$$x = \frac{9}{14} \cdot \frac{4}{15} = \frac{3 \cdot 2}{7 \cdot 5} = \frac{6}{35}$$

SUMMARY

The key to solving an equation of the form

$$a \otimes x = b$$

is to multiply both sides by a^{-1}. The trick, then, is to know whether an inverse exists, and how to find it when it does. If we know that our elements are members of a field, we can be certain that the inverse exists, hence the equation can be solved. Methods for actually finding the inverse have to be developed individually for various fields.

PROBLEM SET 15.4

1. A beginning student in algebra, wishing to solve $x + 3 = -5$, decides to subtract 3 from both sides. He writes

$$x + 3 = -5$$
$$\underline{-3 \quad -3}$$

On the left side he gets x, but on the right side he is momentarily confused. Then he recalls that, to subtract, you change the sign and add. He concludes that $x = -2$. What's wrong? (See Section 2.2.)

2. Solve for x the following equations from elementary algebra.
   (a) $5x = 2(x - 4)$ (b) $-3(x - 2) = 4$
   (c) $\tfrac{2}{3}x - \tfrac{1}{4} = \tfrac{11}{4}$ (d) $ax + b = bx + a$

3. Solve for x the following equations from elementary algebra.
   (a) $2x - 3 = 5x + 2$ (b) $3 - 2(4 - x) = 0$
   (c) $\frac{3}{4}x + \frac{7}{12} = \frac{1}{4}$ (d) $b - ax = a - bx$

4. The following calculations show all the steps in solving two equations in the field of integers modulo 5. Justify each step.

$$2x = 3 \qquad\qquad x + 2 = 1$$
$$3(2x) = 3 \cdot 3 \qquad (x + 2) + 3 = 1 + 3$$
$$(3 \cdot 2)x = 3 \cdot 3 \qquad x + (2 + 3) = 1 + 3$$
$$1x = 4 \qquad\qquad x + 0 = 4$$
$$x = 4 \qquad\qquad x = 4$$

5. Solve the equation $2x + 1 = 6$ in the field of integers modulo 7. Justify every step (see Problem 4).

6. Solve the equation $3x + 5 = 2$ in the field of integers modulo 11.

7. Solve the matrix equations

   (a) $\begin{bmatrix} 3 & 1 \\ 0 & 2 \end{bmatrix}\begin{bmatrix} x_1 \\ x_2 \end{bmatrix} = \begin{bmatrix} 1 \\ 4 \end{bmatrix}$ (b) $\begin{bmatrix} 3 & 1 \\ 0 & 2 \end{bmatrix}\begin{bmatrix} x_1 & x_2 \\ x_3 & x_4 \end{bmatrix} = \begin{bmatrix} 2 & 4 \\ 3 & 2 \end{bmatrix}$

8. Solve the matrix equations

   (a) $\begin{bmatrix} 6 & 2 \\ 2 & 1 \end{bmatrix}\begin{bmatrix} x_1 \\ x_2 \end{bmatrix} = \begin{bmatrix} 3 \\ 4 \end{bmatrix}$ (b) $\begin{bmatrix} 6 & 2 \\ 2 & 1 \end{bmatrix}\begin{bmatrix} x_1 & x_2 \\ x_3 & x_4 \end{bmatrix} = \begin{bmatrix} -2 & -3 \\ 5 & 1 \end{bmatrix}$

9. We can describe the process for finding the inverse of a 2×2 matrix in language that allows the entries to come from any field. To find the multiplication inverse of

   $$\begin{bmatrix} a & b \\ c & d \end{bmatrix}$$

   a. Find $D = ad - bc$.
   b. Replace b and c by their additive inverses.
   c. Interchange a and d.
   d. Multiply each entry by D^{-1}, the multiplicative inverse of D.

   Use this procedure to find the inverse of

   $$\begin{bmatrix} 2 & 1 \\ 6 & 4 \end{bmatrix}$$

   in the field of integers modulo 11. Then solve

   $$\begin{bmatrix} 2 & 1 \\ 6 & 4 \end{bmatrix}\begin{bmatrix} x_1 & x_2 \\ x_3 & x_4 \end{bmatrix} = \begin{bmatrix} 9 & 3 \\ 6 & 2 \end{bmatrix}$$

   in this system.

10. Using the method described in Problem 9, solve the problem that introduced this section.

11. The g.c.d. of 15 and 26 is 1. According to the Euclidean Algorithm (see Problem 22 in Problem Set 15.3), there are integers r and s satisfying

$$1 = r(26) + s(15)$$

Check that $r = -4$ and $s = 7$ work. Thus

$$7(15) - 1 = 4(26)$$

which means that

$$7(15) \equiv 1 \bmod 26$$

(a) What is the multiplicative inverse of 15 in the integers modulo 26?
(b) Solve $15x \equiv 12 \bmod 26$.

12. Check that

$$1 = 9(37) - 4(83)$$

Use what you learned in Problem 11 to
(a) Find the multiplicative inverse of 37 mod 83.
(b) Solve $37x \equiv 16 \bmod 83$.

Answers

PROBLEM SET 1.1

1. 12.
3. 1 hour and 20 minutes is 80 minutes.
5. Give the fifth apple and the basket to the fifth girl.
7. Never; the ship rises with the tide.
9. Three.
11. 30,000.
13. Call the slices: A, B, and C. Put A and B in the pan. After 30 seconds turn A over, take B out, and put C in. After 1 minute, remove A, put B back in the pan on its unbrowned side, and turn C over.
15.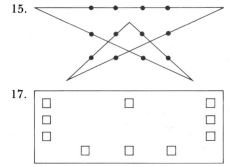
17.
19. The North Pole is one solution; there are infinitely many others. Consider a circle of circumference 1 mile around the South Pole. Start from any point 1 mile north of this circle. There are still other solutions. Find them.
21. Open rings 1, 2, and 3. Use ring 1 to link ring 6 to ring 7. Then use ring 2 to link ring 9 to ring 10. Finally, use ring 3 to link ring 12 to ring 13.

PROBLEM SET 1.2

1. 91 feet.
3. 11, counting the 2 it meets in the bus depots.
5. July 13.
7. 35.
9. Send 2 boys across, send 1 boy back, send 1 soldier across, and send a second boy back. Now repeat this process until all 10 soldiers are across.
11. Split the contents of the beaker between the two test tubes. Return the contents of one (i.e., 24 milliliters) to the beaker. Now split the contents of the remaining test tube between the two test tubes and pour one of them (i.e., 12 milliliters) into the beaker. Finally, split the contents of the remaining test tube between them and pour one of them (6 milliliters) into the beaker.
13. 10.
15. $\sqrt{108}$ miles.
17. 65.
19. Alphabetically by name.

21. 43.
23. 31,832.

PROBLEM SET 1.3

1. 10 pigs and 3 chickens.
3. 14.
5. The largest monument that can be built consists of $4 \times 4 \times 4 = 64$ bricks standing on a plaza of $8 \times 8 = 64$ bricks.
7. $16061 - 15951 = 110$ kilometers.
9. Adding 6 feet to the circumference adds about 1 foot to the radius. You could slip through.
11. Running half of the time is more running than running half of the distance. Homer will get there first.
13. A beat B, 2 to 0; A beat C, 5 to 1; B and C tied, 2 to 2.
15.

3 regions   6 regions

Four regions is impossible.

17. I am a male nurse.

19. Suppose there are m people at the college. Put a label on each person corresponding to the number of friends he or she has. This label will be a number between 1 and $m - 1$. There are m labels, but at most $m - 1$ values. At least two must agree.
21. Three weighings suffice. A complete description would take too much space. Here are some hints. Number the coins 1 to 12. Weigh coins 1 through 4 against 5 through 8. If the pans balance, the bad coin comes from 9 through 12. It can be determined in two more weighings. If the pans do not balance in the first step, coins 9 through 12 are good. Suppose the side with coins 5 through 8 goes up. Then weigh 1, 2, 3, and 8 against 4, 9, 10, and 11. You finish.
23. Take 1 coin from pile 1, 2 coins from pile 2, 3 coins from pile 3, etc., or a total of 55 coins, and put them on the scale. If all are good, they will weigh 308 grams. They will weigh less than this by a number of grams equal to the number of bad coins. This is also the number of the bad pile. Thus one weighing suffices.
25. Over 70 feet.

PROBLEM SET 1.4

1. 88.
3. 24, 25.
5. Lay the ruler across the board at an angle so the ends of the ruler coincide with the edges of the board. Mark at 2-inch intervals.
7. 100 miles.
9. Horace and Homer meet in 4 hours. Therefore Trot runs 120 miles.
11. He drove a post in the ground directly across the river from the house and equally far from it. Then he sighted from the post to the barn. Where his line of sight intersected the river is the desired point.

13. No. The central cube has six sides. It requires six cuts to slice its sides.

15. A domino covers a black and a white square. The removed squares are of the same color. Therefore the mutilated chessboard cannot be covered with dominoes.

17. I noted my clock's time when I left and when I returned, so I knew how long I was gone. From this I subtracted the time I was at my friend's house and divided by 2. This gave me the time required to walk home, which I added to the time showing on my friend's clock when I left his home.

19. Here is one argument. Suppose the string lies along the real line so that each point of it has a coordinate x between 0 and b. After stretching, each point has a new coordinate, $y = f(x)$. Graph $y = f(x)$ in the coordinate plane. The graph must cross the line $y = x$. The x coordinate of the crossing point is also the x coordinate of an unmoved point.

PROBLEM SET 2.1

1. 357.
3. $\frac{252}{27} = 9\frac{1}{3}$.
5. (a) 24 (b) 45 (c) 8 (d) 16
   (e) 15 (f) 62 (g) 123 (h) 104.
7. (a) $xy + wz$ (b) $x(y + z) - w$ (c) $x(y + z - w)$.
9. 1035.
11. \$14.49.
13. (a) $F = 96.8°$ at 11:00 A.M.
    (b) $F = 74.8°$ at noon
    (c) $C = 23\frac{7}{9}°$ at noon.

PROBLEM SET 2.2

1. (a) 4 (b) -15 (c) -9 (d) 25.
3. (a) 20 (b) 3 (c) -19 (d) -9.
5. (a) 176 (b) 216 (c) 224 (d) 0 (e) -136 (f) -304.
7. (a) The ball is thrown downward.
   (b) The ball is thrown from a point below ground level.
9. (a) 49 (b) -148 (c) -2641.

PROBLEM SET 2.3

1. (a) $\frac{3}{4}$ (b) $\frac{3}{4}$ (c) $\frac{8}{9}$ (d) $\frac{4}{9}$ (e) $\frac{4}{3}$ (f) $-\frac{7}{9}$ (g) $\frac{3}{4}$ (h) $\frac{21}{11}$.
3. (a) $\frac{15}{16}$ (b) 2 (c) $\frac{3}{5}$ (d) $\frac{16}{9}$ (e) $\frac{97}{27}$ (f) $\frac{3}{4}$ (g) $\frac{4}{3}$ (h) $\frac{9}{2}$
   (i) $\frac{37}{20}$ (j) $\frac{17}{36}$ (k) $\frac{19}{20}$ (l) $\frac{17}{24}$ (m) $\frac{13}{22}$ (n) $\frac{11}{13}$ (o) $\frac{3}{11}$ (p) $-\frac{7}{9}$ (q) $\frac{1}{24}$.
5. $\frac{12}{5} = 2\frac{2}{5}$ hours
7. \$10,175.
9. \$135.
11. \$277.
13. $\frac{1755}{16} = 109\frac{11}{16}$ inches.
15. $\frac{79}{240}$.

PROBLEM SET 2.4

1. (a) $\frac{3}{10}$ (b) $\frac{45}{100} = \frac{9}{20}$ (c) $\frac{689}{1000}$ (d) $\frac{135}{100} = \frac{27}{20}$ (e) $\frac{1212}{100} = \frac{303}{25}$ (f) $\frac{39}{100}$.
3. (a) 0.24 (b) 0.7143 (c) 0.5556 (d) 2.0794 (e) 1.4167.
5. (a) 34.407 (b) 11.248 (c) 2.164 (d) 6.615 (e) 0.551 (f) 0.039.
7. (a) 0.36 (b) 0.085 (c) 0.092 (d) 0.1341 (e) 1.51 (f) 0.0775.
9. (a) 0.0715 (b) 5.6847 (c) −3.2842 (d) 0.1989 (e) −1.6065 (f) 0.7303.
11. $490.06.
13. $101.19.
15. $1.41.
17. $2713.38.
19. 2.4, 4.44.
21. 2,232,000,000 inches = 186,000,000 feet = 35227.3 miles.
23. $2806.

PROBLEM SET 2.5

1. (a) 16 (b) 16 (c) −16 (d) $\frac{1}{16}$ (e) $\frac{1}{16}$ (f) $\frac{1}{16}$ (g) 16 (h) 128
   (i) 256 (j) 2 (k) 16 (l) $\frac{1}{2}$ (m) −96 (n) 13 (o) 144.
3. (a) $\frac{3}{32}$ (b) 3.
5. (a) 4096 (b) 19,683 (c) 40,960 (d) 708,588 (e) 6149 (f) 3402
   (g) 171.09 (h) 705.83 (i) 0.70138 (j) 315.07.
7. (a) 4.325×10^3 (b) 5.12×10^6 (c) 5.13×10^{-3} (d) 2.341×10^{-1}.
9. (a) 395,680,000,000 (b) 108,181,000.
11. 34328.1 cubic inches.
13. 4671.2 cubic inches.
15. $5600(0.81)^{10} = \$680.83$.
17. $18,000; $36,000; $18,432,000.
19. $1569.22.

PROBLEM SET 3.1

1. (a) 4 (b) 6 (c) $\frac{5}{3}$ (d) $-\frac{5}{2}$ (e) 6 (f) 4 (g) $-\frac{7}{5}$.
3. (a) 4 (b) 3 (c) $\frac{71}{15} = 4.73$ (d) 3 (e) 12 (f) $\frac{19}{6}$.
5. (a) $\frac{1}{2}(x + x^2)$ (b) $3 + 2x$ (c) $x + (x + 2) + (x + 4)$ or $3x + 6$
   (d) $4x$ (e) $x(2x - 3)$ (f) $x^2(\frac{1}{2}x + 3)$ (g) $5x + 30x + 3x - 4$.
7. (a) x, $x + 1$, $x + 2$ (b) $x + (x + 1) + (x + 2) = 636$
   (c) The three numbers: 211, 212, 213.
9. (a) s, $s + 1$, $s + 2$ (b) $s + (s + 1) + (s + 2) = s + 63$
   (c) The three numbers: 30, 31, 32.
11. (a) length $= 2w + 4$ (b) $2w + 2(2w + 4) = 200$
    (c) Dimensions: 32 feet by 68 feet.
13. 20 meters by 22 meters.
15. 5.54930, 13.54930.

PROBLEM SET 3.2

1. (a) $6x + 10$, $8x - 12$ (b) $6x + 10 = 8x - 12$ (c) 11.
3. Equation: $\dfrac{61 + 77 + s}{3} = 75$; Answer: 87.

5. (a) $3x$ (b) $x + 15$, $3x + 15$, $3x + 15 = 2(x + 15)$ (c) 15.
7. 47.
9. 19 ounces.
11. (b) $(10 - x)(0.08)$ (c) $(10 - x)(0.08) = 10(.05)$ (d) 3.75 gallons.
13. 20 gallons.
15. 9 feet.
17. (a) time downstream $= 6 - t$ (b) $3t = 7(6 - t)$
    (c) 4.2 hours after starting, or at 10:12 P.M.
19. 6.96 hours after starting, or at about 6:57 P.M.
21. $23.25.
23. 25.
25. 9 minutes to get to the front, and 6 minutes to return.
    Total: 15 minutes.
27. $1:05\frac{5}{11}$ P.M. or about 5 minutes and 27 seconds past 1:00 P.M.
29. $1\frac{5}{7}$ days.
31. 30 years old.

PROBLEM SET 3.3

1. (a) $x = 3$, $y = 2$ (b) $x = 5$, $y = 2$ (c) $x = 3$, $y = -4$ (d) $x = -1$, $y = 1$
   (e) $x = -2$, $y = 10$ (f) $x = \frac{40}{27}$, $y = -\frac{5}{9}$.
3. (a) $x = 1.2$, $y = 1.32$ (b) $x = 3$, $y = \frac{1}{4}$.
5. (a) $x + y = 17$ (b) $\frac{2}{3}x - y = 8$ (c) $x = 15$, $y = 2$.
7. (a) value of nickels is $5n$;
   value of quarters is $25q$;
   $5n + 25q = 600$
   (b) $n = 3q$ (c) $n = 45$; $q = 15$.
9. (a) $x + y = 10$ (b) $264x + 384y = 3180$
   (c) 4.5 pounds of Aromatic; 5.5 pounds of Caffineo.
11. (a) width $= w + 8$; length $= l - 10$ (b) $l - 10 = w + 8$
    (c) $lw = (w + 8)(l - 10)$ (d) $l = 50$; $w = 32$.
13. (a) rate downstream is $x + y$;
    rate upstream is $x - y$
    (b) $16 = x + y$
        $16 = 2(x - y)$
    (c) $x = 12$, $y = 4$.
15. 50 dresses, 28 coats.
17. 9 children each getting $9000.

PROBLEM SET 4.1

1. (a) 15, 18 (b) 9, 6 (c) $\frac{1}{16}$, $\frac{1}{32}$.
3. (a) 20, 32 (b) $\frac{1}{18}$, $\frac{1}{40}$ (c) $\frac{1}{8}$, $\frac{1}{64}$ (d) 3, -9.
5. (a) $a_n = 3n$ (b) $b_n = 21 - 3(n - 1) = 24 - 3n$ (c) $c_n = (\frac{1}{2})^{n-1}$.
7. (a) 81 (b) 14 (c) $\frac{1}{4}$ (d) 5.
9. (a) $a_n = a_{n-1} + 3$ (b) $b_n = b_{n-1} - 3$ (c) $c_n = c_{n-1}/2$.
11. (a) 486, 1458 (b) 7, 1 (c) $\frac{6}{7}$, $\frac{7}{8}$ (d) 1/216, 1/343 (e) $\frac{3}{2}$, $\frac{3}{16}$ (f) 16, 26
    (g) 48, 88 (h) 19, 23 (i) 28, 36.
13. (a) 126 (b) 33 (c) 0 (d) 70 (e) 15.75 (f) 0.

15. $\frac{1}{7} = 0.142857142857\ldots$, so $a_1 = 1$, $a_2 = 4$, $a_3 = 2, \ldots, a_{53} = 5$.
$\frac{5}{13} = 0.384615384615\ldots$, so $a_1 = 3$, $a_2 = 8$, $a_3 = 4, \ldots, a_{53} = 1$.

17. $a_n = n^2$.

19. $a_n = 3 + 2^{n-1}$.

21. 1225; 41,616; 1,413,721; 48,024,900.

23. For $A = 20$, $a_2 = 6$, $a_3 = 4.67$, $a_4 = 4.4763$, $a_5 = 4.472138$.
For $A = 312$, $a_2 = 79$, $a_3 = 41.5$, $a_4 = 24.51$, $a_5 = 18.62$,
$a_6 = 17.69$, $a_7 = 17.663$.

PROBLEM SET 4.2

1. (a) 9, 11 (b) 17, 20 (c) 84, 80.

3. (a) $d = 2$, $a_{40} = 1 + (39)2 = 79$ (b) $d = 3$, $b_{40} = 122$ (c) $d = -4$,
$c_{40} = -56$.

5. (a) $A_{40} = \frac{40}{2}(1 + 79) = 1600$ (b) $B_{40} = 2540$ (c) $C_{40} = 880$.

7. (a) 2550 (b) 2500 (c) 1683.

9. $16 + 9(32) = 304$ feet, $16 + 48 + 80 + \cdots + 304 = 1600$ feet.

11. $\frac{21}{2}(40 + 60) = 1050$ centimeters.

13. $7 + 14 + \cdots + 350 - \frac{50}{2}(7 + 350) = 8925$ cents $- \$89.25$.

15. 19,900.

17. 5244.

19. $1^3 + 2^3 + 3^3 + \cdots + n^3 = n^2(n + 1)^2/4$.

PROBLEM SET 4.3

1. (a) 512, 2048 (b) $\frac{1}{3}$, $\frac{1}{9}$ (c) 3, -3.

3. $a_{40} = 2(4)^{39}$, $b_{40} = 27(\frac{1}{3})^{39}$, $c_{40} = 3(-1)^{39}$.

5. (a) $A_5 = \dfrac{2(1 - 4^5)}{1 - 4} = \dfrac{2(1 - 1024)}{-3} = 682$ (b) 121/3 (c) 3.

7. 1,024,000.

9. $1 + 7 + 7^2 + 7^3 + 7^4 = 2801$.

11. 2^{40} square inches $\approx 10^{12}$ square inches $\approx 10^{10}$ square feet ≈ 300
square miles.

13. $a_{40} = 2.48328$, $A_{40} = 65.9737$.

15. \$194,530.50.

17. The latter. $1 + 2 + 2^2 + \cdots + 2^{19}$ cents equals \$10,485.75.

19. $(\frac{2}{3})^5 = 32/243$, 12 strokes.

21. (a) $\frac{1}{3}$ (b) $\frac{7}{9}$ (c)1.

23. \$4 billion.

PROBLEM SET 4.4

1. (a) 11, 13 (b) 240, 250 (c) 10.125, 15.1875 (d) 32, 64.

3. (a) Linear (b) linear (c) exponential (d) exponential.

5. 1500.

7. 4000; 8000; 1,024,000.

9. Neither.

11. This is eight half-lives; $200(\frac{1}{2})^8 \approx 0.8$ gram.

13. 200.

15. About 2085.
17. 19,664,010.
19. (a) Town A: 3000, 4000, 5000, 6000
       Town B: 1814, 2443, 3291, 4432.
    (b) 140 years.
21. About 650 years.

PROBLEM SET 4.5

1. (a) 2.69158803 (b) 5.47356576 (c) 1.99900463 (d) 466.095714.
3. (a) $300 (b) $684.85 (c) $724.46.
5. $270.48.
7. (a) 8 years and 4 months (b) between 6 and 7 years (c) 70 months
   (d) a little under 70 months.
9. Slightly over 12%.
11. (a) $225.37 (b) 12.68%.
13. (a) 453.29 (b) 530.22.
15. (a) $689.98 (b) $697.10 (c) $697.77.
17. $1064.43.
19. $1341.21.
21. $774.34.
23. (a) Most of the time Homer will be using much less than $20, since
    he is steadily paying off his debt. (b) 18% compounded monthly.
25. 15% compounded monthly.

PROBLEM SET 4.6

1. 2584, 4181, 6765.
3. 7, 12, 20, 33, 54, 88.
5. 1, 3, 8, 21, f_{2k}.
7. (a) 0, 1, 1, 2, 3, 5, . . . (b) 1, 1, 2, 3, 5, 8, ...
9. 9, 25, 64, 169, 441; 34, 89, 233, 610, 1597; 34, 89, 233, 610,
   1597; $(f_n)^2 + (f_{n+1})^2 = f_{2n+1}$.
11. 1, -1, 1; $f_{n+1}f_{n-1} - (f_n)^2 = (-1)^n$ for $n = 2, 3, 4, ...$
13. (a) a, b, $a + b$, $a + 2b$, $2a + 3b$, $3a + 5b$, $5a + 8b$, $8a + 13b$,
    $13a + 21b$, $21a + 34b$.
15. (a) 3, 5, 9, 17, 31 (b) $t_n = t_{n-1} + t_{n-2} + t_{n-3}$

PROBLEM SET 5.1

1. (a) 44 (b) 100100 (c) XXXVI.
3. (a) 19 (b) 51 (c) 668 (d) 21 (e) 61 (f) 69.
5. (a) 51 (b) 67 (c) 552 (d) 631.
7. (a) 101001 (b) 110111 (c) 101101010 (d) 110011001.
9. Change each base-eight digit to a three-digit base-two number.
   (a) 101111 (b) 110011 (c) 11010100.
11. (a) 165 (b) 122 (c) 42 (d) 167 (e) 1666 (f) 5406.
13. (a) 18B (b) 98A (c) B09.
15. (a)   9,567 (b)   850 (c) 24,765
         1,085   850   24,765
        10,652   29,786   49,530
                      31,486

PROBLEM SET 5.2

9. (a) 10010010 (b) 01110011 (c) 00100111.
11. (a) 11011110 (b) 10010100 (c) 10100011.
13. (a) 10010001 (b) 01110010 (c) 00100110.
15. Take the two's complement of the subtrahend and add; discard the overflow; finally add 1.

PROBLEM SET 5.3

1.   3.

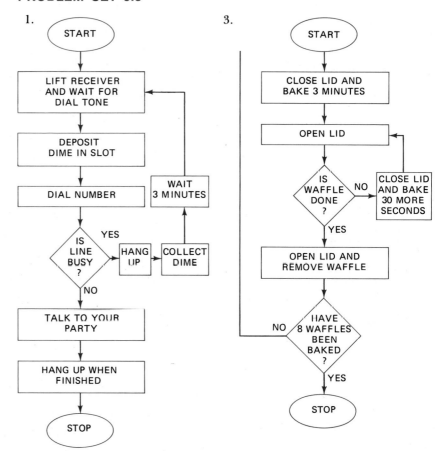

5. (a) *C* (b) *D* (c) *B*.
7. (a) 187, 171, 203 (b) 2, 4, $\frac{13}{5}$ (c) 1, 1, 1.
9. You go in circles.

PROBLEM SET 5.4

3.
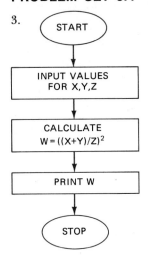

5. 2, 4, 8, 16, 32, 64.

7.
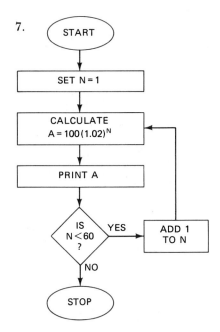

9.
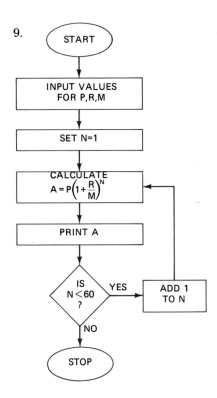

11. 220.

13. $2^2 + 4^2 + 6^2 + \cdots + (2N)^2.$

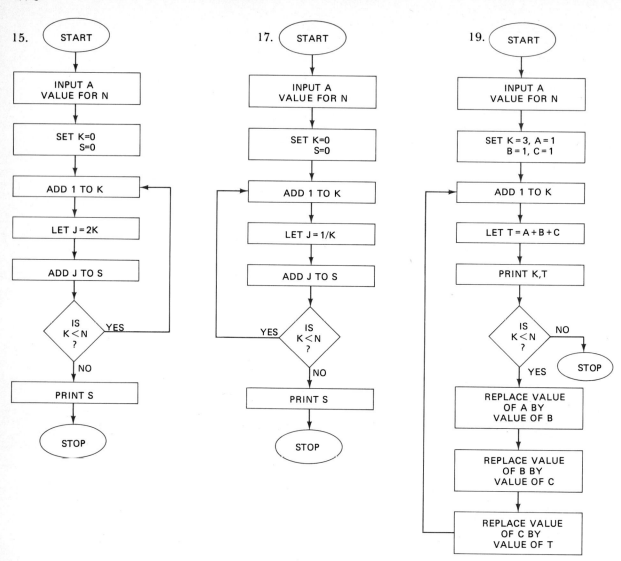

PROBLEM SET 5.5

Usually there are several possible correct BASIC programs correspond-ing to a given flowchart. Yours may be correct even if it doesn't match the one given below. Also, we mention that there is more than one version of BASIC. For example, one version does not use slashes to distinguish zero from the letter O. Another omits the comma in the IF-THEN command. If you actually try to run these programs on a computer, you may need to revise them slightly to fit your local version.

1. (a) $3*X \uparrow 2 - 2*X + 7$ (b) $6*(X+Y) \uparrow 2$ (c) $(X \uparrow 2+3)/5$
   (d) $(X+Y) \uparrow 3/(Z+3)$.

3.
```
1∅    INPUT X
2∅    INPUT Y
3∅    INPUT Z
4∅    LET W=((X+Y)/Z)↑2
5∅    PRINT W
6∅    END
```

7.
```
1∅    LET N=1
2∅    LET A=1∅∅*1.∅2↑N
3∅    PRINT A
4∅    IF N<6∅, THEN 6∅
5∅    STOP
6∅    LET N=N+1
7∅    GO TO 2∅
8∅    END
```

5.
```
1∅    LET N=∅
2∅    LET N=N+1
3∅    LET A=2↑N
4∅    PRINT A
5∅    IF N<6, THEN 2∅
6∅    END
```

9.
```
1∅    INPUT P
2∅    INPUT R
3∅    INPUT M
4∅    LET N=1
5∅    LET A=P*((1+R/M)↑N)
6∅    PRINT A
7∅    IF N<6∅, THEN 9∅
8∅    STOP
9∅    LET N=N+1
1∅∅   GO TO 5∅
11∅   END
```

11.
```
1∅    INPUT N
2∅    LET K=∅
3∅    LET S=∅
4∅    LET K=K+1
5∅    LET J=(2*K)↑2
6∅    LET S=S+J
7∅    IF K<N, THEN 4∅
8∅    PRINT S
9∅    END
```

13.
```
1∅    INPUT N
2∅    LET S=∅
3∅    FOR K=1 TO N
4∅    LET J=(2*K)↑2
5∅    LET S=S+J
6∅    NEXT K
7∅    PRINT S
8∅    END
```

15.
```
1∅    INPUT N
2∅    LET K=∅
3∅    LET S=∅
4∅    LET K=K+1
5∅    LET J=1/K
6∅    LET S=S+J
7∅    IF K<N, THEN 4∅
8∅    PRINT S
9∅    END
```

17.
```
1∅    INPUT N
2∅    LET K=∅
3∅    LET S=∅
4∅    LET K=K+1
5∅    LET J=2*K
6∅    LET S=S+J
7∅    IF K<N, THEN 4∅
8∅    PRINT S
9∅    END
```

19.
```
1∅    INPUT N
2∅    LET K=3
3∅    LET A=1
4∅    LET B=1
5∅    LET C=1
6∅    LET K=K+1
7∅    LET T=A+B+C
8∅    PRINT K, T
9∅    IF K<N, THEN 11∅
1∅∅   STOP
11∅   LET A=B
12∅   LET B=C
13∅   LET C=T
14∅   GO TO 6∅
15∅   END
```

PROBLEM SET 6.1

1. $3 \cdot 10 \cdot 4 = 120.$ 3. $4 \cdot 5 = 20.$
5. $(6 + 8)4 = 56.$ 7. $5 \cdot 5 = 25, 5 \cdot 4 = 20.$
9. $4 + 4 \cdot 3 + 4 \cdot 3 \cdot 2 + 4 \cdot 3 \cdot 2 \cdot 1 = 64.$
11. $3 \cdot 9 \cdot 5 \cdot 4 \cdot 3 \cdot 2 \cdot 5 \cdot 4 \cdot 3.$
13. $(10)^7, 9(10)^6.$ 15. $9 \cdot 10 \cdot 10, 9 \cdot 9 \cdot 8.$
17. $6 \cdot 10 \cdot 3 = 180, 1 \cdot 7 \cdot 3 + 5 \cdot 10 \cdot 3 = 171.$
19. (a) $7 \cdot 6 \cdot 5 \cdot 4 \cdot 3 = 2520$ (b) $8 \cdot 7 \cdot 6 \cdot 5 = 1680$ (c) 35 (d) 756.
21. (a) n (b) (n) (n) $(n - 1) = n^3 - n^2.$
23. (a) 1680 (b) 18,150 (c) $3^9 = 19,683.$
25. $n!.$
27. $(n/2)^n.$

PROBLEM SET 6.2

1. (a) $4! = 24$ (b) $5! = 120$ (c) $7!.$
3. $8 \cdot 7 \cdot 6 \cdot 5, 8 \cdot 7 \cdot 6 \cdot 5 \cdot 4.$
5. (a) $5!/2! = 60$ (b) $6!/2! = 360$ (c) $7!/(2!\ 2!)$ (d) $10!/3!.$
7. $_6P_6 + {}_6P_5 + {}_6P_4 + {}_6P_3 + {}_6P_2 + {}_6P_1 = 1956.$
9. (a) $6!/3! = 120$ (b) $4^0.$
11. $10 \cdot 9 \cdot 8 = 720.$
13. $2^{10} = 1024.$
15. $9!/(4!\ 3!\ 2!).$
17. 70.
19. $26^{10} + 26^9 + 26^8 + \cdots + 26^2 + 26 \approx 1.4681 \times 10^{14}.$

PROBLEM SET 6.3

1. (a) 120 (b) 126 (c) 455 (d) 4950 (e) 56 (f) $8 \cdot 7 \cdot 6 \cdot 5 \cdot 4 = 6720.$
3. $_{50}C_3 = 19600.$
5. (a) $_5C_3 = 10$ (b) $_5C_4 = 5$ (c) $10 + 5 + 1 = 16.$
7. $_{13}C_5, {}_{12}C_4.$
9. (a) $_9C_3 \cdot {}_6C_2 = 84 \cdot 15 = 1260$ (b) $_6C_3 \cdot {}_9C_2 = 20 \cdot 36 = 720$ (c) $_9C_5 + {}_6C_5 = 126 + 6 = 132$ (d) $_9C_3 \cdot {}_6C_2 + {}_9C_4 \cdot {}_6C_1 + {}_9C_5 = 2142.$
11. $_{12}C_4 \cdot {}_8C_4.$
13. $_7C_3 = 35.$
15. (a) 10^5 (b) $10 \cdot 9 \cdot 8 \cdot 7 \cdot 6$ (c) $_{10}C_5 = 252.$
17. $b_n = a_n + a_{n-9} + a_{n-19} + a_{n-29} + \cdots .$
19. (a) $_{52}C_{13} \approx 6.3501 \times 10^{11}$ (b) $_4C_3 \cdot {}_{48}C_{10} \approx 2.6163 \times 10^{10}$
    (c) $_{48}C_{13} \approx 1.9293 \times 10^{11}$ (d) $4 \cdot {}_{39}C_{13} \approx 3.249 \times 10^{10}$
    (e) $_{16}C_{13} = 560$ (f) $4^{13} = 67,108,900$
    (g) 4.

PROBLEM SET 6.4

1. (a) 10 (b) 84 (c) 210 (d) 495.
3. There are $_5C_3 = 10$ such words.
5. (a) $_9C_3 = 84$ (b) $_{12}C_3 = 220$ (c) $_{13}C_7.$
7. (a) 55 (b) 56.
11. Fibonacci Sequence.

13. (a) $a^4 + 4a^3b + 6a^2b^2 + 4ab^3 + b^4$ (b) $c^5 - 5c^4d + 10c^3d^2 - 10c^2d^3 + 5cd^4$
$- d^5$ (c) $u^6 + 6u^5(2v) + 15u^4(2v)^2 + 20u^3(2v)^3 + 15u^2(2v)^4 + 6u(2v)^5$
$+ (2v)^6$ or $u^6 + 12u^5v + 60u^4v^2 + 160u^3v^3 + 240u^2v^4 + 192uv^5 + 32v^6$.

21. $\left(1 + \dfrac{0.12}{n}\right)^n = 1 + n\left(\dfrac{0.12}{n}\right) + \dfrac{n(n-1)}{2 \cdot 1}\left(\dfrac{0.12}{n}\right)^2 + \dfrac{n(n-1)(n-2)}{3 \cdot 2 \cdot 1}\left(\dfrac{0.12}{n}\right)^3$
$+ \cdots \approx 1 + 0.12 + \frac{1}{2}(0.12)^2 + \frac{1}{6}(0.12)^3 = 1.127488.$

PROBLEM SET 7.1

1. (a) $\frac{1}{6}$ (b) $\frac{1}{2}$ (c) $\frac{1}{3}$ (d) $\frac{1}{2}$ (e) $\frac{1}{2}$.
3. (a) $\frac{1}{8}$ (b) $\frac{3}{8}$ (c) $\frac{1}{2}$.
5. (a) $\frac{1}{6}$ (b) $\frac{1}{6}$ (c) $\frac{20}{36} = \frac{5}{9}$.
7. (a) 16 (b) $\frac{2}{16}$ (c) $\frac{13}{16}$.
9. (a) States vary in population; the equal likelihood assumption fails.
(b) A person may both drink and smoke; the disjointness assumption fails. (c) The two numbers should add to one; they don't.
(d) Football frequently results in ties.
11. (a) $\frac{16}{50}$ (b) $\frac{100}{240}$ (c) $\frac{120}{300}$ (d) $\frac{32}{300}$.
13. $\frac{1}{4!} = \frac{1}{24}$.
15. (a) $\frac{1}{2}$ (b) $\frac{1}{4}$ (c) $\frac{1}{13}$.
17. (a) $_{26}C_3/_{52}C_3$ (b) $_{13}C_3/_{52}C_3$ (c) $_4C_1 \cdot {}_{48}C_2/_{52}C_3$ (d) $_4C_3/_{52}C_3$.
19. (a) $\frac{1}{4} + \frac{1}{12} + \frac{1}{12} = \frac{5}{12}$ (b) $\frac{1}{3} + \frac{1}{4} + \frac{1}{6} + \frac{1}{12} = \frac{10}{12}$ (c) $\frac{11}{12}$.
23. $4(12/_{52}C_5)$.
25. .0004952.
27. $_8C_5/_{52}C_5 \approx .0000215$.
29. $_{34}C_5/_{52}C_5 \approx .10706$.

PROBLEM SET 7.2

1. $1/216$.
3. (a) $\frac{1}{4} \cdot \frac{1}{2} = \frac{1}{8}$ (b) $\frac{3}{4} \cdot \frac{1}{2} = \frac{3}{8}$ (c) $\frac{1}{4} \cdot \frac{1}{2} = \frac{1}{8}$ (d) $\frac{1}{4} \cdot \frac{5}{6} = \frac{5}{24}$ (e) $\frac{1}{2} \cdot \frac{2}{3} + \frac{1}{2} \cdot \frac{1}{3}$
$= \frac{1}{2}$.
5. (a) $\frac{5}{12} \cdot \frac{5}{12} = \frac{25}{144}$ (b) $\frac{5}{12} \cdot \frac{4}{11} = \frac{5}{33}$.
7. (a) 0.9 (b) 0.5 (c) 0.8.
11. $\frac{9}{10} \cdot \frac{9}{10} \cdot \frac{9}{10}$.
13. $\frac{1}{2} \cdot \frac{4}{100} + \frac{1}{2} \cdot \frac{1}{100} = \frac{1}{40}$.
15. $\frac{6}{10} \cdot \frac{5}{9} \cdot \frac{4}{8} = \frac{1}{6}$.
17. (a) $\frac{2}{3} \cdot \frac{2}{3} \cdot \frac{2}{3} \cdot \frac{2}{3} = \frac{16}{81}$ (b) $\frac{1}{3} \cdot \frac{1}{3} \cdot \frac{1}{3} \cdot \frac{1}{3} = \frac{1}{81}$ (c) $\frac{17}{81}$ (d) $4 \cdot \frac{1}{3} \cdot \frac{2}{3} \cdot \frac{2}{3} \cdot \frac{2}{3}$
$\cdot \frac{2}{3} = \frac{64}{243}$ (e) $\frac{64}{243} + \frac{8}{243} = \frac{8}{27}$.
19. $\frac{6}{15} = 40\%$.

PROBLEM SET 7.3

1. (a) 6 (b) 15 (c) 20.
3. (a) $(\frac{1}{2})^6 = \frac{1}{64}$ (b) $_6C_1(\frac{1}{2})^6 = \frac{6}{64}$ (c) $_6C_2(\frac{1}{2})^6 = \frac{15}{64}$ (d) $_6C_3(\frac{1}{2})^6 = \frac{20}{64}$
(e) $1 - (\frac{1}{64} + \frac{6}{64} + \frac{15}{64} + \frac{20}{64}) = \frac{22}{64}$.
5. (a) $_{12}C_4(\frac{2}{3})^4(\frac{1}{3})^8$ (b) $_{12}C_6(\frac{2}{3})^6(\frac{1}{3})^6$.
7. (a) .0162 (b) .2522.
9. $_{10}C_0(\frac{1}{4})^0(\frac{3}{4})^{10} + {}_{10}C_1(\frac{1}{4})^1(\frac{3}{4})^9 + {}_{10}C_2(\frac{1}{4})^2(\frac{3}{4})^8 + {}_{10}C_3(\frac{1}{4})^3(\frac{3}{4})^7 = .0563$
$+ .1877 + .2816 + .2503 = .7759$.

11. $_{10}C_3(.35)^3(.65)^7 + _{10}C_4(.35)^4(.65)^6 + \cdots + _{10}C_{10}(.35)^{10}(.65)^0 = .7383.$

13. (a) $6(\frac{1}{6})^5$ (b) $6 \cdot {_5}C_4(\frac{1}{6})^4(\frac{5}{6})^1$ (c) $6 \cdot {_5}C_3(\frac{1}{6})^3(\frac{5}{6})^2$ (d) $6 \cdot 5 \cdot {_5}C_2(\frac{1}{6})^5$
    (e) $2 \cdot 5! \cdot (\frac{1}{6})^5$.

15. $(\frac{5}{6})^{20} + _{20}C_1(\frac{5}{6})^{19}(\frac{1}{6}) + _{20}C_2(\frac{5}{6})^{18}(\frac{1}{6})^2 + _{20}C_3(\frac{5}{6})^{17}(\frac{1}{6})^3 \approx .5665.$

17. $1 - [(.9876)^{100} + _{100}C_1(.9876)^{99}(.0124) + _{100}C_2(.9876)^{98}(.0124)^2] \approx .1282.$

PROBLEM SET 7.4

1. (b) 9 (c) $\frac{9}{24}$ (d) $\frac{9}{24}$.

3. $1 - Q_{20} = 1 - .41 = .59.$

5. $\frac{99}{100} \cdot \frac{98}{100} \cdot \frac{97}{100} \cdots \frac{77}{100}.$

7. Paul is right. Peter's four cases are not equally likely.

11. (a) $1 - (.001)(.001) = .999999$
    (b) $1 - [(.001)^4 + 4(.001)^3(.999)] = .999999996.$

13. $\frac{1}{2} \cdot 1 + \frac{1}{4} \cdot 2 + \frac{1}{8} \cdot 4 + \frac{1}{16} \cdot 8 + \cdots$, which is infinite.

PROBLEM SET 8.1

1.

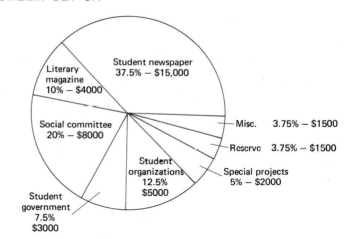

Student congress budget
total $40,000

3.

Class Range	Tally	Frequency
24–29	II	2
30–35	III	3
36–41	I	1
42–47	IIII	4
48–53	ЖH II	7
54–59	ЖH I	6
60–65	III	3
66–71	II	2
72–77	II	2
78–83	III	3
84–89	II	2
90–95	I	1

5.

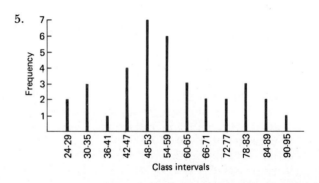

7. Probably scores on a physics exam. Exam scores tend to cluster in the middle and thin out at the ends. It seems unlikely that the age of welfare recipients would cluster around 50.

9. Make a trend line and a comparative bar graph.

11. It will be four times as large. Doubling the dimensions of a cube will make the volume eight times what it was.

13. (a) Invalid. There are more male drivers and males probably drive more. (b) Invalid. Ohio is much bigger and has more drivers.
(c) Invalid. The foreign manufacturer probably began exporting to the United States recently. (d) Possibly invalid. It may have been an exceptionally hard test.

PROBLEM SET 8.2

1. (a) Mean 6.56, median 7, mode 8 (b) mean 16.22, median 16, mode 16.

3.

Word Length	Tally	Frequency
13	I	1
12		0
11	I	1
10	I	1
9	II	2
8	I	1
7	I	1
6	II	2
5	II	2
4	⊞ ⊞ ⊞ I	16
3	⊞ ⊞ II	12
2	⊞ ⊞ II	12
1	II	2

5. 142.2 pounds.

7. Mean $22,000; median $12,000.

9. Not necessarily. The temperature may fluctuate wildly and still have a mean of 25°C.

11. 3.31.

13. 5.47 feet from the end.

15. 150 mph. It can't be done.

17. All students get the same score.

19. 392.3.

21. 2.602.

PROBLEM SET 8.3

1. Mean 7, standard deviation 1.90.

3.

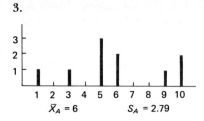 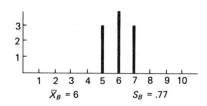

5. $-2.38, -0.63, 0.75, 1.88$

7. German -0.36, Math $-1.30+$ she did better in German.

9. Mean 12, standard deviation 1.90; you add the number to the mean, but the standard deviation is unchanged.

11. 4%, 16%.

13.

Word Length	Tally	Frequency
11	I	1
10		0
9	II	2
8	II	2
7	III	3
6	IIII	4
5	IHI I	6
4	IHI III	8
3	IHI IHI III	13
2	IHI IHI	10
1	III	3

Mean 4.12. Standard deviation 2.22

15. Zero. If the standard deviation is 0,
$(x_1 - \bar{x})^2 + (x_2 - \bar{x})^2 + \cdots + (x_n - \bar{x})^2 = 0$. This implies that $x_i = \bar{x}$ for each i.

17. $\bar{x} = 12.5$, $s_x = 1.3$.

19. $\bar{z} = 12.8$, $s_z = 2.5$.

PROBLEM SET 8.4

1. (a) 72 (b) 80 (c) 55 (d) $\frac{25}{12}$.

3. (a) $\sum\limits_{i=1}^{10} i$ (b) $\sum\limits_{i=1}^{10} i^2$ (c) $\sum\limits_{i=1}^{37} y_i$ (d) $\sum\limits_{i=1}^{n} (y_i - 3)^2$.

5. (a) 4.31 (b) 1293 (c) $3 \sum\limits_{i=1}^{100} x_i + \sum\limits_{i=1}^{100} 1 = 3(431) + 100 = 1393$.

7. $\bar{x} = 2$, $s = 6$.

9. $\bar{x} = 7$, $s = 1.90$.

PROBLEM SET 8.5

1.

$\bar{x} = 4$; $\bar{y} = 6$; $s_x = 2.45$; $s_y = 2.10$; $r = .95$.

3. (a) High negative correlation; y decreases as x increases.
   (b) Low negative correlation; y decreases as x increases.
   (c) Very low (and probably insignificant) correlation.
   (d) Quite high positive correlation; y increases as x increases.
5. Authors' guesses are: (a) high negative (b) high positive (c) low positive (d) high negative (e) low positive (f) low negative (g) low positive (h) high positive (i) low positive.
11. $r_{xy} \approx -0.237$. This slight negative correlation is probably not significant.
13. $r \approx 0.976$. This is a very high positive correlation and indicates a strong linear relationship. Of course, the data lie on the graph of $y = \sqrt{x}$ which is definitely not a straight line. However, if you graph $y = \sqrt{x}$, you will note a linear tendency.

PROBLEM SET 9.1

1. (a) No (b) no (c) yes (d) yes (e) yes.
3. (a), (b), (c), and (e) have Hamiltonian circuits.
5. (a) Yes (b) no.
7. (a) No (b) yes (c) yes.
9. A network has an Euler path if and only if it has either zero or two odd vertices.
11. (a) Yes (b) yes (c) yes.
15. 5.

PROBLEM SET 9.2

1. $d_1 = 27$, $d_2 = 0$, $d_3 = 6$, $d_4 = 3$, $d_5 = 3$, $d_6 = 1$; $1 \cdot 6 + 2 \cdot 3 + 3 \cdot 3 + 4 \cdot 1 = 27 - 2$.
3. $V = 40$, $E = V - 1 = 39$.

5.

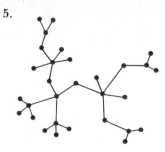

7. 50.

9.

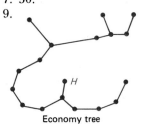

Economy tree

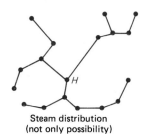

Steam distribution
(not only possibility)

PROBLEM SET 9.3

1. $F = 14$, $V = 15$, $E = 27$.
3. $F = 6$, $V = 5$, $E = 9$. The same number of faces does not meet at each vertex.
5. $E = 9$.

7. Suppose it did, that is, suppose $V = E$. Use Euler's Formula to arrive at the conclusion $F = 2$, which is impossible.
9. Tetrahedron, no; cube, no; octahedron, yes; dodecahedron, no; icosahedron, no.
11. $F = 8$, $V = 12$, $E = 18$; yes.
13. Assume the polyhedron is flexible. Push out the dents. This doesn't change F, V, or E.

PROBLEM SET 9.5

1. (a) Three colors (b) two colors (c) three colors (d) two colors.
3. Four colors.
7. 77.
9. (a) 4 (b) 3 (c) 2 (d) 4 (e) 3.

PROBLEM SET 10.1

1. Obviously, there is no single set of answers. We have in mind such things as the following:
    (a) There was an earthquake last week in ____.
       Certain insulating materials are fire-resistant.
       Company X makes the best calculators.
       The use of toothpaste reduces cavities.
    (b) A certain road will break up every spring.
       If it rains, be it ever so gentle, traffic jams will build up at the bridge.
       The Wonder Department Store will have a January sale.
    (c) Two aspirin will relieve a headache.
       A certain spot in your yard is not a good place to grow tomatoes.
       Brand X ballpoint pens perform better than brand Y ballpoint pens.

5. The numbers through $k = 27$ are classified as:
    Odd type: 2, 3, 5, 7, 8, 11, 12, 13, 17, 18, 19, 20, 23, 27
    Even type: 4, 6, 9, 10, 14, 15, 16, 21, 22, 24, 25, 26
    The odds lead the evens, 14 to 12.

7. (a) The numbers on the left are the binomial coefficients; hence $(x + y)^3 = 1x^3 + 3x^2y + 3xy^2 + 1y^3$. Set $x = y = 1$. The natural and correct conjecture comes from the expansion of $(x + y)^n$.
    (b) The natural conjecture is shattered by circling primes in the next line.

9. (a) The intersections of AB with ab, AC with ac, and BC with bc will be in a straight line. This is Desargues' Two-Triangle Theorem.
    (b) The intersections of the three pairs of external tangents are colinear.

PROBLEM SET 10.2

1. (a) Not valid.   3. (a) Not valid.
   (b) Valid.   (b) Not valid.
   (c) Not valid.   (c) Not valid.
                            (d) Valid.

5.
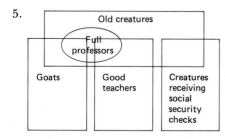

   (a) Valid.
   (b) Not valid.
   (c) Not valid.
   (d) Not valid.

7.

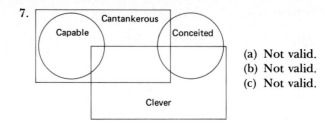

(a) Not valid.
(b) Not valid.
(c) Not valid.

9.

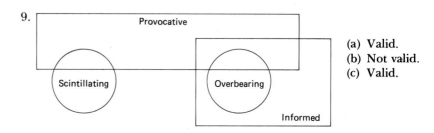

(a) Valid.
(b) Not valid.
(c) Valid.

11. (a) This is not valid. (We need the added assumption that Hugo follows the best tactic.)
    (b) Horace is wrong. (He assumes the converse of the proposition.)
    (c) Valid.

13. (a) Not valid.
    (b) Not valid.
    (c) Not valid.
    (d) Valid.

15. (a) *Hypotheses:* My home state collects taxes on income made in the state.
                        I conduct my business in another state.
        Conclusion: I won't have to pay income tax in my state.
    (b) *Hypotheses:* Dogs can't pull a sled that is too heavy.
                        If you take all that gear, the sled will be too heavy.
        Conclusion: The dogs won't be able to pull the sled.
    (c) *Hypotheses:* A sensitive person does not continually refuse offers of help.
                        Homer accepts any offer of help.
        Conclusion: Homer is a sensitive person.

17.

Dissatisfied with both
39

Satisfied with food service 72 | Satisfied with dorm rooms 61 | 97

TOTAL STUDENTS	
Satisfied with food service only	11
Satisfied with dorm rooms only	36
Satisfied with both	61
Satisfied with neither	39
	147

19. (a) *Hypotheses:* Moon rocks are made of blue cheese.
　　　　　　　　　　No one has touched blue cheese from the moon.
　　　Conclusion: No one has touched a rock from the moon.
　(b) *Hypotheses:* Moon rocks are made of green cheese.
　　　　　　　　　　No one has eaten blue cheese from the moon.
　　　Conclusion: No one has eaten moon rocks.
　(c) If the hypotheses are true and the argument valid, the conclusion must be true.
　(d) *Hypotheses:* Moon rocks can only be obtained by going to the
　　　　　　　　　　moon.
　　　　　　　　　　We have moon rocks.
　　　Conclusion: We have been to the moon.

PROBLEM SET 10.3

1. (a) If the true place for a just man is in prison, then a government imprisons any person unjustly.
　(b) If a man is alone, then he is thinking or working.
　(c) If there are many things a man can afford to leave alone, then he is truly rich.
　(d) If a man hears a different drummer, then he does not keep pace with his companions.
　(e) If a man sweats easier than I do, then he must earn his living by the sweat of his brow.

3. (a) If the true place for a just man is not in prison, then a government does not imprison any man unjustly.
　(b) If a man is not alone, then he is not thinking or working.
　(c) If there is nothing that one can afford to leave alone, then he is not truly rich.
　(d) If a man does not hear a different drummer, then he keeps pace with his companions.
　(e) If a man does not sweat easier than I do, then he will not have to earn his living by the sweat of his brow.

5. (a) The meaning of the word "depression" has shifted.
　(b) If A implies B, but A doesn't happen, no conclusion can be drawn about B.
　(c) If A implies B and B is true, it is not necessarily true that A is true.

7. At a combined selling price of two-fifths of a dollar for one gizmo, the one who had been selling them for one-third of a dollar apiece gained $30(\frac{2}{5} - \frac{1}{3}) = \$2$, so his share of the sales is his expected earnings of \$10 plus \$2. The other exhibitor, however, lost $30(\frac{1}{2} - \frac{2}{5}) = \$3$, so his share is \$15 minus \$3. Nothing is missing.

9. The farmer, in willing $\frac{1}{2} + \frac{1}{3} + \frac{1}{9} = \frac{17}{18}$ of his possessions, didn't provide for giving away all that he had, so the additional contributed horse was left over.

11. (a) Both balls arrive simultaneously. Imagine a 15-pound weight falling. Will it fall faster (or slower) if it has been sawed into two parts?
　(b) Try it.

13. If it is true that every rule has an exception, so must the rule in part (a); hence some rule must have no exception. A similar difficulty is encountered with parts (b) and (c).

PROBLEM SET 10.4

1. $(2n)^2 = 4n^2 = 2(2n^2)$.
3. $(3n + 1)^2 = 9n^2 + 6n + 1 = 3(3n^2 + 2n) + 1$.
   $(3n + 2)^2 = 9n^2 + 12n + 4 = 3(3n^2 + 4n + 1) + 1$.
5. $2n(2n + 1) = 2(2n^2 + n)$.
7. Let the odd number be $2n + 1$. Its square is $4n^2 + 4n + 1$.
   If n is even, $n = 2k$.
   $4n^2 + 4n + 1 = 4(2k)^2 + 4(2k) + 1 = 8[2k^2 + k] + 1$
   If n is odd, $n = 2k + 1$ and
   $4n^2 + 4n + 1 = 4[4k^2 + 4k + 1] + 4[2k + 1] + 1$
   $= 8[2k^2 + 3k + 1] + 1$
9. Suppose an arbitrary angle has been divided into six equal subangles, using only a compass and straightedge. Then, taken in adjacent pairs, they would be trisectors, violating the fact that one cannot so construct angle trisectors.
11. Suppose n is not odd, so that $n = 2k$. Then $n^2 = 2(2k^2)$ is not odd, contradicting what we know about n^2.
13. Imagine each maple tagged with a slip showing the number of leaves. Let n be the maximum number on a tag. By hypotheses, there are more than n trees, hence more than n tags. By the pigeonhole principle, some tags must have the same number written on them.
15. Yes. Again the argument is by the pigeonhole principle.

PROBLEM SET 11.1

11. Gwen loves Alan.
13. The Greater Glory Party.

PROBLEM SET 11.2

1.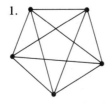

3.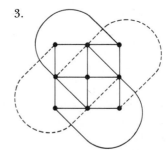

PROBLEM SET 11.3

5. (a) This course is not useful education.
   (b) This course can be useful in preparing for certain jobs.

7. (a) Those who wish to improve themselves go to school.
   (b) Homer wishes to improve himself.

PROBLEM SET 11.4

1. (a) My arm from wrist to elbow measures 28 centimeters, and it appears to have a radius of about 4 centimeters. With a cylinder as a model, the volume appears to be $\pi 4^2 \cdot 28 = \frac{22}{7}(4^2)(28) = 22(4^3) = 1408$ cubic centimeters.
   (b) With a sphere as a model and approximating the radius of my head to be $r = 12$ centimeters, the volume appears to be $\frac{4}{3}\pi \cdot 12^3 = 7235$ cubic centimeters.

PROBLEM SET 12.4

1. There should be a correlation between aptitude and grade; a person with low aptitude should expect a low grade.
3. Since my folks are poor correspondents, then I am excused for not writing to them.
5. Change the second hypothesis to: A liberal arts education should also prepare students for a job.
7. Homer wishes to improve himself.
9. Draw the picture carefully; we encountered the same problem in our proof that all triangles are isosceles.

PROBLEM SET 13.1

1. (a) g.c.d. = 2, l.c.m. = 240 (b) g.c.d. = 3, l.c.m. = 90 (c) g.c.d. = 1, l.c.m. = 1800.
3. (a) $2 \cdot 2 \cdot 2 \cdot 5$ (b) $2 \cdot 2 \cdot 13$ (c) $2 \cdot 3 \cdot 13$ (d) $2 \cdot 2 \cdot 3 \cdot 5$ (e) $2 \cdot 5 \cdot 7$ (f) $2 \cdot 3 \cdot 3 \cdot 7$ (g) $2 \cdot 2 \cdot 3 \cdot 3 \cdot 7$.
5. (a) g.c.d. = 2, l.c.m. = 1560 (b) g.c.d. = 2, l.c.m. = 1260.
7. (a) $\frac{19}{24}$ (b) $\frac{53}{48}$ (c) $\frac{127}{105}$ (d) $\frac{143}{360}$.
9. (a) $2^4 \cdot 3^3 \cdot 5^2$ (b) $2 \cdot 3^4 \cdot 5^2 \cdot 13$ (c) $5^2 \cdot 3 \cdot 11 \cdot 47$.
11. 24.
13. $n = 4$.
15. 4 inches or 9 inches.

PROBLEM SET 13.2

1. (a) 12.615 (b) 11.865 (c) 0.05625 (d) 2.5.
3. $0.\overline{285714}$.
5. (a) $0.375\overline{0}$ (b) $2.375\overline{0}$ (c) $0.4\overline{5}$ (d) $3.91\overline{6}$.
7. $\frac{49}{133}$.
9. (a) $\frac{19}{33}$ (b) $\frac{17}{66}$.
11. (a) 1 2 11 2 111 2 1111 2 11111 2 . . . . (b) This is an irrational number.
13. Yes.
15. Note that $r - \frac{3}{4}$ is rational.
17. Let $r = \frac{3}{4}\sqrt{2}$ and note that if r is rational so is $\frac{4}{3}r$.

19. (a) irrational (b) irrational (c) irrational (d) rational (e) rational (f) irrational (g) irrational (h) rational.
21. 0.0000001/2 will do. There is no smallest rational.
23. If r and s are rational, so is their average $(r + s)/2$.
25. The decimal expansion of π is nonrepeating.

PROBLEM SET 13.3

1. (a) 6 (b) 5 (c) 4 (d) 2.
3. Friday, Sunday.
5. (a) 3 (b) 1 (c) 4 (d) 3 (e) 2 (f) 4 (g) 6 (h) 1.
7. (a) 5 (b) 3, 7, 11 (c) 3 (d) 4.
9. (a) 4 (b) 8 (c) 6 (d) 5.
11. (a) 16,896 (b) okay (c) 302736 (d) 312193.
13. See Problem 21.
15. If $a = b + km$ and $b = c + rm$, then $a = c + (k + r)m$.
17. $a = b + 7k$ and $c = d + 7m$, so $ac = bd + 7[bm + kd + 7mk]$.
19. (a) 4 (b) 5 (c) 2 (d) 6.

PROBLEM SET 13.4

1. (a) $x = 1 \bmod 5$
   $y = 0 \bmod 2$

x	$\cdots$	-4	1	6	11	$\cdots$
y	$\cdots$	2	0	-2	-4	$\cdots$

  (b) $x = -1 \bmod 16$
   $y = 2 \bmod 15$

x	$\cdots$	-17	-1	15	$\cdots$
y	$\cdots$	17	2	-13	$\cdots$

  (c) $x = 9 \bmod 25$
   $y = -13 \bmod 37$

x	$\cdots$	-41	-16	9	$\cdots$
y	$\cdots$	61	24	-13	$\cdots$

  (d) No integer solutions exist.

  (e) $x = 55 \bmod 86$
   $y = -73 \bmod 117$

x	$\cdots$	-31	55	141	$\cdots$
y	$\cdots$	44	-73	-190	$\cdots$

3.

x	$\cdots$	-1	1	3	$\cdots$
y	$\cdots$	2	15	28	$\cdots$
t	$\cdots$	5	12	19	$\cdots$

5. Homer gets $4(17) = \$68$; Horatio gets $25(3) = \$75$.
7. 47 tables, 311 chairs.
9. Coconuts $= 3121 \bmod 15,625$; *i.e.*, 18,746; 34,371; etc.

PROBLEM SET 14.1

1. $AB = \begin{bmatrix} 7 & 10 \\ 9 & 13 \end{bmatrix}$ $BA = \begin{bmatrix} 12 & 19 \\ 5 & 8 \end{bmatrix}$; $BC = \begin{bmatrix} -20 & 9 \\ -9 & 4 \end{bmatrix}$

   $CB = \begin{bmatrix} 13 & 18 \\ -21 & -29 \end{bmatrix}$.

3. $MC = \begin{bmatrix} -29 & 13 \\ -38 & 17 \end{bmatrix} = AN$.

5. $(BC)D = \begin{bmatrix} -20 & 9 \\ -9 & 4 \end{bmatrix}\begin{bmatrix} 5 & -3 \\ -3 & 2 \end{bmatrix} = \begin{bmatrix} -127 & 78 \\ -57 & 35 \end{bmatrix} = \begin{bmatrix} 5 & 7 \\ 2 & 3 \end{bmatrix}\begin{bmatrix} 18 & -11 \\ -31 & 19 \end{bmatrix}$

   $= B(CD).$

7. $A(B + C) = \begin{bmatrix} 1 & 1 \\ 1 & 2 \end{bmatrix}\begin{bmatrix} 8 & 6 \\ -3 & 5 \end{bmatrix} = \begin{bmatrix} 5 & 11 \\ 2 & 16 \end{bmatrix} = \begin{bmatrix} 7 & 10 \\ 9 & 13 \end{bmatrix} + \begin{bmatrix} -2 & 1 \\ -7 & 3 \end{bmatrix}$

   $= AB + AC.$

9. $A, A, D, D.$

11. (a) I (b) $I.$

13. (a) $\begin{bmatrix} 3 & -4 \\ -5 & 7 \end{bmatrix}$ (b) $\begin{bmatrix} -3 & 5 \\ 5 & -8 \end{bmatrix}$ (c) $\begin{bmatrix} 2 & -\frac{7}{2} \\ -1 & 2 \end{bmatrix}$ (d) no inverse.

15. (a) $\begin{bmatrix} 6 & 7 & 3 \\ 6 & 2 & 0 \\ 13 & 6 & 1 \end{bmatrix}$ (b) $\begin{bmatrix} 5 & -2 & 6 \\ 2 & 0 & 2 \end{bmatrix}.$

17. $1 = m$; dimensions $k \times n.$

19. $\begin{bmatrix} 25{,}875 & 10{,}125 \\ 28{,}475 & 14{,}450 \end{bmatrix}.$

PROBLEM SET 14.2

1. $(AB)C = \begin{bmatrix} -7 & -2 \\ -11 & -3 \end{bmatrix}\begin{bmatrix} 2 & -1 \\ 1 & 4 \end{bmatrix} = \begin{bmatrix} -16 & -1 \\ -25 & -1 \end{bmatrix} = \begin{bmatrix} 1 & -2 \\ 2 & -3 \end{bmatrix}\begin{bmatrix} -2 & 1 \\ 7 & 1 \end{bmatrix}$
   $= A(BC).$

3. $AB = \begin{bmatrix} -7 & -2 \\ -11 & -3 \end{bmatrix}$ $BA = \begin{bmatrix} -1 & 2 \\ 5 & -9 \end{bmatrix}.$

5. $A(B + C) = \begin{bmatrix} 1 & -2 \\ 2 & -3 \end{bmatrix}\begin{bmatrix} 1 & -1 \\ 4 & 5 \end{bmatrix} = \begin{bmatrix} -7 & -11 \\ -10 & -17 \end{bmatrix}$

   $= \begin{bmatrix} -7 & -2 \\ -11 & -3 \end{bmatrix} + \begin{bmatrix} 0 & -9 \\ 1 & -14 \end{bmatrix} = AB + AC.$

7. (a) $A^{-1} = \begin{bmatrix} -3 & 2 \\ -2 & 1 \end{bmatrix}$ (b) $C^{-1} = \begin{bmatrix} \frac{4}{9} & \frac{1}{9} \\ -\frac{1}{9} & \frac{2}{9} \end{bmatrix}$

   (c) $(AC)^{-1} = \begin{bmatrix} -\frac{14}{9} & 1 \\ -\frac{1}{9} & 0 \end{bmatrix}$; $A^{-1}C^{-1} = \begin{bmatrix} -\frac{14}{9} & \frac{1}{9} \\ -1 & 0 \end{bmatrix}.$

9. $C^{-1}A^{-1} = \begin{bmatrix} -\frac{14}{9} & 1 \\ -\frac{1}{9} & 0 \end{bmatrix} = (AC)^{-1}$ $(CA)^{-1} = \begin{bmatrix} -\frac{14}{9} & \frac{1}{9} \\ -1 & 0 \end{bmatrix} = A^{-1}C^{-1}.$

11. $(\det A)(\det B) = 1(-1) = \det AB.$

17. $\begin{bmatrix} 0 & 0 \\ 0 & 0 \end{bmatrix}.$

19. No.

PROBLEM SET 14.3

1. (a) $x = 3$, $y = -2$ (b) $x = -2$, $y = \frac{5}{2}$ (c) $x = -1$, $y = 2$.
3. (a) $x = -67$, $y = -17$, $z = 12$ (b) $x = -65$, $y = -18$, $z = 11$.

5. $\begin{bmatrix} 195 & 12 \\ 239 & 16 \\ 288 & 19 \end{bmatrix} \begin{bmatrix} 9 \\ 43 \end{bmatrix} = \begin{bmatrix} 2271 \\ 2839 \\ 3409 \end{bmatrix}$.

7. $\begin{bmatrix} 195 & 12 \\ 239 & 16 \\ 288 & 19 \end{bmatrix} \begin{bmatrix} 40 & 20 & 25 & 60 \\ 300 & 150 & 200 & 400 \end{bmatrix} = \begin{bmatrix} 11,400 & 5700 & 7275 & 16,500 \\ 14,360 & 7180 & 9175 & 20,740 \\ 17,320 & 8610 & 11,000 & 24,880 \end{bmatrix}$.

9.
	Plywood	Time
Style 1	5	12
Style 2	3	8
Style 3	2	6

$\begin{bmatrix} 15 & 12 \\ 22 & 17 \end{bmatrix}$ Plywood / Labor

$= \begin{bmatrix} 339 & 264 \\ 221 & 172 \\ 162 & 126 \end{bmatrix}$ Style 1 / Style 2 / Style 3 (A B).

PROBLEM SET 15.1

1. (a) No; $4 \div 2 \neq 2 \div 4$
   (b) no; $(48 \div 12) \div 2 \neq 48 \div (12 \div 2)$.
3. (a) -41 (b) -41 (c) yes (d) yes.
5. Yes.
7. (a) yes (b) yes (c) no; $2 * 3 = \frac{5}{2}$ is not an integer.
   (d) yes.
9. (a) is determinative; (c) is questionable, since different-sized tires may be on the same car; (b) and (d) are matters of judgment.
11. (a) R, S
    (b) S
    (c) S
    (d) R, T
    (e) R, S
    (f) R, S.
13. "Separated from" is symmetric, not reflexive nor transitive.
15. "Is at least as far north as."

PROBLEM SET 15.2

1. (a) $\bar{x} = 2$ (b) $\bar{y} = 2$ (c) $\bar{x}\bar{y} = 1$ (d) $xy = 1$.
3. (a) Doesn't exist (b) 1 (c) 3 (d) 2 (e) 4.
5. (a) 9 (b) doesn't exist (c) 7 (d) 5 (e) 1 (f) 11.
7. (a) $\begin{bmatrix} -1 & -2 \\ -1 & -3 \end{bmatrix}$ (b) $\begin{bmatrix} 3 & -2 \\ -1 & 1 \end{bmatrix}$ (c) $\begin{bmatrix} -5 & -5 \\ -1 & -1 \end{bmatrix}$ (d) doesn't exist.

9. There is no additive identity, so it doesn't even make sense to ask if each element has an additive inverse.

11. $(a + b)(a - b) = a^2 - ab + ba - b^2$
$$= a^2 - b^2 \quad \text{(if and only if } ab = ba)$$
The result depends on multiplication being commutative.

PROBLEM SET 15.3

1. $\overline{0} = 0, \overline{1} = 4, \overline{2} = 3, \overline{3} = 2, \overline{4} = 1; 1^{-1} = 1, 2^{-1} = 3, 3^{-1} = 2, 4^{-1} = 4.$

3. (a) $\overline{4} = 9$ (b) $4^{-1} = 10$ (c) $\overline{5} = 8$ (d) $5^{-1} = 8.$

13. (a) The universal set U acts as multiplicative identity. (b) The empty set $\emptyset$ acts as the additive identity.

15. $\begin{bmatrix} a & -b \\ b & a \end{bmatrix} \begin{bmatrix} c & -d \\ d & c \end{bmatrix} = \begin{bmatrix} R & -S \\ S & R \end{bmatrix}$

where $\begin{matrix} R = ac - bd \\ S = ad + bc \end{matrix}$

19. Yes.

21. It is not true that any odd value of m will work ($m = 9$ is a counter-example). Guess again. See 23.

PROBLEM SET 15.4

1. To subtract a (positive) 3 from -5, write either

$$\text{subtract} \quad \begin{matrix} -5 \\ \underline{3} \end{matrix} \quad \text{or} \quad \begin{matrix} -5 \\ \underline{-3} \end{matrix}$$

Either way, the answer is -8.

3. (a) $-\frac{5}{3}$ (b) $\frac{5}{2}$ (c) $-\frac{4}{9}$ (d) -1.

5. 6.

7. (a) $\begin{matrix} x_1 = -\frac{1}{3} \\ x_2 = 2 \end{matrix}$ (b) $\begin{bmatrix} \frac{1}{6} & 1 \\ \frac{3}{2} & 1 \end{bmatrix}.$

9. $\begin{bmatrix} 2 & 1 \\ 6 & 4 \end{bmatrix}^{-1} = \begin{bmatrix} 2 & 5 \\ 8 & 1 \end{bmatrix} \quad \begin{bmatrix} x_1 & x_2 \\ x_3 & x_4 \end{bmatrix} = \begin{bmatrix} 4 & 5 \\ 1 & 4 \end{bmatrix}.$

11. (a) 7 (b) $x = 6 \bmod 26$.

Names and Faces Index

Boldface page numbers indicate that a picture appears on that page.

Abbot, E. A., 377
Appel, K., 281
Aristotle, 134, 356

Bach, J. S., 108
Bell, E. T., 101
Bergamini, David, 133
Blake, William, 263
Bolyai, John, 365, 368
Bolyai, Wolfgang, 365, 368

Carroll, Lewis, 38, 309
Chinn, William, 436
Cicero, 192
Courant, Richard, 384
Coxeter, H. S. M., 130

D'Alembert, Jean de Rond, 365
Davis, Philip, 386, 436
DeMorgan, Augustus, 281, 356
Descartes, René, **74,** 181, 270, 290, 308
Diophantus, 412
Dirichlet, P. G. L., 393
Disraeli, 225
Dodgson, Charles, 309

Einstein, Albert, 67, 94, 288, **289,** 347, 351, 376, 385
Eliot, Charles, 380
Emmet, E. R., 331
Eratosthenes, 361, 393
Escher, M. C., 268, 274
Euclid, 356
Euler, Leonhard, **91,** 97, 255, 270

Fermat, Pierre de, 203, 292, **391**
Fibonacci, Leonardo, **127**

Franklin, Benjamin, 311

Galileo, 292, **314**
Gardner, Martin, 133, 277
Gauss, Carl, **101,** 335, 368
Graves, Robert, 330

Hadamard, J., 341
Haken, W., 281
Hamilton, W. R., **258,** 281
Hardy, G. H., 254, 336, **430**
Hilbert, David, 321, 334, **380,** 444
Hodge, Alan, 336
Hooke, Robert, 351
Howe, Edgar, 305
Hurley, James F., 322, 330

Infeld, L., 351

Jacobi, C. G. J., 417
Jacoby, Oswald, 330
Jefferson, Thomas, 308

Kac, M., 317
Kelvin, Lord, 220
Kemeny, John, 164
Kepler, Johannes, 90, 268
Keynes, John M., 342
Kline, Morris, 381
Kordemsky, Boris, 11
Kurtz, Thomas, 164

Lambert, J. H., 367
Laplace, Pierre Simon de, 210
Lobatchevsky, N., 368

Maki, P., 350
Malthus, Thomas, 115

Meadows, D. H., 119
Méré, Chevalier de, 203
Methuselah, 120
Milne, A. A., 316
Montmart, M. de, 214

Newman, M. H. A., 166
Noether, Emmy, **385,** 444

O'Henry, 221

Pascal, Blaise, 143, **186,** 203
Pearson, Karl, 249
Plato, 314
Playfair, John, 358
Poincaré, H., 341, 355
Polya, George, 2, **3,** 4, 24, 74, 82, 186, 292, 293
Pythagoras, 361

Quine, W. V., 313

Riemann, B., 369
Rochefoucald, F. de la, 304, 308
Rogers, W., 340
Russell, B., 334, 340, 378

Saccheri, G., 366
St. Vincent Millay, Edna, 356
Samuelson, Paul, 302
Schattschneider, D., 277
Schweikart, C. F., 367
Shakespeare, William, 264
Siu, 345
Snow, C. P., 113
Stein, Sherman, 282
Steinhaus, H., 110
Synge, J. L., 92

Taurinus, F. A., 367
Thompson, M., 350
Thoreau, H., 311
Twain, Mark, 226, 281
Tuckerman, Bryant, 392

Ulam, S., 317

Van Tassel, Dennie, 149
Von Neumann, John, **144**
Von Watterhouser, S., 335

Weaver, Warren, 191, 193
Wells, H. G., 220
Whitehead, Alfred North, 29
Wiener, N., 346
Williams, B., 416

Subject Index

Abacus, 143
Absolute geometry, 366
Absolute value, 39
Addition, 39, 46, 54, 421
Addition principle, 168
Additive inverse, 409, 445
Algebraic logic, 35
Algorithms, 56, 136
Alphametric, 140
Arithmetic mean, 229
Arithmetic sequence, 101
Associative, 33, 425, 444
Average, 229
Axioms, 333, 340

Bar graphs, 223
Base-eight, 137
Base-ten, 135
Base-two, 138
BASIC, 159
BASIC commands, 163
Bicycle problem, 23
Binary notation, 138
Binary operation, 438
Binomial distribution, 208
Binomial formula, 187
Birthday problem, 214

Calculator, 35, 143
Canceling, 45, 450
Celsius, 30, 32
Chessboard problem, 28, 107
Clock arithmetic, 403
Closed, 439
Coconut problem, 416
Combination, 179
Combination of n things taken r at
  a time, 179
Common difference, 101
Common ratio, 106

Commutative, 33, 425, 444
Comparative bar graph, 225
Complement, 195
Complex fraction, 48
Complex number, 399
Compound interest, 120
Computer, 143
Computer components, 144
Contradiction, 316
Contrapositive, 311
Converse, 307
Correlation coefficient, 249
Counting numbers, 38, 387, 397

Decimal, 52
Deduction, 298, 314
Degree of vertex, 256
Denominator, 45
Dependent events, 202
Determinant, 427
Determinative, 439
Diophantine problems, 411
Distributive, 33, 427
Division, 44, 47, 54
Divisors, 388
Doubling time, 116
Doughnuts, 285

Economy tree, 265
Edge, 256
Elements, 195
Elimination of one unknown, 84
Empty set, 195
Equal likelihood, 192
Equal tempering, 108
Equation, 68, 431, 456
Equivalence relation, 439
Estimating, 55
Euler circuit, 256
Euler path, 261

Euler's formula, 270
Event, 193
Experiment, 293
Explicit formula, 95
Exponent, 60
Exponential growth, 114

Factor, 388
Factorial symbol, 170
Fahrenheit, 30, 32
Fibonacci sequence, 127, 128
Field, 451
Finite geometry, 333
First moment, 231
Flowchart, 149
Fractions, 43
Frequency chart, 223
Fundamental theorem of
  algebra, 399
  arithmetic, 389

Geometric sequence, 106
Golden ratio, 129
Golden rectangle, 129
Greatest common divisor (gcd), 388

Hairy problem, 317
Half-life, 116
Hamiltonian circuit, 257
Hatcheck problem, 213
Hexstat, 209
Hindu-Arabic notation, 135
House payments, 123

Identity
  additive, 444
  multiplicative, 426, 445
Imaginary numbers, 399, 400
Impossible event, 194
Independent events, 202

Induction, 292
Installment buying, 126
Integer, 38, 395
Interest
  compound, 120
  simple, 120
Intersection, 169, 195
Inverse
  additive, 409, 445
  multiplicative, 409, 445
Irrational numbers, 398

Königsberg bridge problem, 255

Least common denominator, 46
Least common multiple (lcm), 388
Line, 333
Linear growth, 113
Loop, 150

Map coloring, 281
Matrix, 419
Mean, 229
Median, 229
Members, 195
Metric system, 33
Mode, 229
Modulo numbers, 402
Monk problem, 25
Mosaic, 274
Multiple, 388
Multiplication, 41, 47, 54, 419
Multiplication principle, 168
Multiplicative inverse, 409
Music, 108

Negative numbers, 38
Network, 256
Nim, 141
Normal curve, 208, 237
Number line, 39, 396, 397
Number sequence, 94
Numerator, 45

Order, 34, 181

Paradox, 308
Parentheses, 34
Pascal's triangle, 185

Path, 256
Pentagonal numbers, 99
Percent, 56
Permutation, 173
Permutation of n things taken r at
    a time, 174
Phyllotaxis, 130
Pie graphs, 222
Pigeonhole principle, 316
Place value, 136
Platonic solids, 269
Point, 333, 340, 357
Poker, 195
Polygon, 269
Polyhedron, 269
Population growth, 115
Prime, 389
Principal, 120
Principal representative, 404
Probability, 193
  of B given A, 201
  properties of, 194
Pythagorean theorem, 361

Quasiregular mosaic, 276

Rabbit problem, 127
Radioactive decay, 116
Range, 235
Rational number, 45, 396
Real line, 397
Real number, 397
Recursion formula, 95
Reduced form, 45
Reductio ad absurdum, 316
Reflexive, 439
Regular mosaic, 275
Regular polygon, 270
Regular polyhedron, 270
Repeating decimal, 397
Repetitions, 181
Reverse Polish logic, 35
Ring, 446
River-crossing problems, 14
Rounding off, 54

Sample correlation coefficient, 249
Scatter diagram, 249

Scientific notation, 63
Semi-regular mosaic, 280
Semi-regular polyhedron, 273
Sequence
  arithmetic, 101
  doubling, 106
  Fibonacci, 127, 128
  geometric, 106
  number, 94
Set language, 195
Sieve of Eratosthenes, 393
Sigma notation, 243
Simple interest, 120
Signed number, 38
Square numbers, 96
Standard deviation, 235, 236
Standardized score, 239
Statistics, 221
Subtraction, 40, 46, 54
Sure event, 194
Symmetric, 439

Tesselation, 274
Theorem, 333, 341
Three-color maps, 284
Tile, 274
Time-sharing, 144
Tournament problem, 24
Transitive, 439
Tree, 263
Trend line, 225
Triangular numbers, 96
Two-color maps, 282
Two's complement, 148

Union, 169, 195
Universe, 195

Valid, 299
Venn diagram, 299
Vertex (vertices), 256

Well defined, 440

Yahtzee, 206

Z-score, 239
Zeno's paradox, 309